Chemical Analysis and Material Characterization by Spectrophotometry

Chemical Analysis and Material Characterization by Spectrophotometry

Bhim Prasad Kafle

Department of Chemical Science & Engineering,
School of Engineering,
Kathmandu University,
Dhulikhel, Kavre, Nepal

ELSEVIER

Elsevier
Radarweg 29, PO Box 211, 1000 AE Amsterdam, Netherlands
The Boulevard, Langford Lane, Kidlington, Oxford OX5 1GB, United Kingdom
50 Hampshire Street, 5th Floor, Cambridge, MA 02139, United States

Notices
Knowledge and best practice in this field are constantly changing. As new research and experience broaden our understanding, changes in research methods, professional practices, or medical treatment may become necessary.

Practitioners and researchers must always rely on their own experience and knowledge in evaluating and using any information, methods, compounds, or experiments described herein. In using such information or methods they should be mindful of their own safety and the safety of others, including parties for whom they have a professional responsibility.

To the fullest extent of the law, neither the Publisher nor the authors, contributors, or editors, assume any liability for any injury and/or damage to persons or property as a matter of products liability, negligence or otherwise, or from any use or operation of any methods, products, instructions, or ideas contained in the material herein.

Library of Congress Cataloging-in-Publication Data
A catalog record for this book is available from the Library of Congress

British Library Cataloguing-in-Publication Data
A catalogue record for this book is available from the British Library

ISBN: 978-0-12-814866-2

For information on all Elsevier publications visit our website
at https://www.elsevier.com/books-and-journals

Publisher: Susan Dennis
Acquisition Editor: Kathryn Eryilmaz
Editorial Project Manager: Sara Pianavilla
Production Project Manager: Kamesh Ramajogi
Cover Designer: Mark Rogers

Typeset by TNQ Technologies

Working together to grow libraries in developing countries

www.elsevier.com • www.bookaid.org

Contents

Spectrophotometry and its application in chemical analysis

1.1 Spectroscopy and applications (overview)

Spectroscopy is a branch of science (analytical chemistry) which deals with the study of the interaction of electromagnetic radiation with matter. In fact, traditionally, the interactions of analyte were between matter and electromagnetic radiation, but now spectroscopy has been broadened to include interactions between matter and other forms of energy. Such examples include beams of particles such as ions and electrons. These kinds of analytical methods that are considered to be one of the most powerful tools available for the study of materials' fundamental properties (e.g., atomic and molecular structure, optical properties) and also used in quantifying the wide range of chemical species prevailing in a given sample. In this method, an analyst carries out measurements of light (or light-induced charged particles) that is absorbed, emitted, reflected or scattered by an analyte chemical or a material. Then these measured data are correlated to identify and quantify the chemical species present in that analyte. Ideally, a spectrometer makes measurements either by scanning a spectrum (point by point) or by simultaneous monitoring several positions in a spectrum; the quantity that is measured is a function of radiant power.

Specifically, over all the other analytical methods, the spectroscopic techniques possess the following advantages:

1. These techniques are less time consuming and much more rapid.
2. They require a very small amount (at mg and μg levels) of the compound and even this amount can be recovered at the end of evaluation in many cases.
3. The structural information received from the spectroscopic analysis is much more accurate and reliable.
4. They are much more selective and sensitive and are extremely valuable in the analysis of highly complex mixtures and in the detection of even trace amounts of impurities.
5. Controlled Analysis can be performed on a computer, and therefore, continuous operation is possible which is often required in industrial applications.

A wide array of different spectroscopic techniques can be applied in virtually every domain of scientific research - from environmental analysis, biomedical sciences and material science to space exploration endeavors. In other words, any application that deals with chemical substances or materials can use this technique: Spectro-chemical methods have provided perhaps the most widely used tools for the elucidation of molecular structure as well as the quantitative and qualitative determination of both inorganic and organic compounds. For example, in biochemistry; it is used to determine enzyme-catalyzed reactions. In clinical applications, it is used to examine blood or tissues for clinical diagnosis.

Chemical Analysis and Material Characterization by Spectrophotometry. https://doi.org/10.1016/B978-0-12-814866-2.00001-4

A chemist routinely employs spectroscopic techniques for determination of molecular structure (e.g., NMR Spectroscopy), molecular weight, molecular formula and decomposition to simpler compounds or conversion into a derivative (MS Spectroscopy) and presence or absence of certain functional groups (IR Spectroscopy). Also, there are tremendous efforts in improving (e.g., instruments' resolution, detection limits, etc) and expanding this branch of the analytical method for quantitative analysis in various fields such as chemistry, physics, biochemistry, material and chemical engineering, clinical applications and industrial applications.

This book aims to cover chemical analysis and material characterization with this technique, this chapter aims to build a foundation for the book by providing properties of EM and the processes which arise after interaction with matter.

1.2 Classification of spectroscopic techniques

Methods of spectroscopy can be classified according to the type of analytes they are being analyzed or type of light that they employ. For stance, on the basis of type of analyte (elemental or molecular), it is divided into the following two heads:

1. Atomic spectroscopy: This kind of spectroscopy is concerned with the interaction of electromagnetic radiation with atoms which are commonly in their lowest energy state, called the ground state.
2. Molecular spectroscopy: This spectroscopy deals with the interaction of electromagnetic radiation with molecules. The interaction process results in a transition between rotational and vibrational energy levels in addition to electronic transitions. The spectra of molecules are much more complicated than those of atoms, as molecules undergo rotations and vibrations besides electronic transitions. Molecular spectroscopy is of great importance nowadays due to the fact that the number of molecules is extremely large as compared with free atoms.

Alternatively, the spectroscopic techniques are also classified according to the type of radiation they employ and the way in which this radiation interacts with matter. These methods include those that use from radio wave to Gamma-rays and causes to change from nuclear spin to change in nuclear configuration (see Table 1.1). On this basis, spectroscopic methods are listed below.

- (i) **Gamma-ray emission spectroscopy:** Uses light over the Gamma-ray range (0.005−1.4 Å) of electromagnetic radiation spectrum
- (ii) **X-Ray absorption/emission/fluorescence/diffraction spectroscopy:** Uses light over the X-ray range (0.1−100 Å)
- (iii) **Vacuum ultraviolet absorption spectroscopy:** Uses light over the vacuum ultraviolet range of (10−180 nm)
- (iv) **Ultraviolet−visible absorption/emission/fluorescence spectroscopy**: Uses light over the ultraviolet range (180−400 nm) and visible range (400−780 nm).
- (v) **Infra-red absorption spectrophotometry:** Uses light over the infrared range (0.78−300 μm).
- (vi) **FT-IR spectroscopy:** (0.78−300 μm)
- (vii) **Raman scattering spectroscopy:** (0.78−300 μm)
- (viii) **Microwave absorption spectroscopy:** Uses light over the infrared range of (0.75−375 mm)
- (ix) **Electron spin resonance spectroscopy:** Uses the light of (3 cm)
- (x) **Nuclear magnetic resonance spectroscopy:** Uses light over the infrared range (0.6−10 m)

Table 1.1 Regions of the electromagnetic spectrum and the most important atomic or molecular transitions pertinent to the successive regions.

	Region	Limits	Wave number limits (cm^{-1})	Frequency limit (Hz)	Molecular transitions (process)
1	X - rays	0.01−100 Å		10^{20}−10^{16}	K and L shell electrons
2	Far UV	10−200 nm		10^{16}−10^{15}	Inner (middle) shell electrons
3	Near UV	200−400 nm		10^{15} - $7.5 * 10^{14}$	Valence electrons
4	Visible	400−750 nm	25000−13000	$7.5*10^{14}$−$4.0*10^{14}$	Valence electrons
5	Near IR	0.75−2.25 μm	13000−4000	$4*10^{14}$−$1.2*10^{14}$	Molecular vibrations
6	Mid IR	2.5−50 μm	4000−200	$1.2*10^{14}$−$6.0*10^{12}$	Molecular vibrations
7	Far IR	50−1000 μm	200−10	$6.0*10^{12}$ - 10^{11}	Low lying vibrations and molecular rotations
8	Microwaves	1−1100 cm	10−0.01	10^{11}−10^{8}	Molecular rotations
9	Radio waves	1−1000 m		10^{8}−10^{5}	Nuclear spin

Each of these instruments consists of at least three essential components: (1) a source of electromagnetic radiation in the proper energy region, (2) a cell that is highly transparent to the radiation and that can hold the sample, (3) Grating: A holographic grating that disperses the radiation allowing a very precise selection of wavelengths and (4) a detector that can accurately measure the intensity of the radiation after it has passed through the analyte in a sample cell (The beam is focused on the center of the sample compartment to allow maximum light throughput and reduce noise).

The spectroscopic techniques of type (iv) and (v); Ultraviolet−visible absorption/emission/fluorescence spectroscopy and Infra-red absorption spectrophotometry are also named (sub-classified) as *spectrophotometry*. As each compound uniquely absorbs, transmits, or reflects light over a certain range of wavelength, the spectrophotometric method is mainly used to measure how much a chemical substance absorbs, transmits or emits radiation and to correlate the absorbed or emitted radiation with the quantity of an analyte of interest. Therefore, spectrophotometry is a spectro-analytical method for both the qualitative and quantitative measurement of the transmission (or absorption), reflection and emission properties of a chemical species (or material) as a function of wavelength.

1.3 Introduction to electromagnetic radiation
1.3.1 Fundamental properties of EM

Electromagnetic radiation (EM) is composed of a stream of mass-less particles (called photons) each traveling in a wave-like pattern at the speed of light. Each photon of EM possesses a certain amount of energy. The type of radiation is defined by the amount of energy found in the photons and exhibits properties of both the particle and wave, known as the wave-particle duality, and comprises electric and magnetic fields. EM spectrum comprises radiation, ranging from radio waves to gamma-rays (see Fig. 1.1): Radio waves have photons with low energies, microwave photons have a little more energy than radio waves, infrared photons have still more, then visible, ultraviolet, X-rays, and, the most energetic of all, gamma-rays. The wavelength range of each kind of radiation and the process they can initiate after interaction with material is given in Table 1.1.

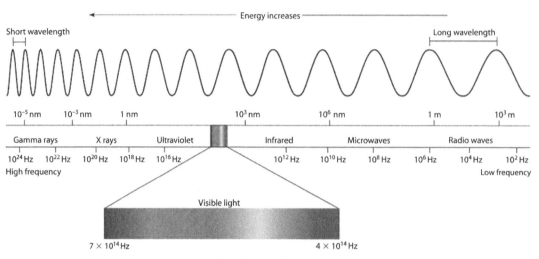

FIG. 1.1

Electromagnetic radiations of different wavelengths (upper section) and the effects after interaction of a photon of certain energy with a molecule (bottom section). For example, when photons of γ-rays interact with an atom or molecule, they can excite K-shell electrons.

In the wave model, electromagnetic radiation is characterized by its frequency, v, wavelength, λ, and velocity, c. These three values are related by the relationship

$$c = v\lambda. \tag{1.1}$$

The value of c is constant in a given medium (e.g., $c = 2.99\,9 \times 10^8$ ms^{-1} in vacuum), while the frequency and wavelength of light are inversely proportional to one another. The SI units for wavelength and frequency are the meter (m) and the hertz (Hz), respectively. Traditionally, spectroscopists also define electromagnetic radiation by the unit wave numbers as:

$$\underline{v} = \frac{1}{\lambda}\,[cm^{-1}], \tag{1.2}$$

where λ denotes the wavelength in centimeters.

The energy of a photon (quantum of electromagnetic radiation) depends solely on its frequency (or wavelength) and is defined as

$$E = hv = h\frac{c}{\lambda} = hc\bar{v}, \tag{1.3}$$

where h is Planck's constant ($h = 6.63\,9 \times 10^{-34}$ J). Note that energy is directly proportional to frequency and wave number, and inversely proportional to wavelength.

Example 1.1. Calculate the frequency of radiation whose wavelength is 600 nm. Express this wavelength in wave number.

Solution: Wavelength (λ) = 600 nm $\times$ 10 Å = 600 nm $\times$ 10 $\times$ 10^{-8} cm = 6 $\times$ 10^{-5} cm.

Now frequency $\vartheta = \frac{c}{\lambda} = \frac{3 \times 10^{10} \text{ cm sec}^{-1}}{6 \times 10^{-5} \text{ cm}} = 5 \times 10^{14}$ cycles per sec.

$$\bar{\vartheta} = \frac{1}{\lambda} = \frac{1}{6 \times 10^{-5} \text{ cm}} = 16666.67 \text{ per cm.}$$

Exercise 1.2. Calculate the wave number of the radiation if the frequency is 2.06×10^{14} Hz. (Given: $c = 3 \times 10^{10}$ cm per sec.)

$$\bar{\vartheta} = \frac{\vartheta}{c} = \frac{2.06 \times 10^{14}}{3 \times 10^{10} \text{ cm}} = 6866 \text{ per cm.}$$

1.3.2 Light-matter interaction

What happens when light meets matter? When light meets matter, there is always an interaction: For example, light is refracted when it enters the glass, reflected off the surface of water or ice, partially absorbed and partially reflected by a green leaf, and generates photo-current by exciting the semi-conductor of a solar cell. The details depend on the structure of the matter and on the wavelength of the light. Additional phenomena are refraction, diffraction and fluorescence. In the following consecutive chapters, we will discuss in detail some of these processes and their manifestations. For example, the fluorescence spectroscopy makes use of light that is released by matter (analyte), with a detector examining how this radiation is released by chemicals in the analyte sample.

1.3.2.1 Absorption of light

As mentioned above, a way in which matter can interact with light is through absorption. Absorption is the process in which energy transfer from a photon of EM radiation to the analyte's atoms or molecules takes place. The general processes which occur during light absorption and emission are shown in (Fig. 1.2). Chemical species that is at low energy state move to a higher energy state by absorption of light.

The general process of absorption can be understood as follows. Atoms and molecules contain electrons. It is often useful to think of these electrons as being attached to the atoms by springs. The electrons and their attached springs have a tendency to vibrate at specific frequencies. Similar to a tuning

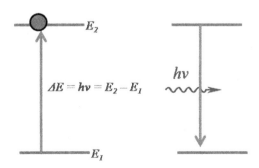

FIG. 1.2

Pictorial demonstration of fundamental concepts related to absorption and emission of light.

fork or even a musical instrument, the electrons of atoms have a natural frequency at which they tend to vibrate. When a light wave with that same natural frequency impinges upon an atom, then the electrons of that atom will be set into vibrational motion. If a light wave of a given frequency strikes a material with electrons having the same vibrational frequencies, then those electrons will absorb the energy of the light wave and transform it into vibrational motion. During its vibration, the electrons interact with neighboring atoms in such a manner as to convert its vibrational energy into thermal energy. Subsequently, the light wave with that given frequency is absorbed by the object, never again to be released in the form of light. So the selective absorption of light by a particular material occurs because the selected frequency of the light wave matches the frequency at which electrons in the atoms of that material vibrate. Since different atoms and molecules have different natural frequencies of vibration, they will selectively absorb different frequencies of visible light. This statement is illustrated with the help of the absorption spectrum of hydrogen gas. As shown in Fig. 1.3, when exposed to a photon of electromagnetic radiation, hydrogen atom absorbs it and is in what is called an "excited" state. As this is not the natural state of an atom or molecule, the electron will eventually drop back down to the lower energy (ground state). However, the atom has to lose energy to do this, and so it releases a photon of the same energy as the one is absorbed. This process is called emission because a photon of radiation is emitted by the excited atoms, molecules or solids again at a very specific wavelength.

For atoms excited by a high-temperature energy source this light emission is commonly called *atomic* or *optical emission* and for atoms excited with light it is called *atomic fluorescence* (see fluorescence spectroscopy). For molecules, it is called *fluorescence* if the transition is between states of

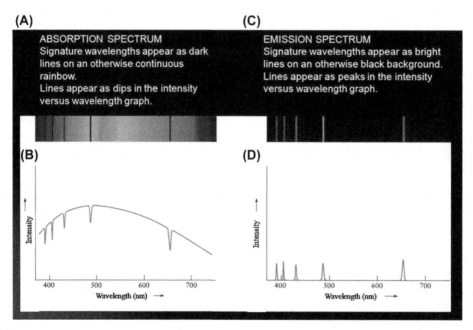

FIG. 1.3

Absorption and emission spectra of neon gas. Also, for a particular analyte, the emission intensity of an emitting substance is linearly proportional to analyte concentration at low concentrations. Atomic emission and molecular fluorescence are therefore useful for quantifying emitting species.

the same spin and *phosphorescence* if the transition occurs between states of a different spin (see in the chapter 9 for details).

We have been discussing one specific transition or "energy jump" in one atom, but of course, in any physical system, there are many atoms. In a hydrogen gas, for example, all of the separate atoms could be absorbing and emitting photons corresponding to the whole group of "allowed" transitions between the various energy levels, each of which would absorb (or emit) at the specific wavelengths corresponding to the energy difference between the energy levels. This pattern of absorptions (or emissions) is unique to hydrogen (see Fig. 1.3): No other element can have the same pattern and causes a recognizable pattern of absorption (or emission) *lines* in a spectrum.

Extending this a bit, it should become clear that since every chemical element has its own unique set of allowed energy levels, each element also has its own distinctive pattern of spectral absorption (and emission) lines! (See diagram below (See Fig. 1.3) for hydrogen). It is this spectral "fingerprint" that astronomers use to identify the presence of the various chemical elements in astronomical objects. Spectral lines are what allow us, from a "spectrum," to derive so much information about the object being observed!

Exercise: 1.3. **(a)** In the absorption spectrum of KMnO$_4$ (shown in Fig. 1.4), what wavelength of light is most strongly absorbed by KMnO$_4$? What wavelengths are the most easily transmitted? Which wavelengths would you select if you wished to use light absorption to measure the KMnO$_4$? What type of light at this wavelength (UV, visible or IR)?

2. The interactions of light with chlorophyll are used in remote sensing to examine the plant and algae content of the land and sea.

(a) In the absorption spectrum for chlorophyll, what wavelength of light is most strongly absorbed by chlorophyll *a* and *b*? What wavelengths are the most easily transmitted?

(b) Which wavelengths would you select if you wished to use light absorption to measure the chlorophyll?

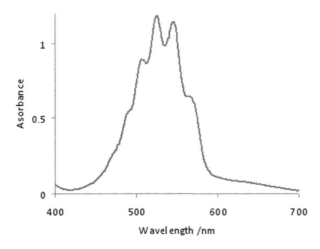

FIG. 1.4

The absorption spectrum for KMnO4.

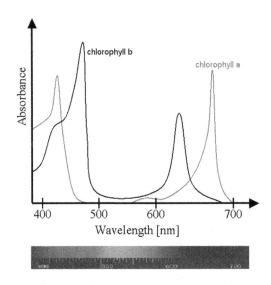

Absorbance as a function of radiation wavelength for chlorophyll 'a' and chlorophyll 'b'.

Solution:

The strongest absorption of light for chlorophyll 'a' occurs at about 450 nm and 660 nm. The strongest absorption of light for chlorophyll 'b' takes place at roughly 460 nm and 635 nm. Wavelengths between 500 and 600 nm have the greatest degree of transmittance by both chlorophylls 'a' and 'b'. Small differences in these wavelength ranges are present for these two types of chlorophyll because of their different chemical structures, which create slight differences in their energy levels and in the types of light they can absorb.

(a) The measurement of chlorophyll a and b would be best performed by using the wavelengths at which these pigments have their strongest absorption of light 435 nm or 660 nm for chlorophyll a, and 460 nm or 635 nm for chlorophyll b, which represent visible light.

If you look closely at the spectra of chlorophyll a and b (Fig. 1.5), we will notice that it is the light that is not absorbed by chlorophyll a and b between 500 and 600 nm that gives these pigments and leaves their green and yellow color: the color of the absorbing species is determined by the remaining types of light that are transmitted (or reflected) by the object. For instance, the passage of white light through a blue solution of copper sulfate indicates that blue light is being transmitted while the complementary (orange in this case) is absorbed.

1.3.2.2 Transmission and reflection (visible light)

Transmission: As a result of the absorption, the intensity of this light, after passing through the sample will be lower than its original value at the energy that was absorbed by the sample. The remaining light that leaves through the sample is said to have undergone transmission. In other words, the transmission is defined as the passage of radiation through matter with no change in energy taking place. The amount of light that is transmitted by a sample plus the amount of light that is reflected (or scattered) and absorbed by the sample will be equal to the total amount of light that is originally entered the sample. A plot of the intensity of the light that is transmitted by a sample at various wavelengths, frequencies, or energies is called a transmittance spectrum.

The mechanism for transmission of the incident light to the other side of a transparent material is understood as follows: Transmission (or reflection) of light waves occurs because the frequencies of the light waves do not match the natural frequencies of vibration of the objects. When light waves of these frequencies strike an object, the electrons in the atoms of the object begin vibrating. But instead of vibrating in resonance at large amplitude, the electrons vibrate for brief periods of time with small amplitude of vibration; then the energy is reemitted as a light wave. If the object is transparent, then the vibrations of the electrons are passed on to neighboring atoms through the bulk of the material and reemitted on the opposite side of the object. Such frequencies of light waves are said to be *transmitted* (this process will be discussed in the consecutive chapter in detail). Pure water is a classic example of an almost completely transparent medium for visible light. An eye exhibits several portions of tissue that are more or less transparent, such as the cornea, crystalline lens, aqueous humor, and the vitreous body, as well as the inner layers of the retina. A medium is always transparent only to a certain rather than the whole part of the electromagnetic spectrum. For example, water is opaque to radiation in the infrared range, while the cornea blocks radiation in the ultraviolet range.

Reflection: *Reflection* of electromagnetic radiation is the change in direction of a wavefront at an interface between two different media so that the wavefront returns into the medium from which it originated (see Fig. 1.6). In particular, when the waves of radiation encounter a surface or other boundary between two regions that have different refractive indices and bounces the radiation waves back to the medium in which it was originally traveling. Alternatively, it can be defined as follows. If the object is opaque, lightwave of unmatched frequencies with the natural frequencies of an electron, then the vibrations of the electrons are not passed from atom to atom through the bulk of the material. Rather the electrons of atoms on the material's surface vibrate for short periods of time and then reemit the energy as a reflected light wave. Such frequencies of light are said to be *reflected*. Common examples include the reflection of light by a mirror.

When light is reflected, its characteristics and properties may not be the same. How light is affected by matter depends on the strength of the field of the light, its wavelength, and the matter itself. In addition, external influences on the matter, such as temperature, pressure, and other external fields (electrical, magnetic) influence the interaction of light with matter.

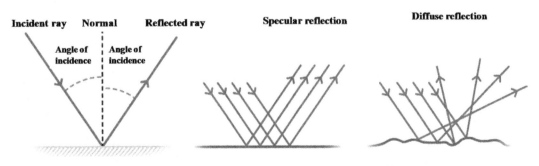

FIG. 1.6

Schematic diagram for reflection from a smooth mirror surface (left) and irregular surface (right).

Similar to absorbed or emitted light reflected light from a sample gives information about it and has been used in some types of spectroscopy in which reflected light from an analyte is detected by the instrument's detector and analyzed. A good example in material science is an evaluation of optical properties such as refractive indices, absorption coefficient and thickness of the partially transparent thin film of materials (e.g., metals and semiconducting material) by analyzing the reflected light from the surface of such materials. Depending on the nature of the surface of the analyte, there are several types of reflections.

Specular Reflection: If the boundary between the two regions that causes the reflection is a flat plane (smooth surface), the reflected light will be in well-defined manner and will retain its original image such type of reflection is called *Specular reflection or regular reflection* (See Fig. 1.6). The fraction of the light that is reflected depends on the angle of incidence, the ratio of the refractive indices of the two media, as well as from the state of polarization of the incident light, but it does not depend on the color in most situations.

An example of such a process is reflection by plane, which mirrors smooth surface of water: In specular reflection, a mirror surface reflects a beam of light so that the angle of reflection is equal to the angle of incidence (e.g., the angle of the incoming light is the same as the angle of the outgoing/reflected light). The diagram shows how the incoming (incident) light is reflected off the mirror at the same angle to the perpendicular line (which in geometry is called the 'normal'). For a perpendicular incidence from air to glass (or for a perpendicular exit out of glass into air), approximately 4 % of the light is reflected. For the transition from air to water, this is approximately 2 %.

As shown in Fig. 1.6 (left), in a perfect reflection, all the light will be reflected from the mirror surface. In real-life, non-perfect situations, the base material of the mirror and the surface of the mirror may:

■ Absorb light: the light is absorbed (the energy of the light is taken up by the material).
■ Imperfectly reflect light: The surface is an imperfect mirror and some of the light is scattered.

Diffuse reflection: Rough surfaces, such as a piece of paper, reflect light back in all directions. This also occurs when sunlight strikes the wall of a house or the green leaf of a plant. Thanks to the diffuse character of the reflection, we see the illuminated object from every angle. Diffuse reflection occurs when an incident ray of light strikes a surface and the light is scattered. As shown in Fig. 1.6 (right), in perfect or ideal diffuse reflection, all the light will be perfectly distributed in a hemisphere of even illumination around the point the light strikes the diffusion surface. The diagram shows the way light is scattered by a diffusion surface. Although the diagram is only two dimensional light scatter forms a hemisphere around a light strike-point.

This general picture will now be made more precise. The most obvious is the phenomenon of the color of the reflecting surface. The wall of a house, being illuminated by the sun, appears white when its paint reflects all wavelengths of the incident light completely. The yellow color of a sunflower arises through the absorption of blue: together, the remaining green and red produce the perception of yellow. If a surface partially absorbs all the spectral portions of the light uniformly (50 % of it, for example), it appears to be gray, that is, without any color.

As indicated in the I[st] paragraph of this section, the degree to which light will be reflected at a boundary will depend on the relative difference in the refractive indices for the two sides of the boundary. The larger this difference, the greater the fraction of the light that will be reflected. This idea is illustrated by Eq. (1.4) (the Fresnel equation), which gives the fraction of light that will be reflected as it inters the boundary at a right angle.

$$\frac{P_r}{P_0} = \frac{(n_2 - n_1)^2}{(n_2 + n_1)^2} \tag{1.4}$$

The symbol P_0 in this Eq. (1.4) represents the incident radiant power (original) of the light (in the units of Watts), which is defined as the energy in a beam of light that strikes a given area per unit time, P_r denotes the radiant power of the reflected light where the ratio $\frac{P_r}{P_0}$ gives the fraction of the original light versus the reflected light. For boundaries that have only a small difference in refractive indexes, such as between the vacuum in space and air, the fraction of reflected light will be small and most of the light will pass through the boundary and into the new medium. If a large difference in refractive index is present, as occurs between air and the silver-coated surface of a mirror, a large fraction of light will be reflected. An example of a completely white surface is provided by snow. Its white color has a simple explanation: the tiny ice crystals reflect the light without any absorption.

Exercise 1.4. Some sensors on the Terra Satellite make use of reflection patterns to map the surface of the Earth.

(a) If a beam of light passes through the air (n = 1.0003) and strikes the smooth surface of the water (n = 1.333) at a right angle, what fraction of this light will be reflected by the water back into the air?

(b) If this beam of light strikes the water at an angle of 65.0, what will be the angle of reflection?

Solution:

(a) $n_1 = 1.0003$, $n_2 = 1.333$

$$\frac{P_r}{P_0} = \frac{(1.333 - 1.0003)^2}{(1.333 + 1.0003)^2} = 0.0203 \text{ (or 2.03\% reflection)}$$

(b) If the light is undergoing perfect regular reflection, it will be reflected at an angle of 65.0 on the other side of the normal from the incoming light. If the surface of the water is rough and diffuse reflectance instead occurs, the light will be reflected at many different angles.

1.3.2.3 Refraction of light

When a beam of light meets a smooth interface between two transparent media that have different refractive indices, both reflection and refraction occur (Fig. 1.7). The refraction of light is the basis for the optical imaging through the crystalline lens, eyeglasses, and optical instruments (e.g., magnifying glasses, microscopes, and refractive telescopes).

The incident ray of light onto a surface, refracted and reflected rays, and the surface normal all lie in the same plane (Fig. 1.7). The amount of light refracted depends on the ratio of the refractive indices of the two media. The relationship between the two angles α *and* β is specified by the law of refraction:

$$\sin \alpha \, / \, \sin \beta \, = \, n_2 \, / \, n_1. \tag{1.5}$$

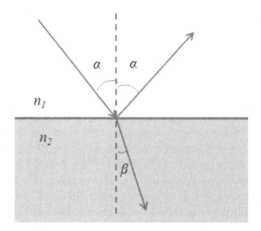

FIG. 1.7

Refraction at the interface of two media. The primary ray is partially reflected and partially refracted. α and β are the angles of incidence and refraction with respect to the surface normal, respectively.

Keeping in mind that light is scattered when it encounters an obstacle, the existence of transparent media such as glass, water, corneas, crystalline lenses, and air seems quite miraculous. Inside these media, interactions between the light and the materials still occur, but it only leads to the light's traveling more slowly than it would in a vacuum (Refraction is a consequence of the differing speeds of light in two media). To understand this, we first note that the frequency of the light vibrations remains the same in both media. Therefore, inside the medium with the slower light speed, the wavelength is smaller since the light moves one wavelength further during one period. This slowing down is quantified as the refractive index n: the velocity of light in the medium amounts to c/n, where c is the velocity of light in vacuum ($c \gg 300{,}000$ km/s). For example, in water, light travels with a velocity of c *is about* 225,000 km/s ($n = 1.33$).

1.3.2.4 Dispersion of light

The refractive index of a transparent medium is slightly dependent on the wavelength and increases with shorter wavelengths. This gives rise to dispersion during refraction, i.e., to a breaking up of white light into various colors, as we mainly know from prisms or crystals. The colors of the rainbow are also based on the dispersion in water droplets. Newton was interested in the chromatic aberration of the human eye. Its focal plane for blue light lies approximately 1 mm in front of that for a red light. It is amazing that we perceive the chromatic error of our eyes only under rare conditions, even though the difference between the refractive power for red and blue light amounts to ca. 1.5 D.

1.3.2.5 Light scattering in media

The term scattering refers to the change in travel of one particle (such as photon) due to its collision with another particle (e.g., an atom or molecule). One common type of scattering is 'Rayleigh scattering'. The mechanism for scattering is understood as: An isolated atom scatters light because the electric field of the incident light wave forces the electrons in the atom to oscillate back and forth about their equilibrium position. By the laws of electromagnetism, when a charge changes its velocity, it emits radiation. Light is emitted uniformly in all directions in the plane perpendicular to oscillation but decreases in amplitude as the viewing angle shifts away from that plane.

An outer surface does not always scatter back all the light that penetrates it. Instead, the scattering can also take place deeper inside the interior of the medium. Among numerous examples, the blue of the sky is the most well known: the air molecules scatter sunlight and mainly the shorter wavelengths. Without the scattering, the sky would appear black to our eyes and light would come into our eyes when looking directly at the sun (The blue portion of the sunlight is approximately six times more strongly scattered than the red portion). A glass of beer absorbs light of short wavelengths and scatters light with longer wavelengths in all directions. The scattering of light by particles depends on the particle size. The scattering due to particles that are considerably smaller than the light wavelength is known as Rayleigh scattering. Rayleigh scattering occurs in all directions. The best-known example is the scattering of sunlight by the molecules of the atmosphere.

For larger particles, e.g., from atmospheric pollution, with diameters on the order of light wavelengths or larger, the scattering takes place mainly in the forward direction, and it is less color-dependent (called Mie scattering). The scattered light loses the blue dominance of the Rayleigh scattering and becomes increasingly whiter with the increasing diameter of the scattering particles. A nice manifestation of the forward direction of the scattering of sunlight on atmospheric particles is the whitish appearance of the sky near the sun: The scattering due to water droplets and atmospheric pollutants (aerosols, salt appearance of heavy clouds) derives from the fact that the light coming from above is mainly scattered and reflected back upward, while only a small part passes through.

1.3.2.6 Diffraction of light

Diffraction is the slight bending of light as it passes around the edge of an object. The amount of bending depends on the relative size of the wavelength of light to the size of the opening. If the opening is much larger than the light's wavelength, the bending will be almost unnoticeable. However, if the two are closer in size or equal, the amount of bending is considerable, and easily seen with the naked eye.

In the atmosphere, as shown in Fig. 1.8 diffracted light is actually bent around atmospheric particles: most commonly, the atmospheric particles are tiny water droplets found in clouds. Diffracted light can produce fringes of light, dark or colored bands. An optical effect that results from the diffraction of light is the silver lining sometimes found around the edges of clouds or coronas surrounding the sun or moon. The illustration in Fig. 1.8 shows how light (from either the sun or the moon) is bent around small droplets in the cloud.

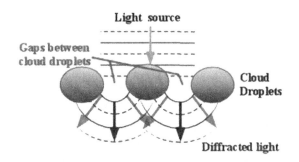

FIG. 1.8

Example of diffraction of light by atmospheric particles.

All forms of waves, including light waves, are diffracted when they encounter an edge or pass through a narrow opening — like water waves when passing through the entrance to the harbor. A diffraction image can also be understood as an effect of interference among the entity of the waves that extend from all the points of the opening. Strictly speaking, diffraction is less the consequence of certain interactions between light and matter and more the expression of an inner property of light: its wave nature. This concept also applies to light waves. When sunlight encounters a cloud droplet, light waves are altered and interact with one another. If there is constructive interference, (the crests of two light waves combining), the light will appear brighter. If there is destructive interference, (the trough of one light wave meeting the crest of another), the light will either appear darker or disappear entirely.

A curious phenomenon of diffraction was the object of controversy when the wave theory of light was being established. In 1818, Poisson pointed out that, as a consequence of the wave theory, a bright spot must appear in the center of a sphere's shadow because the waves originating from all the edges would arrive there in phase, irrespective of the position of the screen. To him, this seemed so absurd that he believed he had therefore disproved the wave theory. However, a short time afterward, the "Poisson's spot" was actually observed and became one of the pillars supporting the wave theory of light. Diffraction also has an influence on image formation in the eye.

Diffraction of X-Rays by a Crystal: It is possible to predict the angles at which constructive interference will be observed for X-Ray diffraction by the atoms in a crystal. This can be achieved by using the Bragg equation,

$$n\lambda = 2d \sin(\theta), \tag{1.6}$$

(Note: Eq. (1.6) can also be written as $n\lambda = 2d(nhkl) \sin(\theta)$, where h, k, and l are lattice parameters).

Where λ is the wavelength of the X-rays passed through the crystal, d is the interplanar distance (or lattice spacing) between atoms in the crystal, and n is the order of diffraction for the observed constructive interference band (e.g., $n = 1$ for first-order interference). The term θ represents the specific angle at which the X-Rays will strike the crystal surface and produce, on the other side of the normal and at the same angle, a diffraction band for the given order of constructive interference.

Exercise 1.5. If X-Rays with a wavelength of 0.711 Å are diffracted by a crystal of sodium chloride $(d = 2.82$ Å), at what angle will first-order constructive interference be observed for this crystal?

Solution:

Rearranging Eq. (1.5), and substituting the given values of n, d, and λ, we can find the term θ.

$$sin(\theta) = \frac{n\lambda}{2d} = \frac{n(1)(0.0711\ nm)}{(2)(0.2820\ nm)}$$

$$sin(\theta) = 0.1260r = 7.24^o$$

The same type of calculation predicts that additional bands of constructive interferences will be seen at angles of 14.6° $(n = 2)$, $(n = 3)$, and so on for higher orders of interference.

1.3.2.7 Emission (fluorescence/phosphorescence)

Fluorescent materials are utilized, for example, in laundry detergents. The visual effect produced from fluorescent material often originates from blue or UV light, which illuminates a material and, in turn, excites the emission of yellow or orange light to which our eyes are especially sensitive. The

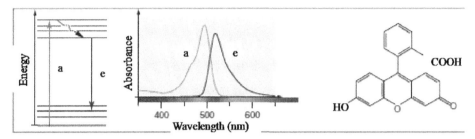

FIG. 1.9

Fluorescein. Left: transition scheme. Middle: absorption spectrum (a) and emission (e) spectrum. An absorption maximum at 485 nm, emission maximum at 514 nm. Right: formula of fluorescein.

conversion always happens toward the longer wavelength and, thus, in the direction of decreasing photon energy because some energy is converted into molecular vibrations (heat). This behavior (absorption of short-wave light, emission of longer-wave light) is termed fluorescence (Fig. 1.9). The name stems from the mineral fluorite. This phenomenon can occur in organic materials and also in minerals. If we irradiate minerals with ultraviolet light, we can ascertain that individual mineral samples shine more or less brightly in various colors. Fluorescent materials often produce light weakly.

The amount of light that is emitted by a sample will be directly related to the concentration of the atoms or molecules in the sample that are creating this emission. We can represent this relationship through the following equation,

$$P_E = kCP_E, \tag{1.7}$$

where P_E represents the radiant power of the emitted light, C is the concentration of species emitting this light, and k is the proportionality constant. Eq. (1.7) can be used to estimate the concentration of the atoms or molecules that are in excited states, after interaction with radiation or other excitation sources (e.g., thermal). Details of methods for such analysis will be discussed in chapter 9.

1.4 Questions
1.4.1 Subjective problems

1. Describe three ways that light can interact with matter.
2. Transmitted light may be refracted or scattered. When does each process occur?
3. Why does matter increase in temperature when it absorbs light?
4. Compare and contrast transparent, translucent, and opaque matter.
5. What wavelengths of light have the most intense emission when given off by the sun? Which wavelengths of light are taken up by (or absorbed) to the greatest extent by earth's atmosphere?

1.4.2 Numerical problems

1. Calculate the wave number of radiation whose wavelength is 4000 nm. $\left[Ans : \vartheta = 2500\ cm^{-1}\right]$
2. Calculate the wavenumber of the lines of frequency 4×10^{14} .

$$\left[Ans: \vartheta = 1.33 \times 10^4\ cm^{-1}\right]$$

3. Radiation with a wavelength of 700 nm is visible red light
   (a) Calculate the wavelength in $\overset{\circ}{A}$
   (b) What is the wavenumber and frequency of this radiation?

$$\left[Ans: \lambda = 7000 \text{ Å}, \quad \underline{\vartheta} = 1.43 \times 10^6 \ m^{-1}, \quad \vartheta = 4.28 \times 10^{14} \ s^{-1}\right]$$

4.
   (a) What is the energy of a photon that has a wavelength of 500 nm, if n = 1.00?
   (b) What is the energy contained in one mole of photons with this wavelength?

$$\left[Ans: (a) Energy \ of \ a \ photon = 3.98 \times 10^{-19} \ J, \quad (b) Energy \ of \ one \ mole \ of \ p = 2.4 \times 10^5 \ J = 240 \ kJ\right.$$

5. One application of emission in remote sensing is in the detection of forest fires. One of the kinds of satellite Landsat has a sensor that detects active fires by examining their emission in the wavelength region of 1.55 μm−1.75 μm.
   (a) What type of light (visible, ultraviolet, etc) is it?
   (b) What are the energies of single photons of light at these wavelengths?

[Hint: IR region. Ans: (b) energies of single photons = 1.2×10^{-19} J]

Further reading

[1] D.A. Skoog, F.J. Holler, S.R. Crouch, Instrumental Analysis, Cengage Learning India Pvt. Ltd., New Delhi, 2012, p. 158.
[2] Guide for use of terms in reporting data: spectroscopy nomenclature, Anal. Chem. 62 (1990) 91−92.
[3] https://www.nist.gov/programs-projects/spectrophotometry, visited on: 7/23/2017.
[4] M.G. Gore, Spectrophotometry and Spectrofluorimetry: A Practical Approach, Oxford University Press, 2000.
[5] G.S. Jeffery, J. Basset, J. Mendham, R.C. Denney, Vogel's Textbook of Quantitative Chemical Analysis, fifth ed., 1991.
[6] B.K. Sharma, Instrumental Methods of Chemical Analysis, Krishna's Education Publishers, 2012.
[7] Ref. For the Absorption Spectrum of Hydrogen Spectrum (Or Simple Absorption and Emission).
[8] Ref. For the Absorption Spectrum of KMnO$_4$.
[9] J. Inczedy, T. Lengyel, A.M. Ure, Compendium of Analytical Nomenclature, third ed., Blackwell Science, Malden, MA, 1997, p. 433. From Hage's Book.
[10] D.S. Hage, J.D. Carr (Eds.), Analytical Chemistry and Quantitative Analysis, International, Pearson Education, Inc., New Jersey, USA, 2011.
[11] http://blair.pha.jhu.edu/spectroscopy/basics.html, Visited: 07/14/2017.
[12] A.T. Young, Rayleigh scattering, Appl. Opt. 20 (1981) 522−535.

Theory and instrumentation of absorption spectroscopy: UV−VIS spectrophotometry and colorimetry

2.1 Absorption (UV−VIS) spectrophotometric measurements: general concept

Traditionally, the light from 185 nm to 760 nm is generally called ultraviolet−visible (UV−VIS) region. However, with the advancement of technology, the modern spectrophotometers come with the extended wavelength range from 185 nm−2100 nm. In order to know what wavelength of light is absorbed by an analyte contained in a given sample, we must know (or determine) the UV−VIS absorption spectrum of the standard sample of which we wish to study.

As mentioned in the previous chapter, most often in spectrophotometric methods, the electromagnetic radiation that is provided by the instrument is absorbed by the analyte, and the amount of absorption (or transmittance) is measured. The radiation power (i.e., the energy, in the form of electromagnetic radiation, transferred across a unit area per unit time) of the incident radiation decreases as it passes through the sample. The amount of decrease in radiation power is proportional to the analyte concentration. In other words, essentially, the amount of absorbed radiation increases with the concentration of the analyte and with the distance through the analyte that the radiation must travel (the cell path length).

Application in quantitative analysis: As radiation is absorbed in the sample, the power of the radiative beam arriving at the photon detector decreases. By measuring the decreased power through a fixed-path-length cell containing the sample, it is possible to determine the concentration of the sample. Because different substances absorb at different wavelengths, the instruments must be capable of controlling the wavelength of the incident electromagnetic radiation. In most instruments, this is accomplished with a monochromator. In other instruments, it is done by use of radiative filters or by use of sources that emit radiation within a narrow wavelength band.

Application in qualitative analysis: Also, because the wavelength at which substances absorb radiation depends on their chemical composition, spectrophotometry can also be used for qualitative analysis. The analyte is placed in the cell, and the wavelength of the incident radiation is scanned throughout a spectral region while the transmission (or absorption) is measured. The resulting plot of transmitted radiation or absorbance as a no instruction function of wavelength or energy of the incident radiation is called transmittance or absorption spectrum. The wavelength(s) at which peak(s) (maximum absorbance) are observed are used to identify components of the analyte.

Chemical Analysis and Material Characterization by Spectrophotometry. https://doi.org/10.1016/B978-0-12-814866-2.00002-6

FIG. 2.1

UV—VIS Spectrophotometer with a double beam in space (Shimadzu: UV-1800).

Within ranges of light, calibrations are needed on the machine using standard samples (samples of analyte with known concentration).

As an example, Fig. 2.1 shows a UV/Visible spectrophotometer (product of Shimadzu company, Japan) which is widely used in academia, research and industrial quality assurance. In particular, it can measure optical absorbance, transmittance or extinction coefficient from ultraviolet to near IR region (190—1100 nm) of electromagnetic radiation with a resolution of 1 nm with user user-friendly features.

The UV—VIS Spectrophotometers available in the market can be used either as a stand-alone instrument or as a PC-controlled instrument, comprising of the following knobs to ease the measurement process.

1. On/Off switch
2. Wavelength selector/Readout
3. Sample chamber
4. Zero transmission adjustment/blank adjustment (or auto Zero) knob
5. Measurement mode: Absorbance, Transmittance, Extinction coefficient
6. Measurement type: (i) at a broad wavelength range (Spectrum), (ii) at a single wavelength
7. Absorbance/Transmittance/Extinction coefficient scale

2.2 Principle of absorption spectroscopic measurements

2.2.1 Absorption and emission processes (concept from quantum physics)

In atomic or molecular absorption spectroscopy, following terms and mathematical relations are used to quantify the analyte concentration, to study about the structural features, optical properties, and the functional groups present in a molecule of interest, and so on.

Any atomic or molecular system possesses energy only in certain specific amounts or quanta. These are referred to as energy levels of system. A set of quantized energy levels is shown in Fig. 2.2A.

From Fig. 2.2, it can be noticed that in order that system may go from energy level E_1 to E_2, it will require the absorption of an amount of energy equal to

$$\Delta E = E_2 - E_1 \tag{2.1}$$

This energy can be provided by electromagnetic radiation of the proper frequency if the radiation can interact with the system. Thus the energy of the radiation absorbed by the system is related to the energy difference between quantized energy levels or,

$$\Delta E = h\vartheta; Bohr\ condition. \tag{2.2}$$

However, if the system were to undergo a change from state E_4 to E_3 then there would be an emission of energy equal to

$$\Delta E' = E_4 - E_3. \tag{2.3}$$

And the frequency of the radiation emitted would be given by

$$\vartheta' = \frac{\Delta E'}{h} = \frac{E_4 - E_3}{h} = \frac{E_{excited\ state} - E_{ground\ state}}{h} \tag{2.4}$$

The emission of radiation occurs when a molecule or an atom in higher energy state (called excited state) returns to a lower energy state (or ground state).

However, in order to observe and interpret a spectrum we must keep three fundamental conditions in mind:

1. *The Bohr Quantum Condition:* When a system makes a transition between quantized energy levels, the frequency of the radiation absorbed or emitted is given by the conditions (see Eqs. 2.1 and 2.4).
2. *Electric dipolar interaction:* In an atom or molecule, if absorption or emission of one photon of radiation, the dipole moment is either created or destroyed due to electron charge redistribution. In such a situation, the electronic transitions take place. On the other hand, if there is no net change in the dipole moment on redistribution of electronic charge (i.e., charge distribution remains symmetric), the transition does not take place and it is said to be forbidden transition. Thus, on the absorption of light, a dipole is either created or destroyed in its excited state. In case of molecules a g-state (g: gerade) changes to u-state (u: ungerade) or vice versa. This means, in the case of a molecule, transition from the ground state to an excited state is allowed only when the transition is from bonding molecular orbital to antibonding molecular orbital and vice versa. That is, $u \leftrightarrow g$; *bonding MO $\leftrightarrow$ antibonding MO.*
3. *Selection rules:* The creation or destruction of dipoles restrict the transitions that occur between energy levels. These selection rules are a consequence of the symmetric properties of the wave function in two energy states. For example, the atomic wave functions or orbitals *s, p, d, f* are alternatively symmetric anti-symmetric with respect to the inversion about the origin of the system. The selection rules for allowed electronic transitions for light diatomic molecules:

$$\wedge = 0, \pm1, \Delta \sum = 0.$$

[Note: $\Delta\Sigma = 0$ mean spin state. All transitions are singlet, doublets and so on. Inter combination singlet-Triplet and similar transitions are forbidden].

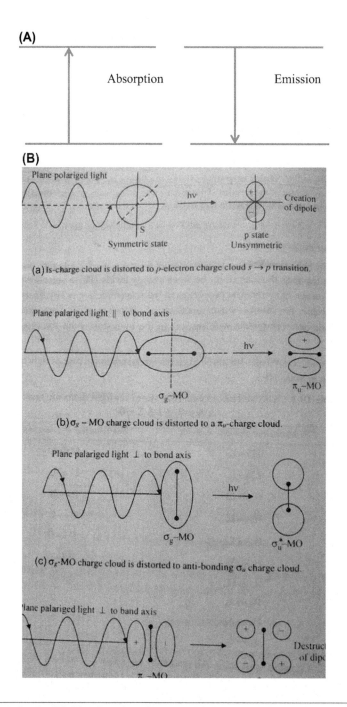

(A)

Absorption

Emission

(B)

Plane polariged light

hν

Creation of dipole

S

p state
Unsymmetric

Symmetric state

(a) Is-charge cloud is distorted to p-electron charge cloud $s \rightarrow p$ transition.

Plane palariged light ‖ to bond axis

hν

π_u–MO

σ_g–MO

(b) σ_g – MO charge cloud is distorted to a π_u-charge cloud.

Plane palariged light ⊥ to bond axis

hν

σ_g–MO

σ_u^*–MO

(c) σ_g-MO charge cloud is distorted to anti-bonding σ_u charge cloud.

Plane palariged light ⊥ to band axis

Destruct of dipc

π –MO

FIG. 2.2

(A): Quantized energy levels. (B): The distortion of charge cloud due to radiation.

2.2.2 Molecular orbital picture of excitation due to absorption of light

After absorption or light by an atom or molecule, the electron charge cloud around an atom or molecule is disturbed. The distortion of charge cloud produces a dipole in the direction of the incident radiation. This phenomena is pictorially demonstrated in Fig. 2.2B.

2.2.3 Terms employed in absorption spectroscopy

In absorption spectroscopy, among the processes after impinging light on an object, the following are measured: Transmission, absorption, reflection, scattering, and refraction. Below we discuss terms associated with absorption spectroscopy.

Transmittance: Transmittance, T, is the fraction of incident light (electromagnetic radiation) at a specified wavelength that passes through a sample. The transmittance of a beam of light as it passes through a cuvette filled with analyte solution with concentration c is shown in Fig. 2.3.

The transmittance of a sample is sometimes given as a percentage, as given in Eq. (2.5). In this equation, scattering and reflection are considered to be close to zero or otherwise accounted for.

$$T = \frac{P_{solution}}{P_{solvent}} = \frac{P}{p0}, T(\%) = \frac{P}{P_0} \times 100\%, \tag{2.5}$$

where P_o is the power of the incident radiation and P is the power of the radiation coming out of the sample.

Absorbance: Absorbance, or absorption factor, is the fraction of radiation absorbed by a sample at a specified wavelength. For liquids, the transmittance is related to absorbance (A) as

$$A = -\log_{10} = -\log_{10}\frac{P}{P_0}. \tag{2.6}$$

In the case of gases, it is customary to use natural logarithms instead.

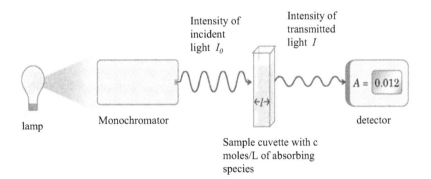

Intensity of incident light I_0

Intensity of transmitted light I

$A = 0.012$

lamp   Monochromator   detector

Sample cuvette with c moles/L of absorbing species

FIG. 2.3

Scheme for absorption measurement of a given sample (Diagram of Beer-Lambert Law). Light sources provide radiation with various wavelengths. Monochromator filters a wavelength of the interest, with intensity P_0. After passing through the sample cell the radiation intensity decreases to P.

The transmittance is directly proportional to the concentration (C) of the analyte and width l of the sample (which is determined by the width of the cuvette) (Notes: details of derivation is given in the consecutive section). That is, A is directly proportional to concentration and path length:

$$A = \alpha C \times l; \tag{2.7}$$

where α is the proportionality constant. When concentration is expressed in a number of moles per centimetre and l is expressed in centimeters, the above equation becomes

$$A = \varepsilon C \times l; \tag{2.8}$$

where ε is the molar absorption coefficient. Above equation is known as *Beer's Law* (detailed derivation for Eq. 2.8 is discussed in the consecutive section). Beer's Law states that molar absorptivity is constant and the absorbance is proportional to concentration for a given substance dissolved in a given solute and measured at a given wavelength. Accordingly, molar absorptivities are commonly called molar extinction coefficients. Since transmittance and absorbance are unitless, the units for molar absorptivity must cancel with units of measure in concentration and light path. Accordingly, molar absorptivities have units of $M^{-1}cm^{-1}$. Standard laboratory spectrophotometers are fitted for use with 1 cm width sample cuvette; hence, the path length is generally assumed to be equal to one and the term (in Eq. 2.7) is dropped altogether in most calculations $A = \varepsilon * C$.

Molar Extinction Coefficients: In chemistry, biochemistry, molecular biology, or microbiology, the mass extinction coefficient and the molar extinction coefficient (also called molar absorptivity) are parameters defining how strongly a substance absorbs light at a given wavelength, per mass density or per molar concentration, respectively. The SI unit of molar attenuation coefficient is the square meter per mole (m^2/mol), but in practice, it is usually taken as the $M^{-1} \cdot cm^{-1}$ or the $L \cdot mol^{-1} \cdot cm^{-1}$. The molar attenuation coefficient is also known as the molar extinction coefficient and molar absorptivity, but the use of these alternative terms has been discouraged by the IUPAC.

Molar absorptivities (= molar extinction coefficients) for many proteins are provided in the Practical Handbook of Biochemistry and Molecular Biology (Gerald D. Fasman, 1992). Expressed in this form, the extinction coefficient allows for estimation of the molar concentration ($c = A/\varepsilon$) of a solution from its measured absorbance.

Extinction coefficients for proteins are generally reported with respect to an absorbance measured at or near a wavelength of 280 nm. Although the absorption maxima for certain proteins may be at other wavelengths, 280 nm is favored because proteins absorb strongly there while other substances commonly in protein solutions do not.

Example 2.1. The absorbance of 5.4×10^{-4} M solution of Fe^{3+} at 530 nm was 0.54 when measured in a cell with a 1 cm path length. Calculate the molar absorption coefficient.

Solution:
From Eq. (2.8),

$$\varepsilon = \frac{A}{c \times l} = \frac{0.54}{5.4 \times 10^{-4} \, M \times 10^{-2} \, cm} = \times 10^5 \, M^{-1} \, cm^{-1}.$$

2.2.4 Beer-Lambert's law: quantitative aspects of absorption measurements

Credit for investigating the change of absorption of light with the thickness of the medium is frequently given to German physicist Johan Lambert, although he really extended concepts originally developed by French scientist Pierre Bouguer. German scientist August Beer later applied similar experiments to solutions of different concentrations and published his results just before those of French scientist F. Bernard. The two separate laws governing absorption are usually known as Lambert's land and Beer's Law. In the combined form they are known as the Beer-Lambert law.

Lambert's Law: Consider a beam of monochromatic light passing through a sample that absorbs some of the light. If P is the radiation power of light incident upon a thin layer of a sample of thickness dl the fractional change in intensity (dP/P) as light passes through the sample is proportional to its thickness,

$$-\frac{dP}{P} = \alpha dl \tag{2.9}$$

where α is the proportionality constant and is called the absorption coefficient. This is called Lambert's Law. Lambert's law states that when monochromatic light passes through a transparent medium, the rate of decrease in intensity with the thickness of the medium is proportional to the intensity of the light. This is equivalent to stating that the fraction of the incident radiation absorbed by a transparent medium is independent of the intensity of incident radiation and that each successive layer of the medium absorbs an equal fraction of incident radiation. The intensity of the emitted light decreases exponentially as the thickness of the absorbing medium increases arithmetically, that any layer of a given thickness of the medium absorbs the same fraction of the light incident upon it. The light intensity transmitted through a sample of finite thickness l will be obtained by integrating Eq. (2.9) as

$$\int_{P_0}^{P} -\frac{dP}{P} = \int_{0}^{l} \alpha dl,$$

Integrating above equation, for radiation power P_0 to P and path length 0−1 cm, yields:
$-\ln\frac{P}{P_0} = \alpha l$

$$-log\frac{P}{P_0} = \frac{\alpha}{2.303}l \tag{2.10}$$

As stated in Eq. (2.6), the term '$-log\frac{P}{P_0} = A'$ is the absorbance (formerly called optical density, O.D.).

Thus a medium with absorbance 1 for a given wavelength transmits 10% of the incident light at that wavelength.

Beer's Law: Lambert's law considers the light absorption and light transmission for monochromatic light as a function of the thickness of the absorbing layer only. For quantitative chemical analysis, however, largely solutions are used. Beer studied the effect of concentration of the colored constituent in solution upon the light transmission or absorption. He found the same relation (Eq. 2.10) between transmission and concentration as Lambert had discovered between transmission and thickness of the layer. That is, Beer noticed that when light passes through the solution, the fractional change in light intensity of a beam of monochromatic light is proportional to the concentration of the absorbing species, c, in addition to the thickness of the optical path (sample container width through which light passes). That is

$$-\frac{dP}{P} = \alpha\, c\, dl. \tag{2.11}$$

Integrating Eq. (2.11), we get,

$$-ln\frac{P}{P_0} = \alpha\,c\,l;$$

$$-log\frac{P}{P_0} = \frac{\alpha}{2.303}cl = \varepsilon cl; \tag{2.12}$$

where $\frac{\alpha}{2.303} = \varepsilon$, as defined before, is the molar absorption coefficient or molar absorptivity (formerly called molar extinction coefficient) and Eq. (2.12) is called Lambert-Beer-Law. Note that to obey Beer's law, the chemical form of the substance should not change with concentration and there should not be association or dissociation of after interaction of the sample with light. Optical arrangement to study the Lambert-Beer law is shown in Fig. 2.3.

2.2.5 Application of Beer's law

One of the most widely used applications of Beer's law is to identify the concentration of an analyte by measuring its absorbance (or transmittance). Specifically, by varying c (and also l) the validity of Beer-Lambert law can be tested, and the value of ε can be evaluated. When the value of ε is known, the concentration of an unknown analyte (c_x) solution can be calculated from Eq. (2.12): $C_x = \frac{logP_0/P}{\varepsilon l}$.

The molar absorption coefficient ε depends upon the wavelength of the incident radiation, the temperature, and the solvent employed. To overcome these issues absorbance is always measured at λ_{max}.

For matched cells (i.e. l constant), the Beer-Lambert law may be written as

$$C\,\alpha log P_0/P$$

$$Or\ C\alpha A.$$

Hence plotting A, or $log(1/T)$, against concentration will produce a straight line, and this will pass through the point $C = 0, A = 0\,(T = 100\%)$. This straight line may then be used to determine unknown concentrations of solutions of the same material after measurement of absorbance.

Beer's law can be used to a solution containing more than one kind of absorbing species, provided that the mixed substances do not interact. For more than one absorbing species, the absorbance, A is $\Sigma\varepsilon_i c_i l$ where ε_i and c_i are the molar absorption coefficient and concentration of ith species. Hence $A_{total} = A_1 + A_2 + \ldots + A_i$

$$A_{total} = \varepsilon_1 c_1 l + \varepsilon_2 c_2 l + \cdots + \varepsilon_n c_n c_n l \tag{2.13}$$

where the subscripts refer to absorbing components $1, 2, \ldots, n$.

Example 5.1. (a) Light of wavelength 400 nm is passed through a cell of 1 mm path length containing 10^{-3} mol dm^{-3} of compound A. If the absorbance of this solution is 0.25, calculate the molar absorption coefficient and the transmittance.

Solution: (a) Using Eqs. (2.8), $A = 0.25, C = 10^{-3}$ moldm^{-3}, $l = 0.1$ cm.
$\varepsilon = 0.25/(10^{-3}$ moldm$^{-3} \times 0.1$ cm$) = 0.25/10^{-4}$ moldm^{-3} cm.
$= 2.5 \times 10^3$ dm^{3}mol^{-1}cm^{-1}

2.2.6 Limitations to Beer's law (deviation from Beer's law)

When the linear relationship between absorbance and concentration with a fixed path length l does not hold good this situation is understood as a limitation to Beer's law. Mostly there are three types of limitation occur which are given as i) real limitation (deviation), ii) chemical deviation and iii) instrumental deviation.

(i) *Real limitation (deviation):* Beer's law describes the nature of absorption of an analyte which contains relatively low concentrations, which is considered a limitation within the law. This kind of deviation occurs due to changes in the refractive index of the absorbing medium to concentrations higher than 0.01 M. At higher concentrations interactions between the molecules or ions of the absorbing species affect the charge distribution and affect the extent of absorption. A similar effect occurs sometimes in the presence of large amounts of electrolytes causing a shift in the maximum absorption and may also change the value of the molar absorption coefficient. The proximity of ions to the absorber alters the molar absorptivity of the latter by electrostatic interactions, the effect is lessened by dilution.

Deviations from Beers's law also arise because the absorption coefficient depends on the refractive index of the medium. Thus if changes in concentration cause significant alterations in the refractive index n of a solution, departures from Beer's law are observed. A correction for this effect can be made by substituting the quantity $\varepsilon*n/(n^2 + 2)$ for ε in Eq. (2.7). However, in general, this correction is rarely significant at concentrations less than 0.01 M.

(ii) *Chemical limitation:* This kind of limitation occurs when the structure of the absorbing species changes with concentration. Discrepancies are usually found when the colored solute ionizes, dissociates or associates, forms complexes and the nature of the species varies with concentration. For example, orange-colored potassium dichromate shows maximum absorbance at 450 nm. On diluting the solution, the yellow-colored potassium chromate with a different λ_{max} and ε is formed:

$$Cr_2O_7^{2-} + H_2O = 2HCrO_4^- = 2CrO_4^- + 2H^+$$

Also, when more than one absorbing species is present in the medium, the law does not hold as the measured absorbance is the sum of the absorbances of the individual species each having its own absorption coefficient value. Therefore, it is very essential to be sure that only one species (only analyte) exists, by keeping proper sample preparation conditions.

(iii) *Instrumental deviation:* Instrumental deviations are of three kinds:

   (a) *Due to Polychromatic radiation:* This kind of limitation (deviation) occurs due to contamination (presence) of other wavelengths than that of λ_{max} during the absorbance or transmittance measurements. As the radiation sources in spectrometers produce lights of a certain range of wavelengths called bandwidth. To filter the desired wavelength for our measurements monochromators are used. However, even the best monochromator (e.g., diffraction grating) does not produce a beam of monochromatic radiation (radiation with a single wavelength).

   The instrumental deviation can be reduced by using radiation source combined with monochromator with narrow bandwidth (range of radiation wavelengths that make up a radiation) and choosing wavelength close to the λ_{max} (where change in the absorption

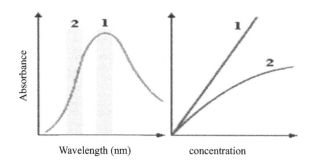

FIG. 2.4

The effect of polychromatic light on Beer's law.

coefficient is negligible for small changes in the incident beam wavelength (For e.g., as shown in Fig. 2.4, one need to measure in the wavelength region, not in the region 2 or below where absorbance is sharply decreasing). Fig. 2.4 (right) demonstrates the effect on absorbance values in region '1' and '2'.

Remediation: When an absorbance measurement is made with radiation composed of both wavelengths, the power of the beam emerging from the solution is the sum of the intensities emerging at the two wavelengths, $P' + P''$. Likewise, the total incident power is the sum $p_o' + p_o''$.

$$A_m = \log \frac{(P_0' + P_0'')}{(P_0' + P_0'') 10^{-\varepsilon' lc}} = \log\left(10^{-\varepsilon' lc}\right) \tag{2.14}$$

But, when the molar absorptivities are the same at the two wavelengths ($\varepsilon' = \varepsilon''$), this equation simplifies to

$$A_m = \varepsilon' * l * c = \varepsilon'' * l * c \tag{2.15}$$

In the spectrum at the left, the absorptivity of the analyte is nearly constant over band (1). Note in the Beer's law, at the right, using wavelength at band 1 gives a linear relationship. Eq. (2.15) also supports this result (as the analyte shows the same absorbance at the radiation at wavelengths lying within the band 1. (e.g., λ_1, λ_2 at which absorption coefficients are ε', and ε'', respectively). However, in the same figure, the curved spectrum (measured with radiation in band 2) corresponds to a region where the absorptivity shows substantial changes: That is, dramatic deviation from Beer's law as shown in the curve (2).

(b) *Instrumental deviations due to the presence of stray light:* Besides, the radiation coming out from monochromator may also be mixed with a small amount of unwanted scattered light which is commonly outside the nominal wavelength band used for the analysis (such radiation is called stray light).

(c) Instrumental deviations due to Mismatched Cells

Besides deviations discussed above, instrumental effects can also take place if the sample cuvettes are dirty or scratch or when the photodetector is faulty.

2.3 Instrumentation: components of UV–VIS spectrophotometry

Most spectroscopic instruments for use in the UV/visible and near IR regions are made up of five components (see Fig. 2.5):

(1) A stable source of radiant energy
(2) A wavelength selector that isolates a limited region of the spectrum for measurement
(3) One or more sample containers
(4) A radiation detector, which converts radiant energy to a measurable electrical signal
(5) A signal processing and readout unit, usually consisting of electronic hardware and, in modern instruments, a computer

A brief description of each component is given below.

2.3.1 The light source

In order to be suitable for spectroscopic studies, a source must generate a beam of radiation with sufficient power for easy detection and measurement and its output power should be stable for reasonable periods. Light sources are of two types.

1. Continuous light source.
2. Line source

2.3.1.1 Continuous light source (continuum sources in the ultraviolet/visible region)

A light source is required which gives the light in the regions at the near ultra-violet and the entire visible spectrum so that we are covering the range from about 200 nm to about 800 nm. This extends slightly into the near infra-red as well. Note: "Near UV" and "near IR" simply means the parts of the UV and IR spectra which are close to the visible spectrum. The most widely used continuum sources in the UV/visible range are listed in Table 2.1. For example, as shown in the table, H_2 and D_2 produces light in the UV-region from 160 nm to 380 nm. Whereas, tungsten lamp (it is also called tungsten/halogen lamp) produces in the visible region from 320 to 2400 nm. Table 2.1 also, clearly indicates that we can't get the range of wavelengths (near UV to near IR) required for UV–VIS spectroscopy from a single lamp, and so a combination of the two is used - a deuterium lamp for the UV part of the spectrum, and a tungsten/halogen lamp for the visible part. In the consecutive paragraph, we learn how these two devices work.

H_2 and D_2 Lamp: An H_2 lamp or deuterium lamp (D_2 lamp) contain H_2 or deuterium gas under low pressure subjected to a high voltage. It contains two electrodes, one a pure metal electrode and another

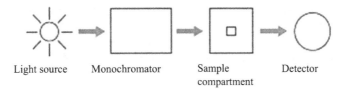

Light source   Monochromator   Sample   Detector
                                 compartment

FIG. 2.5

Basic construction layout of a spectrophotometer.

Table 2.1 Common sources of electromagnetic radiation for spectroscopy.

Source	Wavelength region	Useful for
H_2 and D_2 lamp	Continuum source from 160 to 380 nm	UV molecular absorption
Tungsten lamp	Continuum source from 320 to 2400 nm	Vis molecular absorption
Xe arc lamp	Continuum source from 200 to 1000 nm	Molecular fluorescence
Nernst glower	Continuum source from 0.4 to 20 µm	IR molecular absorption
Globar	Continuum source from 1 to 40 µm	IR molecular absorption
Nichrome wire	Continuum source from 0.75 to 20 µm	IR molecular absorption
Hollow cathode lamp	Line source in UV/Vis	Atomic absorption
Hg vapor lamp	Line source in UV/Vis	Molecular fluorescence
Laser	Line source in UV/Vis	Atomic and molecular absorption, fluorescence and scattering

Abbreviations: IR, *infrared;* UV, *ultraviolet;* VIs, *viable.*

an oxide-coated filament. As shown in Fig. 2.6, D_2 lamp produces a continuous spectrum in the part of the UV spectrum we are interested in (It produces continuous spectrum from 160 nm to the beginning of the visible region, 400 nm, with the maximum intensity at ~ 225 nm). The process by which a continuum spectrum is produced involves the initial formation of an excited molecular species (D_2*) followed by dissociation of the excited molecule to give two atomic species plus a UV photon. The reactions for deuterium are.

$D_2 + E_e \rightarrow D' + D'' + h\nu$, where E_e is the electrical energy absorbed by the molecule. Most modern light sources contain D_2 lamp as it can work in relatively low voltages. Uniquely, H_2 and D_2 Lamps can produce a beam of light of about 1−1.5 mm diam. It is important that the power of the radiation source does not change abruptly over its wavelength range which is satisfied by the D_2 lamp. A power supply with a voltage regulator is required for producing constant radiation intensities.

Precautions: While absorption or emission measurements with H_2 and D_2 Lamps, we must use quartz cuvette, as glass absorbs strongly at wavelengths less than about 350 nm. Although D_2 Lamp produces continuum spectrum as short as 160 nm, the lower limit we can use is up to 190 nm due to absorption by the quartz windows.

Tungsten filament lamp: Tungsten/halogen lamps contain a small amount of iodine (I_2) in a quartz "envelope", besides the tungsten (W) filament. Like all incandescent light bulbs, a halogen lamp produces a continuous spectrum of light, from near-ultraviolet to deep into the infrared (from 350 to 2400 nm). Its intensity reaches a maximum in the near-IR region of the spectrum (~ 1200 nm in this case). Typically, instruments switch from deuterium to tungsten at ~ 350 nm. The combined output of these two light sources (D_2 lamp and tungsten lamp) is focused on to a diffraction grating for selecting radiation of a desired wavelength.

Quartz allows the filament to be operated at a temperature of about 3500 °C, which leads to higher intensities and extends the range of the lamp into the UV region. A halogen lamp, also known as a *tungsten halogen*, quartz-halogen or an incandescent lamp consisting of a tungsten filament sealed into a compact transparent envelope that is filled with a mixture of inert gas and a small amount of a halogen such as iodine or bromine. The halogen gas reacts with gaseous tungsten, formed by sublimation, producing the volatile compound WI_2. When molecules of tungsten iodide (WI_2) hit the

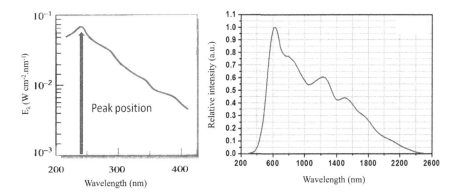

FIG. 2.6

Radiation spectra produced by deuterium lamp (left) and tungsten halogen lamp (right).

filament they decompose, redepositing tungsten back on the filament. This reversible process increases the lamp's life and assures maintaining the clarity (cleanliness) of the envelope. The lifetime of a tungsten/halogen lamp is approximately double that of an ordinary tungsten filament lamp. Therefore, they are used in many modern spectrophotometers. Because of this, a halogen lamp can be operated at a higher temperature than a standard gas-filled lamp of similar power and operating life, producing light of a higher luminous efficacy and color temperature. That is tungsten/halogen lamps are very efficient, and their output extends well into the ultra-violet. The small size of halogen lamps permits their use in compact optical systems for projectors and illumination.

2.3.2 Filter, monochromator and slit

As most of the UV–VIS absorption spectrophotometers use grating due to its narrow bandwidth, here we will discuss only about diffraction gratings in combination with the slit.

(i) *The diffraction grating and the slit:* Similar to the way that a prism splits light into its component colors, a diffraction grating does the same job, but more efficiently. The blue arrows represent the split lights in to the various wavelengths of the incoming light in different directions. The slit only allows light of a very narrow range of wavelengths to pass through the sample container into the spectrometer. By gradually rotating the diffraction grating, as shown in the Fig. 2.7, you can allow light from the whole spectrum (a tiny part of the range at a time) to pass through into the instrument.

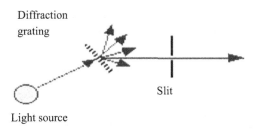

FIG. 2.7

Basic layout on working principle of grating and slit used in a spectrophotometer.

2.3.3 The sample and reference cells

These are small rectangular plastic, glass or quartz containers. They are often designed so that the light beam travels a distance of 1 cm through the contents. The sample cell contains a solution of the substance you are testing - usually very dilute (<0.01 M). The solvent is chosen so that it doesn't absorb any significant amount of light in the wavelength range we are interested in (200−800 nm). The reference cell just contains the pure solvent. However, if you are also adding a reagent(s) to develop a color in the analyte, then in this case, besides solvent one need to add the reagent in the same proportion.

2.3.4 The detector

The light beams that pass through the sample compartment enter the detector, which is the last element in the spectrophotometer. Photomultipliers and silicon photodiodes are typical detectors used with spectrophotometers for the ultraviolet and visible regions. For the near-infrared region, PbS photo-conductive elements have always been used in the past, but recently, instruments incorporating InGaAs photodiodes have also been used. Silicon photodiode array detectors are used, in combination with the back spectroscopy method, for high-speed photometry instruments.

The detector converts the incoming light into an electric current. Therefore, the higher the current, the greater the intensity of the light. For each wavelength of light passing through the spectrometer, the intensity of the light passing through the reference cell is measured. As mentioned in the beginning of this chapter, this is usually referred to as P_o. The intensity of the light passing through the sample cell is also measured for that wavelength - given the symbol, P. If P is less than P_o, then obviously the sample has absorbed some of the light. A simple bit of mathematics is then done in the computer (using transducer function) to convert this into the absorbance A of the sample Two general classes of transducers are used for optical spectroscopy, photon transducers and thermal transducers. The conventional photo-detectors, their range of wavelength that could be measured and type of output signals are given in table 2.2.

(i) *Photon Transducers:* As mentioned above, photomultipliers and silicon photodiodes are the common detectors. Phototubes and photomultipliers contain a photosensitive surface that absorbs radiation in the ultraviolet, visible, and near infrared (IR), producing an electric current proportional to the number of photons reaching the transducer. Below we learn about their working principle and spectral sensitivity.

Table 2.2 Characteristics of Transducers for optical Spectroscopy.

Detector	Class	Wavelength Range	Output Signal
Phototube	Photon	200−1000 nm	Current
Photomultiplier	Photon	110−1000 nm	Current
Si photodiode	Photon	250−1000 nm	Current
Photoconductor	Photon	750−6000 nm	Current
Photovoltaic cell	Photon	400−5000 nm	Current
Thermocouple	Thermal	0.8−40 μm	Voltage
Thermistor	Thermal	0.8−40 μm	Change in resistance

Reproduced with permission from David Harvey, Modern Analytical Chemistry, McGraw Hill, 1999.

2.3.4.1 Photomultiplier

A photomultiplier is a detector that uses the fact that photoelectrons are discharged from a photo-electric surface when it is subjected to light (i.e., the external photoelectric effect). The photoelectrons emitted from the photoelectric surface repeatedly cause secondary electron emission in sequentially arranged dynodes, ultimately producing a large output for a relatively small light intensity. The most important feature of a photomultiplier is that it achieves a significantly high level of sensitivity that cannot be obtained with other optical sensors. If there is sufficient light intensity, this feature is not particularly relevant, but as the light intensity decreases, this feature becomes increasingly useful. For this reason, photomultipliers are used in high-grade instruments. The spectral sensitivity character-istics of a photomultiplier are mainly determined by the material of the photoelectric surface. Fig. 2.8 shows an example of the spectral sensitivity characteristics of a multi-alkali photoelectric surface, a type of surface that is often used in spectrophotometers.

2.3.4.2 Silicon photodiode

Other photon detectors use a semiconductor (silicon photodiode) as the photosensitive surface. When the semiconductor absorbs photons, valence electrons move to the semiconductor's conduction band, producing a measurable current. That is, a Si photodiode is a detector that uses the fact that the electrical properties of a detector change when it is exposed to light (i.e., the internal photoelectric effect). One advantage of the Silicon (Si) photodiode is that it is easily miniaturized (made on a smaller size). Groups of photodiodes may be gathered together in a linear array containing from 64 to 4096

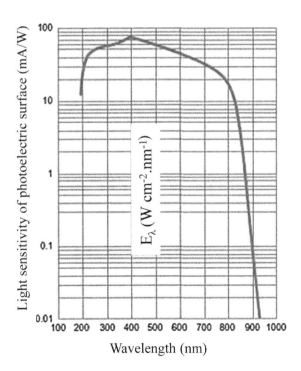

FIG. 2.8

Spectral sensitivity characteristics of a photomultiplier.

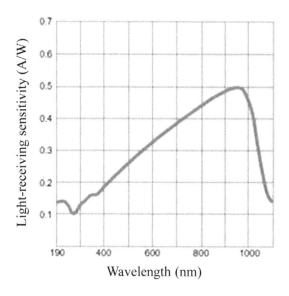

FIG. 2.9

Spectral sensitivity characteristics of a silicon photodiode.

individual photodiodes. With a width of 25 mm per diode, for example, a linear array of 2048 photodiodes requires only 51.2 mm of linear space. By placing a photodiode array along the monochromator's focal plane, it is possible to monitor simultaneously an entire range of wavelengths. The light beams that pass through the sample compartment enter the detector, which is the last element in the spectrophotometer.

Also in comparison with photomultipliers, silicon photodiodes offer advantages such as low cost and the fact that a special power supply is not required. Even regarding sensitivity, if the light intensity is relatively large, they can obtain photometric data that is not inferior to that obtained with photomultipliers. Fig. 2.9 shows an example of the spectral sensitivity characteristics of a silicon photodiode.

Photomultipliers and silicon photodiodes are typical detectors used with spectrophotometers for the ultraviolet and visible regions. For the near-infrared region, PbS photoconductive elements have always been used in the past, but recently, instruments incorporating InGaAs photodiodes have been sold. Silicon photodiode array detectors are used, in combination with the back spectroscopy method, for high-speed photometry instruments.

(ii) Thermal Transducers Infrared radiation generally does not have sufficient energy to produce a measurable current when using a photon transducer. A thermal transducer, therefore, is used for infrared spectroscopy. The absorption of infrared photons by a thermal transducer increases its temperature, changing one or more of its characteristic properties. The pneumatic transducer, for example, consists of a small tube filled with xenon gas equipped with an IR-transparent window at one end, and a flexible membrane at the other end. A blackened surface in the tube absorbs photons, increasing the temperature and, therefore, the pressure of the gas. The greater pressure in the tube causes the flexible membrane to move in and out, and this displacement is monitored to produce an electrical signal.

2.4 Classification (types) of UV–VIS spectrophotometers

UV–VIS spectrophotometers are constructed adopting mainly the following three designs.

2.4.1 Single beam (SB)

Schematic drawing of single-beam instrument is shown in Fig. 2.10. Radiation produced from the light source is wavelength selected by either the filter or monochromator and then passes through either the reference cell (cuvette is filled only with solvent that was used to prepare sample solution) or the cuvette with sample before striking the photo-detector.

Advantages and limitations: It is a simple device to use. However, the accuracy of a single-beam spectrophotometer is limited by the stability of its source and detector.

2.4.2 Double-beam (DB)-in-space

Schematic drawing of a double-beam-in-space instrument is shown in Fig. 2.11. In this kind of in-strument, radiation from the filter or monochromator is split into two beams (with the help of a lens, beam splitter and mirrors) that simultaneously pass through the reference and sample cells before striking two matched photo-detectors.

Advantages and limitations: It overcomes the problem of radiation stability, as equal amount light is split into sample and reference produced by the same light source. However, the radiation intensity is decreased to due to division that reduces the signal strength.

2.4.3 Double-beam (DB)-in-time

As shown in Fig. 2.12, the radiation beam is alternately sent through reference and sample cells before striking a single photo-detector. Only a matter of milliseconds separates the beams as they pass through the two cells. Therefore, this device addresses the problem of low radiation intensity and stability, as it alters very quickly from a reference cell to the sample cell.

IMPORTANT! The color-coding of the light beams through the spectrometer is not to show that some light is red or blue or green. The colors are simply to emphasize the two different paths that light can take through the device. Where the light is shown as a blue line, this is the path that it will *always* take (light from diffraction grating to rotating disc). Where it is shown red or green, it will go *either* one way *or* the other - depending on how it strikes the rotating disc (see in Fig. 2.13).

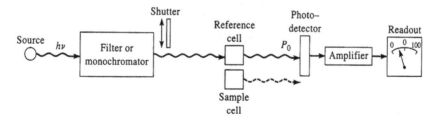

FIG. 2.10

Schematic drawing of single-beam instrument.

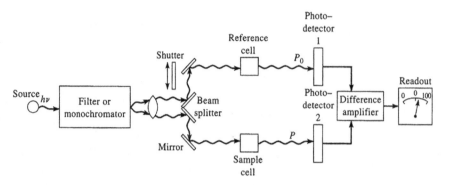

FIG. 2.11

Schematic drawing of double-beam in space instrument.

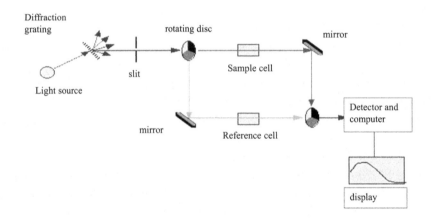

FIG. 2.12

Schematic drawing of double-beam in time instrument.

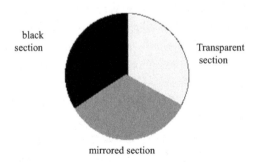

FIG. 2.13

Schematic view of rotating disc used in double-beam in time spectrophotometer.

The light coming from the diffraction grating and slit will hit the rotating disc and one of three things can happen as explained below.

If it hits the transparent section, it will go straight through and pass through the cell containing the sample.

(i) It is then bounced by a mirror onto a second rotating disc.

This disc is rotating such that when the light arrives from the first disc, it meets the mirrored section of the second disc. That bounces it onto the detector.

(ii) If the original beam of light from the slit hits the mirrored section of the first rotating disc, it is bounced down along the green path. After the mirror, it passes through a reference cell (more about that later). Finally the light gets to the second disc which is rotating in such a way that it meets the transparent section. It goes straight through to the detector.

(iii) If the light meets the first disc at the black section, it is blocked - and for a very short while no light passes through the spectrometer. This just allows the computer to make allowance for any current generated by the detector in the absence of any light.

2.5 Method for performance test of a spectrophotometer

For any type of sample, whether it be clinical, pharmaceutical, environmental analysis or research, it is essential that the instrument is performing according to specification. For this following tests are required to be performed.

2.5.1 Wavelength accuracy testing

Wavelength accuracy is normally assessed by using either a sample containing a series of very sharp peaks such as a solution of holmium perchlorate or a holmium oxide and/or didymium doped glass filter or by measuring the emission from a lamp. If the instrument is equipped with a deuterium (D_2) lamp as the UV source, this can be used. The advantage of emission lines is that they are inviolate (i.e. the emission wavelengths don't change over time).

2.5.2 Photometric (absorbance) accuracy testing

As with wavelength accuracy testing, either solutions or glass/quartz filters can be used for absorbance accuracy testing. The most commonly used solution for checking absorbance accuracy is potassium dichromate ($K_2Cr_2O_7$). The original 1988 Ph. Eur. method tests absorbance at four wavelengths. (235, 257, 313 and 350 nm) using between 57.0 and 63.0 mg of $K_2Cr_2O_7$ in 0.005 M sulfuric acid (H_2SO_4) diluted to 1000 mL. Since 2005 a second solution has been added to provide an additional test point at 430 nm. This uses the same amount of $K_2Cr_2O_7$ but is made up to 100 mL (i.e. it is ten times more concentrated than the original solution). Care has to be taken when preparing the solutions as potassium dichromate is hygroscopic and so it is important that this is dried thoroughly otherwise the weight will be incorrect and this could result in an unnecessary failure.

2.5.3 Resolution testing

The European Pharmacopoeia introduced a resolution test as part of the original 1988 methodology. This is based on measuring a solution of 0.02% w/v solution of toluene in hexane, measuring the ratio

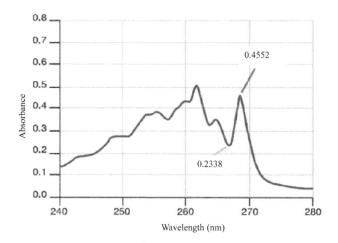

FIG. 2.14

Measurement of absorbance versus radiation wavelength for testing resolution using UV–Visible Spectrophotmeter (Shimadzu: UV-1800) with Toluene/Hexane solution (in this example also ratio of peak at 269 nm to trough at 266 nm is approx 1.9).

of the peak (at 269 nm) and the trough (at 266 nm) and measuring the ratio (see Fig. 2.14). This ratio should be 1.5 or greater *unless prescribed in the manual*. This means that, although it is highly desirable for an instrument to pass this test, it is not essential but most manufacturers regard this to be a part of making an instrument "pharmaceutical compliant" and some carry out the measurement as part of their final test procedure. In practice, an instrument with a bandpass (slit width) of around 1.5 nm should pass this test. For a fixed bandpass instrument, the measurement should give fairly constant results as there is no mechanical movement involved.

2.5.4 Limit of stray light

Stray light may be detected at a given wavelength with suitable filters or solutions; for example, absorbance of a 1.2 % w/v solution of potassium chloride (KCl) in a 1 cm cell should be greater than 2.0 at about 200 nm when compared with water as reference liquid.

The reaction product can be extracted by chloroform, while the original dyestuff is insoluble in this medium, and the intensity of the color in the chloroform layer is proportional to the concentration of the detergent. The method is especially valuable for the determination of small concentrations of detergents and is therefore useful in pollution studies.

2.5.4.1 Problems

1. What is UV–Vis spectroscopy? Explain why ultraviolet or visible light is useful in absorbance measurement for many types of organic compounds.
2. Define the word "chromophore". What are some typical features that are found in a chromophore of an organic molecule that can absorb ultraviolet or visible light?
3. Define the word "Blue shift" and "Red shift".

4. Explain why Beer −Lambert's law is used for the measurement of analytes in UV−VIS spectroscopy.

5. A solution has an analyte concentration of 5.7×10^{-3} M that gives a transmittance of 43.6% at 480 nm and when measured in a 5.0 cm cuvette. Calculate the molar absorptivity of the analyte. What is the expected lower limit of detection for this analyte if the smallest absorbance that can be measured by the instrument is 0.001.

6. What is the upper limit of the linear range for the analyte for the analyte in problem 5 if deviations for Beer-Lambert's law are found to occur at an absorbance of approximately 1.0?

7. Four standard solutions and an unknown sample containing the same compound give the following absorbance reading the same compound give the following absorbance reading at 535 nm when using 1.0 cm cuvettes. What is the concentration of the analyte in the sample?

Solution	Analyte Concentration (M)	Absorbance
Standard #1	0.0	0.005
Standard #2	2.5×10^{-3}	0.085
Standard #3	5×10^{-3}	0.175
Standard #4	25×10^{-3}	0.805
Unknown sample	?	0.465

8. A sample of coffee is analyzed to determine its caffeine concentration. Two portions of this sample are prepared for analysis.The first portion contains 50 mL of brewed coffee to which is added 10 mL water. The second portion contains 50 mL of brewed coffee that has been spiked with 10 mL of an aqueous solution that contains 1.0×10^{-2} M caffeine. The first portion of the sample is found to give a measured absorbance of 243 units, and the second portion gives an absorbance of 387 units. What is the concentration of caffeine in the brewed coffee?

9. The complex of Fe^{2+} with 1,10-phenanthroline has a molar absorptivity of about 11,000 L/mol.cm in water at 510 nm. A 20. mL water sample thought to contain Fe^{2+} is mixed with an excess of 1,10-phenanthroline. Acetic acid, sodium acetate and hydroxylamine are also added to buffer the solution and assure that all the iron is reduced to Fe(II). This mixture is diluted with distilled water to a total volume of 50 mL, and the final solution is found to give an absorbance of 0.762 at 510 nm when using a 1.0 cm cuvette.

   (a) If it is assumed that there are no absorbing species in this solution other than the Fe^{2+} complex with 1,10-phenanthroline, what was the concentration of Fe^{2+} in the original sample?

10. An aqueous sample containing Fe^{2+} is treated 1,10-phenanthroline to form a colored complex for detection. This treated solution gives an absorbance of 0.367 when measured with a 1.0 cm logn cuvette at 510 nm. Next 5, 5 mL of a 0.02 M Fe^{2+} solution is added to 10.0 mL of the unknown sample, treated with 1,10-phenanthroline in the same fashion as the previous sample, and found to give an absorbance of 0.538 at 510 nm. Based on this information, what was the concentration of Fe^{2+} in the original unknown?

11. Describe the basic components of a UV−VIS absorbance spectrophotometer. Give the function of each of these components.

12. What are some requirements for sample holder in UV−VIS spectroscopy?

13. What is the difference between 'single beam' and 'double beam instrument'? What are some common advantages disadvantages for each of these instrument designs?

14. List two types of detectors for light that can be used in UV−VIS spectroscopy and describe how each of these detects the light.

Further reading

[1] IUPAC, Compendium of Chemical Terminology, second ed., 1997. Online corrected version: (2006−) "Extinction".

[2] IUPAC, Compendium of Chemical Terminology, second ed., 1997. Online corrected version: (2006) "Absorptivity".

[3] G.D. Christian, J.E. O'Reilly, Instrumental Analysis, second ed., allyn and Bacon, Boston, 1986.

[4] D. Gerald, Practical Handbook of Biochemistry and Molecular Biology, first ed., CRC Press, 1989.

[5] Hamamatsu Photonics Deuterium Lamp, Photomultiplier, Photodiode Brochure, Hamamatsu Photonics K K., 2019. Iwata city, Sizuoka Pref.

[6] J.G. Speight, Lange's Handbook of Chemistry, sixteenth ed., McGraw-Hill, Inc., New York, 1992. CD&W Inc., Laramie, Wyoming Dean.

[7] R.C. Weast, in: Handbook of Chemistry and Physics, 56th Edition, CRC Press, Cleveland, 1975.

[8] D.G. Fasman (Ed.), Practical Handbook of Biochemistry and Molecular Biology, CRC Press, Boston, 1992.

[9] D.A. Skoog, F.J. Holler, S.R. Crouch, Instrumental Analysis, Cengage Learning India Pvt. Ltd., New Delhi, 2012, p. 158.

[10] Guide for use of terms in reporting data: spectroscopy nomenclature, Anal. Chem. 62 (1990) 91−92.

[11] https://www.nist.gov/programs-projects/spectrophotometry (assessed on: 7/23/2017).

[12] M.G. Gore, Spectrophotometry and Spectrofluorimetry: A Practical Approach, Oxford University Press, 2000.

[13] J. Mendham, R.C. Denney, J.D. Barnes, M. Thomas, B. Shivasankar, Vogel's Textbook of Quantitative Chemical Analysis, sixth ed., Pearson Education, Ltd., 2009.

[14] B.K. Sharma, Instrumental Methods of Chemical Analysis, Krishna's Education Publishers, 2012.

[15] J. Inczedy, T. Lengyel, A.M. Ure, in: Compendium of Analytical Nomenclature, third ed., Blackwell Science, Malden, MA, 1997.

[16] D.S. Hage, J.D. Carr, in: Analytical Chemistry and Quantitative Analysis, International, Pearson Education, Inc., New Jersey, USA, 2011.

[17] http://blair.pha.jhu.edu/spectroscopy/basics.html (Assessed on: 07/14/2017).

[18] T. Young, Rayleigh scattering, Appl. Opt. 20 (1981) 522−535.

Sample preparation methods and choices of reagents

3.1 Sample preparation methods and reagents

The sample preparation usually comprises sample collection, weighting, safe storage, purification and preparation of solution. After we have acquired samples, it is often necessary to treat or prepare these samples, with great care, before making any analytical measurements including spectroscopic studies. Below, we highlight some of the major reasons why careful sample preparation is necessary and discuss procedures for performing good sample preparation.

3.1.1 Why sample preparation is required?

Some of the important reasons why sample preparation is required are highlighted below:

1. One very common reason is that analytes are not present in a matrix suitable for the chosen method (matrix refers to the components of a sample other than the analyte of interest). This factor must be considered because all analytical techniques have some requirements regarding the types of samples that these methods can be used to study. For instance, when you are analyzing heavy metals such as iron, arsenic in soil sample with UV−Visible-, IR- and other spectroscopic methods, it is preferable to have the analyte in the dissolved form with minimal effect of other contaminants. For this, the samples are acid digested to extract analytes and then separated from unwanted contamination by filtration.
2. A second reason for sample pretreatment is to deal with substances present in the sample that interfere with measurement of the analyte. To overcome this problem you can use a procedure that separates the analytes from other components or you can eliminate the problem by converting them into different forms. For instance, if you wish to measure a particular drug in a blood sample, you would use probably pretreatment step to first isolate this drug from the sample. Alternatively, the analytes can be converted into different forms (e.g., conversion into metal complex).
3. A third reason for sample pretreatment is to place the analyte into a chemical form that can be studied by your method. For example, suppose you wanted to determine the total iron content in a water sample collected from a particular sampling site. This iron may be present in various chemical forms (such as Fe^{+2} and Fe^{3+}), which is difficult to analyze. Instead, this sample could be treated so that all of the individual types of iron are converted into single species for measurement: One way of performing this measurement would be converting them into Fe^{+2}.

4. A fourth reason for sample pretreatment is to place the analyte at a level that is suitable for detection. For example, when the analyte is present above the upper limit, dilution process is performed for the analyte to the level where detection can be made reliably.

3.1.2 Selection of sample preparation methods

In analytical chemistry, we need to tackle the wide range of analyte samples which require different approaches (processes) for preparation before chemical analysis. Some of the commonly used processes are listed below: Filtration, weighing, dilution, pH adjustment, evaporation, internal standards addition, concentration, centrifugation, column chromatography, liquid-liquid extraction, reagent addition, drying, mixing, heating, precipitation, digestion, grinding, solvent exchange, cooling, dialysis, ultrafiltration, sonification.

One common feature of these methods is that they all involve the use of either a physical or chemical change in the analyte or its matrix. A physical change might involve diluting a sample solution, weighing a portion of a solid for analysis, filtering a precipitate from a liquid or heating a solution to release a volatile chemical. Example of chemical changes for sample pretreatment includes the addition of reagents to make an analyte easier to detect.

Note that, for sample preparation, it often requires the use of more than one step, such as weighing, dissolving, and filtering, to prepare liquid sample from solids.

3.1.3 Classification of reagents and their choices for sample preparation

Because the purity of a substance of which standard solution to be prepared and the solvent used for solution preparation may greatly influence the accuracy of measurements, one should be very careful when preparing a solution for the analytical method. It is important to use specified grades otherwise errors can arise due to contamination from reagents themselves. Ideally, we do not want to have reagents that may interfere with detection of the analyte and cause the inaccurate result.

To help in this selection, commercial chemicals are often classified according to their purity (i.e., *"Chemical Grade" definitions from highest to lowest purity)* are given below:

Reagent A.C.S. Grade: This designates a high-quality chemical for laboratory use. The abbreviation "A.C.S.," means the chemical meets the specifications of the American Chemical Society (ACS). Such grades are useful for high-quality work.

Guaranteed Reagent (GR): Suitable for use in analytical chemistry, products meet or exceed ACS requirements where applicable.

AR: The standard Mallinckrodt grade of analytical reagents; it is suitable for high precision work, mainly for analytical applications, research and quality control. If the reagent also meets the requirements of the ACS Committee on Analytical Reagent, it will be denoted as an AR (ACS) reagent.

Primary Standard: Analytical reagent of exceptional purity that is specially manufactured for standardizing volumetric solutions and preparing reference standards.

Reagent: The highest quality commercially available for this chemical. The American Chemical Society has not officially set any specifications for this material.

OR: Organic reagents that are suitable for research applications.

Purified: Defines chemicals of good quality where there are no official standards. This grade is usually limited to inorganic chemicals.

Practical: Defines chemicals of good quality where there are no official standards. Suitable for use in general applications. Practical grade organic chemicals may contain small amounts of isomers of intermediates.

Lab Grade: A line of solvents suitable for histology methods and general laboratory applications.

USP: Chemicals manufactured under current "Good Manufacturing Practices" and which meet the requirements of the US Pharmacopeia.

NF: Chemicals that meet the requirements of the National Formulary.

FCC: Products that meet the requirements of the Food Chemical Codex.

CP (**Chemically Pure**): Products of purity suitable for use in general applications.

Technical: A grade suitable for general industrial use.

Solvents for High-Pressure Liquid Chromatography & Spectrophotometry: Chromatography or chromatography combined with spectrophotometry require highly pure solvents and named as HPLC grade and spectroscopy grade solvent.

Besides, there are also other types of reagents targeted for specific applications and are discussed below.

HPLC Grade Solvents: The reagents of product quality is tested for suitability in high-performance liquid chromatography. Products meet ACS requirements for use in HPLC and Ultraviolet spectrophotometry.

Spectroscopy grade: It includes solvents of high purity, low residue on boiling and having absorption blank in the wavelength region of interest. For example this kind of chemicals are required in HPLC and spectroscopic studies.

3.1.4 Classification of acids

Suprapur (E − Merck): High purity grade acids having metallic impurities in ppb range.

Environmental grade (Anachemia): High purity acids refined through sub-boiling distillation.

Environmental grade plus (Anachemia): Produced by additional distillation of environmental grade acids.

As discussed above, chemical reagents with a very high purity are known as analytical grade reagents, *"AR"*. For quantitative analysis, whenever possible, researchers need to use reagents of AR grade. Certain companies (such as Sigma-Aldrich, US and Merck, US) produce chemicals of high purity, and each package of these analyzed chemicals has a label giving the manufacturer's limits of certain impurities. It is often labeled as percentage purity with a list of possible contaminants.

The label on a chemical container is not a foolproof guarantee of the contents. Purity may be compromised for several reasons. For example some impurities may not have been tested for by the manufacturer. The reagent may have been contaminated after its receipt from the manufacturer; the stopper may have been left open for some time, exposing the chemical in open air. Solid reagent may not be sufficiently dried. However, if the chemicals are purchased from the reputed manufacturers, then the possibility of these errors is significantly reduced.

When pouring a liquid reagent, great care should be taken not to contaminate it. Pipette is not recommended to use, as it may contain impurity on its inner or outer walls and great care should be taken to avoid contamination of the stopper of the reagent bottles. During the pouring of chemical from bottle the stopper should be either held by hand or put on the clean watch glass then the return to the bottle immediately after the required chemicals has been taken out. Never keep on the working table or shelves where it can pick up contaminants.

In case there is doubt as to the purity of the chemical used, they should be tested by standard methods for the impurities that might cause errors in the analysis. Where a chemical required for a quantitative analysis is not available in the form of analytical grade, the commercial product should be further purified by known methods which are discussed in the following sections.

(i) Purification of substances

If a reagent of adequate purity for a particular determination is not available, then the purest available product must be purified. For inorganic compounds this is most commonly done by recrystallization from water. In recrystallization technique the chemicals are purified by dissolving both impurities and a compound in an appropriate solvent, either the desired compound or impurities can be removed from the solution, leaving the other behind. It is named for the crystals often formed when the compound precipitates out. For example for inorganic compounds this is most commonly done by recrystallization from water. Using a conical flask, known weight of the solid is dissolved in a volume of water sufficient to give a saturated or nearly saturated solution at the boiling point. Then, as shown in Fig. 3.1, the hot solution is filtered through a filter paper and the filtrate collected in a clean beaker. This process will remove the contaminants. For a successful procedure, one must ensure that the filtration apparatus is hot in order to stop the dissolved compounds crystallizing from solution during filtration, thus forming crystals on the filter paper or funnel.

One way to achieve this is to heat a conical flask containing a small amount of clean solvent on a hot plate. A filter funnel is rested on the mouth, and hot solvent vapors keeps the stem warm. The filter paper is preferably fluted, rather than folded into a quarter; this allows quicker filtration, thus less opportunity for the desired compound to cool and crystallize from the solution.

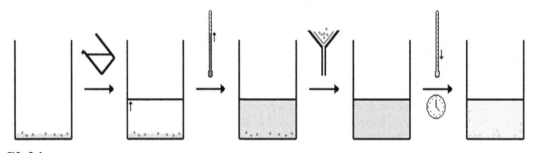

FIG. 3.1

Solvent added (clear) to a mixture of compound (orange [gray in print version]) + insoluble substance (*purple solid spheres* [dark gray in print version]) → Solvent heated to give saturated compound solution (orange [gray in print versions]) + insoluble substance (*purple solid spheres* [dark gray in print version]) → Saturated compound solution (orange [gray in print version]) filtered to remove insoluble substance (*purple solid spheres* [dark gray in print version]) → Saturated compound solution (*orange solid spheres* [light gray in print version]) allowed to cool over time to give crystals (orange [gray in print version]) and a saturated solution (pale-orange [gray in print version]).

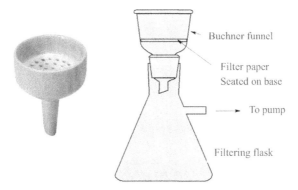

FIG. 3.2

Butchner funnels. Butchner funnel fitted to a filter flask.

Often it is simpler to do the filtration and recrystallization as two independent and separate steps. That is, dissolve "compound A" containing "impurity B" in a suitable solvent at room temperature, filter (to remove insoluble compound/glass), remove the solvent and then recrystallize. The clear hot filtrate is cooled rapidly by immersion in a dish of cold water or in a mixture of ice and water. The solution should be stirred constantly for the formation of small sized crystals. The solid is then separated from the mother liquor by filtration, using funnels shown in Fig. 3.2.

When the liquid has been filtered, the solid is pressed down on the funnel with a wide glass stopper, sucked as dry as possible, then finally washed with small portions of the original solvent to remove adhering mother liquor. Then the recrystallized solid is dried in an oven above the laboratory temperature. The dried solid is preserved in an airtight bottle.

Many organic compounds can be purified by recrystallization from suitable organic solvents. Moreover, liquids can be purified by fractional distillation as shown in Fig. 3.3 (bottom).

(ii) Sublimation

As shown in Fig. 3.3 (top), specific solids on heating directly change from solid to vapor state without passing through a liquid state, such types of substances are known as sublimable and this procedure is called sublimation. The sublimation process is employed for the separation of sublimable volatile compounds from non sublimable impurities. The process is usually employed for the purification of compounds such as camphor, anthracene, naphthalene, benzoic acid NH_4Cl, $HgCl_2$, solid SO_2, Iodine, arsenic oxide and salicylic acid etc, containing non-volatile impurities. The chemical to be separated is gently heated in a porcelain dish, and the vapor produced is condensed on a flask which is kept cool by circulating cold water inside it.

(iii) Zone refining

It a technique for the purification of a crystalline material and especially a metal in which a molten region travels through the material to be refined, picks up impurities at its advancing edge, and then allows the purified part to recrystallize at its opposite edge— called also zone melting.

The molten region melts impure solid at its forward edge and leaves a wake of purer material solidified behind it as it moves through the ingot. The impurities concentrate in the melt, and are moved to one end of the ingot.

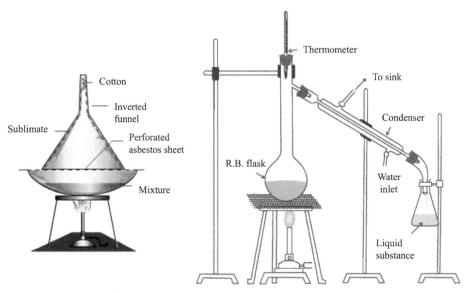

FIG. 3.3

(Top) represents the sublimation process for separation of volatile solid compound from nonvolatile one. Bottom panel represents simple distillation process. The vapor of a substance formed are condensed and the liquid is collected in the conical flask.

In a zone refining apparatus, the substance to be purified is packed into a column of glass or stainless steel, which may vary in length from 15 cm to 1 m. An electric ring heater which heats a narrow band of the column is allowed to fall slowly by a motor controlled drive, from top to the bottom of the column. The heater is set to produce a molten zone of material at a temperature 2−3 °C above the melting point of the substance, and the substance travels slowly down the tube with the heater. Since impurities normally lower the melting point of a substance, the impurities tend to flow down the column in step with the heater, becoming concentrated in the lower part of the tube. The process is usually repeated several times until the required degree of purification is achieved.

3.1.5 Storage of samples and solutions

Before analysis, most samples are dried at 105−110 °C. Also, in any analytical laboratory, it is essential to maintain stocks of various reagents in solution the prepared standard solutions of known concentration need to be stored correctly and precautions need to be taken depending on the type of standard solutions. Below we discuss the suitable apparatus and conditions of storage of the solution.

Solutions which are comparatively stable and unaffected by exposure to air may be stored in 1L or 2.5L bottles. For the measurements which require the highest accuracy, the bottles should be Pyrex, or other resistance glass, and fitted with ground-glass stoppers; this considerably reduces the solvent action of the solution. For alkaline solution, a plastic stopper is preferable to a glass stopper, and a polyethene container may be used instead of glass vessels. But some solutions, e.g., iodine and silver nitrate can be stored only in a glass container, coated with dark brown color. Solutions of polyethene are best stored in a polyethylene container.

Bottles for storing standard solutions should initially be clean and dry. They should be rinsed with a small amount of the standard solution then allowed to drain before the bulk of the solution is poured in bottles which has been washed should be rinsed three times with the standard solution which is to be stored. Then they can be filled with a standard solution. Immediately after the solution has been transferred, the bottles should be labeled with the name of the solution and its concentration, the date of preparation and the initial name of the person who prepared it. For expressing the concentration usually, the molar system is expressed.

Some standard solutions are likely to be affected by air (e.g., alkali hydroxides which absorb carbon dioxide, Iron (II) and Titanium (III) salts which are readily oxidized. Ideally, they should be kept under the inert conditions such as nitrogen.

3.1.6 Methods for weighing the samples and method for standard solution preparation

The procedure for weighing sample depends upon the type of equipment (device) and the standard solutions which are classified as:

1. Reagent solutions which are of approximate concentration.
2. Standard solutions which have a known concentration of some chemical.
3. Standard reference solutions which have a known concentration of a primary standard substance.
4. Standard titrimetric solutions which have a known concentration of a substance other than a primary standard.

For reagent solutions of type (1), it is usually sufficient to weigh approximately the amount of material required, using watch glass, and non-sticky paper and then add to the required volume of solvent which has been measured with a measuring cylinder.

To prepare standard solution use the following method. A suitable amount of the chemical is weighed using a weighing bottle (the bottle is weighed with and without chemical of which standard solution to be made) and transferred to a graduated volumetric flask with the aid of a funnel. Funnel is washed with the stream of solvent so that no chemical remains in it. Then shake and dissolve to make a homogeneous solution and then carefully made up to the mark using a wash bottle connected with a narrow tube.

If a watch glass or non-stick paper is employed to weigh out the sample, the contents are transferred as completely as possible to the funnel, and then pouring distilled water with a wash bottle the last traces of the substance is removed from the watch glass or paper.

In the cases of poor solubility of the substance, transfer the weighted substance in a beaker followed by deionized water. Then the mixture is heated gently with constant stirring until the substance is completely dissolved. Then let the solution cool slightly and transfer to a graduated volumetric flask with short-necked funnel. Rinse the funnel with deionized water and finally, the solution is made up to the mark.

Caution: Never heat the graduated volumetric flask.

3.1.6.1 Weighing techniques (what precautions should be taken for correct weighing of laboratory samples?)

We have to keep in mind that the accuracy and precision of our results are largely dependent on the reliability of our weighing. It is at this stage that we have to take adequate precautions and adhere to the standard operating procedures to ensure the reliability of our reports. Following procedure will ensure the accuracy and precision of the measurement results.

3.1.6.1.1 Balance calibration

Weighing of samples is always against standard reference weights. Ensure that your laboratory has a set of certified standard weights and their calibration is validated as per the prescribed procedure through a recognized national calibration laboratory.

Keep your balance always calibrated using standard calibration procedures against daily, weekly and monthly schedules.

Standard weights should never be touched with hands.

3.1.6.1.2 Environmental control of the balance

1. Horizontal positioning of balance should be daily checked with the help of inbuilt spirit level
2. The balance should be kept on a vibration-free support
3. The balance should be kept in an area with controlled temperature and humidity
4. Silica gel bag should be placed inside the weighing chamber to keep the chamber free from humidity. The bags should be periodically recharged or replaced.
5. The balance should not be exposed to direct sunlight as this can cause temperature variations inside the weighing chamber
6. The balance should not be located next to doors or windows as opening and closing give rise to air drafts.
7. Always weigh samples after closing the weighing chamber doors.
8. Keep the inside of the weighing chamber and surrounding bench space scrupulously clean so as to avoid errors and cross-contamination of samples.

3.1.6.1.3 Handling of weights

1. Weights should not be touched with baer hands as grease or oil from hands can lead to errors. Always use a pair of clean forceps.
2. Place weights gently onto the pan. The weights should be kept in the center of the pan so as to avoid errors due to eccentricity
3. After use weights should not be left outside on the workbench but kept inside the slots in the weight box and keep the box closed so as to avoid minimum environmental exposure

3.1.6.1.4 Procedure for weighing samples

1. Every time use a clean spatula of appropriate size
2. Butter paper weighing can introduce errors. Instead, weigh the required sample quantity in the volumetric flask and after weighing of samples make up the volume with the diluent
3. Allow the reading to stabilize before recording
4. Accurate weighing of samples of small amounts requires additional precautions such as the use of disposable gloves, use of head caps to prevent fall of hair or dandruff and use of face masks to prevent breath from disturbing the reading

5. Reading should be recorded directly into the laboratory notebook. Making note of readings on scrap paper, filter paper or hands should be strictly forbidden
6. Preserve weight slips along with the laboratory records.
7. Weighing of samples is a critical component of any analytical determination. You can eliminate errors by adopting the recommended weighing practices as a regular habit.
8. The procedure for weighing the sample varies slightly depending on the type of balance used which are of two types. The type of balance and operation procedures is explained below.

Two kinds of weighing balance are commonly practiced which are:
(a) Top loader balance and (b) Analytical balance.

3.1.6.1.5 top loader balance
For taking the weight of a given sample, with this kind of balance, two methods are employed.

3.1.6.1.5.1 direct weighing

1. With nothing on the pan, set to zero by pressing the "on" button.
2. Place weighing bottle, beaker, or vial on balance and set to zero again.
3. Use a clean spatula to transfer the sample into container slowly, until you reach the desired mass.

3.1.6.1.5.2 Indirect weighing (weighing by difference)

1. Place enough of the sample in a weighing bottle, put the lid on, and place on the scale. Then, record the mass.
2. Take some sample out and place it in a different container (whatever you will be using for the experiment). Record the new mass. The difference in mass is the mass of the sample transferred.
3. Continue this procedure until you have as much sample as you need.
4. It is best to transfer small amounts at a time, so you do not take more than you need. You should not put the excess sample back into the weighing bottle (Fig. 3.4).

(A)   **(B)**

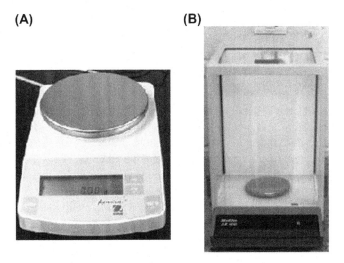

FIG. 3.4

(A) A top-loading balance. (B) An analytical balance.

3.1.6.1.6 Analytical balance

Use the same procedure as with a top loader, remembering these additional points:

1. Close all the doors before taking measurements.
2. Remember the number of significant figures. It is higher than on a regular Top loader balance.

Precautions: Make sure the sample is completely cooled when weighing. If a sample is still warm, it will weigh less because of buoyancy due to the upward circulation of hot air. For example, a 50 mL beaker 3 min after removal from a 110-degree oven weighs 27.0271 g. At room temperature, it weighs 27.0410 g.

3.1.6.2 Solution preparation technique

To prepare a standard solution use the following procedure. A short-stemmed funnel is inserted into the neck of a graduated flask of the appropriate size. A suitable amount of the chemical is placed on a non-stick paper or watch glass or weighing bottle which is weighed its weight need to be subtracted from the total weight. After weighing the required amount of substance, it is then transferred to the funnel, taking care that no particles are lost. The substance in the funnel washed down with a stream of the liquid. Then the contents of the flask are dissolved, if necessary, by shaking or swirling the liquid, and then made up to the mark. For the final adjustment of volume, use of dropping tube drawn out (connected at the tip of a plastic water bottle) to form a very fine jet.

If the substance is not readily soluble in water, it is suggested to transfer the solution from the weighing apparatus to a beaker and heat gently stirring with a magnetic stirrer until the solid has dissolved. After allowing cooling the resulting concentrated solution, it is transferred to the graduated volumetric flask. The beaker need to be rinsed with distilled water and added to the contents of the flask, finally, the solution is made up to the mark (Caution never heat the graduated flask).

It is also possible to prepare the standard solution by using commercial solutions supplied in sealed ampoules. All they require is dilution in a graduated flask to produce a standard solution.

Preparation of standard solutions from the commercial volumetric solution: We can also prepare the standard solution by using one of the commercial volumetric solutions supplied in sealed ampoules; all they require is dilution in a graduated flask to produce a standard solution.

3.2 Standard terminology in describing samples and reagent concentration

Most analytical techniques start with a mass or volume measurement. For example, the analysis of river water quality requires the collection of water samples with specified volumes, and determining the composition of iron might first involve weighing an iron sample. We will learn now how masses and volumes are used to describe the content of samples and reagents. Then we will discuss some issues to consider when preparing a sample and reagents for analysis.

Solution: A chemical mixture that has a uniform distribution for all its components, this mixture is called a solution.

Solvent: The most abundant components of the solution or the component that is used to dissolve and contain the other chemicals) is known as the solvent.

Solutes: All other substances, except solvent in the mixture, are called solutes. As an example, when a small amount of sodium chloride is dissolved in water, this salt dissociates into sodium ions and chloride ions (the solutes) in water (the solvent). The solutions have the same compositions throughout, so their overall content will be identical for any reasonably sized portion of the solution. Therefore, we can describe the content of a solution in concentration.

Here, we will now examine several ways of reporting the concentration of a solution.

Concentration: The amount of substance that is present within a given volume or mass of the solution is known as concentration and usually expressed in molality (m) or molarity (M).

Molality and molarity: Weight and volume ratios are important when dealing with the measures of masses or volumes of chemicals. It is even more important to know the actual number of a given type of molecules, atom, or ion that might be present in solution. To relate the number of moles of a particular molecule to its mass, a chemist uses that substance's molar mass. The molar mass for any substance is defined as the number of grams that are contained in one mole of that substance. For molecular compounds, the molar mass is commonly referred to as the molecular weight (MW), while for ionic compounds and elements the molar mass is also called the formula weight or atomic weight or atomic weight (or formula mass and atomic mass), respectively. Another concentration unit that measures the amount of solute in moles is molality: It is also called molal concentration, is a measure of the concentration of a solute in a solution in terms of the amount of substance in a specified amount of mass of the solvent. A commonly used unit for molality in chemistry is mol/kg. A solution of concentration 1 mol/kg is also sometimes denoted as 1 molal. This can be formulated as;

Molality (m): Molality (m) is defined as,

$$m = \frac{Moles\ of\ solute}{mass\ of\ solvent\ (kg)}.$$

This contrasts with the definition of molarity which is based on a specified volume of solution. That is

$$M = \frac{Moles\ of\ solute}{Volume\ of\ solution\ (in\ L)}.$$

Weight and Volume ratios: One method for describing the composition of any chemical mixture is to use the masses or volumes of the various components that are present in the mixture. One way this can be done is by using a weight −per-weight in percent (w/w %),

Weight percentage (w/w %) = (Mass of chemical / Mass of mixture)∗100.

For example, if 10 kg of iron ore extracted from an iron mining contains 1 kg of iron, then weight percentage of iron would be (1/10)∗100 = 10 %.

But when you are working with minor or trace components of a mixture, other multiplying factors besides 100 can be used with weight-per-weight ratios. Using 1000 instead of 100 would give a result in parts per thousand. If even a lower amount of a substance, it is expressed in parts per million. This is a way of expressing very dilute concentrations of substances. Just as percent means out of a hundred, parts per million or ppm (means out of a million). One ppm is equivalent to 1 mg of something per liter of water (mg/L) or 1 mg of something per kilogram soil (mg/kg). It is usually used to describe the concentration of something in water or soil.

This idea can be illustrated using rare earth metals like lanthanum being found in bastnasite. If 100 g of a sample of bastnasite-(Ce) contained 2 mg of lanthanum would be $(2*10^{-3}$ g La/100 g Bastnasite $-(Ce)) = 20$ ppm La sample$)*10^6 = 20$ ppm La or 20 parts-per-million.

A second way of describing chemical composition is to use a volume-per-volume (v/v) ratio. This type of ratio is employed when you are working with a mixture of liquid of liquids or gases for which volumes are easier to measure than masses. Similar to masses, these ratios are expressed as v/v % as:

$$\text{Volume percentage (v/v \%)} = (\text{volume of chemical / volume of mixture})*100.$$

One common example of an application of these units is to describe mixtures of alcohols with water. For example, a container labeled "25% ethanol (v/v)", should contain 25 mL of methanol for every 100 mL of solution.

But it can also be given in parts per thousand and parts per million or parts per billion.

$$\text{v/v (in ppm)} = (\text{volume of chemical / volume of mixture})*10^6.$$

Exercises

1. What is the difference between "Analytical" grade chemicals and Pharmaceutical grade chemicals?
2. One source of cerium is the mineral bastnasite-(Ce), which has the formula $Ce(CO_3)F$. If you have one mole of pure bastnasite-Ce, what would be the % w/w of cerium in this ore?
3. Calculate the molar concentration of 0.025 mol of iron (II) chloride dissolved in 0.5 kg water.
4. A 2.50 g portion of sodium of NaOH is placed into a 250.0 mL Volumetric flask and diluted to the mark with water. It is later found that the added mass of water was 497.2 g. What is the final concentration of the NaOH in units of molarity and molality?
5. A solution with a total volume of 500 mL that contains 1.0 g glucose (5.56×10^{-3}). What is the molar concentration of the solution?

***

Further reading

[1] IUPAC, Compendium of Chemical Terminology, the "Gold Book", second ed., Online corrected version, 1997 (1989) "Matrix (in analysis)".
[2] http://lab-training.com/2014/06/24/correct-weighing-samples/ (accessed on: 7/27/2019).
[3] https://study.com/academy/lesson/sampling-methods-in-analytical-chemistry.html (accessed on: 7/27/2019).
[4] D.S. Hage, J.D. Carr, Analytical Chemistry and Quantitative Analysis, International ed., Pearson Education, Inc., Publishing as Prentice Hall, Upper Saddle River, New Jersey, 2011 (chapter 18, pg. 450).
[5] D.A. Skoog, F. J Holler, S.R. Crouch, Instrumental Analysis, 11th Indian Reprint, Brooks/Cole, Cengage Learning India Pvt. Ltd., Delhi, 2012 (chap. 15, pg. 443).
[6] D.C. Harris, Exploring Chemical Analysis, second ed., Freeman and Company, USA, 2001 (chap. 18, pg. 401).

The chemical analysis process

4.1 Sampling

As an analytical chemist, we might imagine to monitor the earth's surface and provide a complete analysis of every part of the environment. This is, of course, almost impossible. Instead, we must collect a small portion of the substance or material for analysis. This small representative portion which is used for analysis is called 'sample'. For example, it is obviously impossible to analyze all the water in a lake, so portions of the water must be collected and analyzed to determine the true concentrations of the chemical species in the lake. Similarly, to study chemical contamination in a river around an industrial area, numerous water samples are needed to quantify the extent of the pollution.

Let us also have a concept on sampling from a survey (research) of social science. For example, a researcher may want to study characteristics of female smokers in Nepal. This would be the 'population' being analyzed in the study (A population is a group of individual units with some commonality), but it would be impossible to collect information from all female smokers of the whole country. Therefore, the researcher would select individuals from which to collect the data during their sampling process. The group from which the data is drawn is a representative sample of the population; then the results of the study can be generalized to the population as a whole.

The sample will be representative of the population if the researcher uses a random selection procedure to choose participants. The group of units or individuals who have a legitimate chance of being selected are sometimes referred to as the 'sampling frame'.

From the concept given above, now you understood that the sample is the source of information about the environment. The first step, sample collection, should provide a truly representative subsection of the whole sample, in a fairly small specimen. Even a perfect analytical procedure cannot be rectify the problems created by faulty sample collection (in other words, if it is not collected properly, if it does not represent the system we are trying to analyze, then all our careful laboratory work is useless). A good sampling plan will ensure that the samples obtained will, on average, closely represent the bulk composition of the environment being measured. In addition, the sample must be collected and handled in a such a way that its chemical composition does not change by the time it is analyzed.

Finally, the sampling must be done with the requirements of the analytical method in mind. Care must be taken to avoid the introduction of bias or error (the methods for statistical analysis will be discussed in a later section).

Optimization of a sampling plan increases the reliability of the analytical process which might at the end have beneficial effects on the economics of chemical analysis.

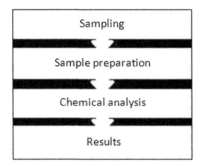

FIG. 4.1

Chemical analysis flowchart.

An overview of the chemical analysis process is presented in Fig. 4.1. The process of chemical analysis consists of several steps, beginning with the sample collection and followed by sample preparation. The sample is subjected to the analytical method and we arrive at a result, as indicated in Fig. 4.1. The actual spectrometric instrumental analysis is only one part of the process. Equally important are sampling, sample preparation, and analysis of the results.

We can think of sample handling as a two-step process, as in the flowchart in Fig. 4.1: sample collection (sampling) followed by sample preparation. The sample is then introduced to a spectrometer (or other analytical tool) in order to determine the analyte (or analytes) of interest.

4.2 Steps and important factors for sampling

The steps involved in environmental sampling are:

- Development of a sampling plan, including where and when samples will be collected and the number of samples required.
- Collection of the samples.
- Preservation of samples during transportation and storage.

In other words, the sampling plan should be able to clarify the following five questions.

1. From where and when, within the target population, should we collect samples?
2. What type of samples should we collect?
3. What is the minimum amount of sample for each analysis?
4. How many samples should we analyze?
5. How can we minimize the overall variance for the analysis?

To carry out a successful experimental study (including environmental study), it is necessary to have a 'plan of action', a sampling plan. If the content of heavy metals in a river is being studied, for example, the purpose might be to examine the effect of these metals on fish, or it might be to monitor the content because the river is a drinking water source. The sampling plan will be different for each of these purposes. The first step is to clearly define the problem being studied and identify the environmental "population" of interest. Some of the major steps involved in the development of a successful study (research) plan are as follows:

- Clearly outline the goal (objective) of the study. Different objectives require different sampling strategies. Decide what hypothesis is to be tested and what data should be generated to obtain statistically significant information. For example, if the objective is to measure the total release of heavy metals into a river by an industry, a 24 h integrated sample may be taken. However, if the goal is to monitor for accidental releases, then sampling and analysis may have to be done almost continuously.
- Identify the area of interest: Such as 'The pattern and variability of environmental contamination': The number of samples to be collected in space and time depends upon the variability in the concentrations to be measured. For example, pollutant levels in air can vary significantly depending upon meteorological conditions, or traffic patterns. In general, if the spatial or temporal variability is high, a larger number of samples need to be analyzed.
- Obtain information about the physical environment: For example, geographical features, locality (industrial, agricultural field, etc) are important if water samples are to be taken.
- Research the site history.
- Carry out a literature search and examine data from similar studies previously carried out. This can provide information about trends and variability in the data. In the absence of previous data, a pilot study may be necessary to generate preliminary information on which to base a more detailed study.
- Identify the measurement procedures to be used, because these affect the way samples are collected and handled.
- Develop an appropriate field sampling design. Decide how many samples are to be collected and delimit the time and area to be covered by the study.
- Determine the frequency of samples to be taken, both in temporal (time) and spatial (area, volume), depending upon the project objectives. Decide if, for example, 24 h integrated samples will be collected or individual samples will be taken every few hours.
- Develop a plan to insure and document the quality of each of the processes involved in the study: sampling, laboratory analysis, contamination control, etc.
- *Estimate the cost of the study:* If more samples are analyzed, the information obtained will have higher precision and accuracy. However, more samples also require more money, time, and resources. So, it is necessary to design an effective sampling plan within the available resources.
- Other factors such as convenience, site accessibility, limitation of sampling equipment and regulatory requirements often play important roles in developing a sampling plan, as well. A well designed strategy is needed to obtain the maximum amount of information from the number of samples.
- Once the sampling and analysis are complete, assess the uncertainty of the measurements.
- Perform statistical analysis on the data. Determine mean concentrations, variability, and trends with time and location.
- Evaluate whether study objectives have been achieved. If not, additional work may be necessary to provide the needed information.

4.3 Sampling methods (approaches or strategies)

There are several approaches (methods) to sampling: Systematic, random, judgmental (non-statistical), stratified, and haphazard. More than one of these may be applied at the same time. Very often, not much is known about the environmental area to be studied. A statistical approach is taken to increase the accuracy and decrease bias. It would be expected that the concentration of the pollutants present in the wastewater outfall are at maximum near the discharge point.

A systematic sampling plan would divide the water surface into a grid, and take samples in a regular pattern.

Sampling a few of the grid blocks chosen in a genuinely random way constitutes random sampling.

Judgmental sampling would concentrate on the area around the outfall.

Taking a few samples at locations chosen by the person doing the sampling would be termed haphazard sampling.

Finally, a *continuous monitor* may eliminate the time factor by giving real-time measurements all the time. This is still a sampling process, however, as the location of the sensor must serve as a typical location to give information about a larger area. More description of each kind is given below.

4.3.1 Systematic sampling

In this kind of sampling, measurements are taken at locations and/or times according to a predetermined pattern. That is, the area to be analyzed may be divided by a grid (see Fig. 4.2A), and a sample taken at each point of the grid. For example, a systematic sampling plan would divide the water surface of the lake or river into a grid, and take samples in a regular pattern. For air pollution studies, an

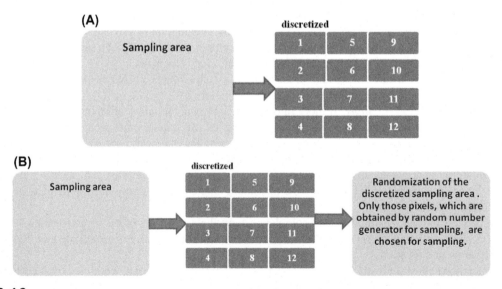

FIG. 4.2

(A): An example how a sampling area or volume may be divided into pixels and numbered. (B): An example how a sampling area or volume may be divided into pixels and numbered, according to Random sampling.

air sample might be taken at fixed intervals of time, say every 3 h. This approach does not require the prior knowledge of pollutant distribution, is easy to implement, and should produce unbiased samples. However, systematic sampling may require more samples to be taken than some of the other methods.

4.3.2 Random sampling

Sampling a few of the grid blocks chosen in a genuinely random way constitutes random sampling. The basis of random sampling is that each population unit has equal probability of being selected. The process for this kind of sampling are shown in see Fig. 4.2B. Random methods are good if the population does not have any obvious trends or patterns. Typically, in this kind of sampling method, the area to be sampled is divided into triangular or rectangular areas with a grid. Three dimensional grids are used if the variation in depth (or height) also needs to be studied. The grid blocks are given numbers. A random number generator or a random number table is then used to select the grid points at which samples should be collected. If a waste site contains numerous containers of unknown wastes and it is not possible to analyze every container, a fraction of the containers are selected at random for analysis.

If a system varies with time, as a stream might, we must sample at a variety of times, so that any time has an equal chance of being chosen. If the system varies with location within it, as a landfill would, we have to sample across the surface and down into it (horizontally and vertically), so that any point in the three dimensional space of the landfill has an equal chance of being chosen.

4.3.3 Judgmental sampling

Judgmental sampling would concentrate on the particular location (for example area around the origin or outfall of a river). This is a non-statistical sampling procedure. Here, the prior knowledge of spatial and temporal variation of the (parameters to be studied (e.g., pollutants in water) is used to determine the location or time for sampling. In the lake example, samples might be collected just around the outfall point. This type of judgmental sampling introduces a certain degree of bias into the measurement. For example, it would be wrong to conclude that the average concentration at these clustered sampling points is a measure of the concentration of the entire lake. However, it is the point which best characterizes the content of the waste stream. In many instances, this may be the method of choice, especially when the purpose of the analysis is simply to identify the pollutants present. Judgmental sampling usually requires fewer samples than statistical methods, but the analyst needs to be aware of the limitations of the samples collected by this method.

4.3.4 Stratified sampling

When a system contains several distinctly different areas (e.g., ground water resources remained in different areas such as industrial areas, residential areas and agricultural areas), these may be sampled separately, in a stratified sampling scheme. The target population is divided into different regions or strata. The strata are selected so that they do not overlap each other. Random sampling is done within each stratum. For example, in a pond or lagoon where oily waste floats over water and sediment settles to the bottom, the strata can be selected as a function of depth, and random sampling can be done within each stratum.

The strata in a stratified scheme do not necessarily have to be obviously different. The area may be divided into arbitrary subareas. Then a set of these are selected randomly. Each of these units is then sampled randomly. For example, a hazardous waste site can be divided into different regions or units. Then, the soil samples are collected at random within each region or within randomly selected regions. Stratification can reduce the number of samples required to characterize an environmental system, in comparison to fully random sampling.

4.3.5 Haphazard sampling

Taking a few samples at locations chosen by the person doing the sampling would be termed haphazard sampling. That is, in this kind of sampling, sampling location or sampling time is chosen arbitrarily. This type of sampling is reasonable for a homogeneous system. Since most environmental systems have significant spatial or temporal variability, haphazard sampling often leads to biased results. However, this approach may be used as a preliminary screening technique to identify a possible problem before a full scale sampling is done.

4.3.6 Continuous monitoring

Continuous monitor gives real-time measurements all the time. This is still a sampling process, however, as the location of the sensor must serve as a typical location to give information about a larger area. An ideal approach for some environmental measurements is the installation of instrumentation to monitor levels of pollutants continuously. These real-time measurements provide the most detailed information about temporal variability. If an industrial waste water discharge is monitored continuously, an accidental discharge will be identified immediately and corrective actions can be implemented while it is still possible to minimize the damage. A grab sample would have provided information about the accidental release only if a sample happened to be taken at the time the release was taking place, and that might well not have been when the problem began. A sample composited frequently enough could have identified the accidental release, but the time for preventive action would likely have passed.

Continuous monitoring is often applied to industrial stack emissions. Combustion sources, such as incinerators, often have CO monitors installed. A high CO concentration implies a problem in the combustion process, with incomplete combustion and high emissions. Corrective action can be triggered immediately. Continuous monitoring devices are often used in workplaces to give early warnings of toxic vapor releases. Such monitors can be lifesaving, if they prevent or minimize chemical accidents such as the one which occurred in Bhopal, India.

At present, a limited number of continuous monitoring devices are available. Monitors are available for gases such as CO, NO_2, and SO_2 in stack gases, and for monitoring some metals and total organic carbon in water. These automated methods are often less expensive than laboratory-analyzed samples, because they require minimal operator attention. However, most of them do not have the sensitivity required for trace level determinations.

4.4 Types of samples

(i) **Grab sample:** A grab sample is a *discrete* sample which is collected at a specific location at a certain point in time (a grid or pixel of the whole sampling area). If the environmental medium varies spatially or temporally, then a single grab sample is not representative and more samples need to be collected.

(ii) Composite sample: A composite sample is made by thoroughly mixing several grab samples. The whole composite may be measured or random samples from the composites may be withdrawn and measured.

A composite sample may be made up of samples taken at different locations, or at different points in time. Composite samples represent an average of several measurements and no information about the variability among the original samples is obtained. A composite of samples which all contain about the same concentration of analyte can give a result which is not different from that obtained with a composite made up of samples containing both much higher and much lower concentrations. During compositing, information about the variability, patterns, and trends is lost. When these factors are not critical, compositing can be quite effective. When the sampling medium is very heterogeneous, a composite sample is more representative than a single grab sample. For example, in a study of exposure to tobacco smoke in an indoor environment, a several hour composite sample will provide more reliable information than several grab samples.

Composite samples may be used to reduce the analytical cost by reducing the number of samples. A composite of several separate samples may be analyzed and if the pollutant of interest is detected, then the individual samples may be analyzed individually. This approach can be useful for screening many samples. A common practice, for example, in clinical laboratories screening samples for drug abuse among athletes is to analyze a composite of about ten samples. If the composite produces a positive result, then the individual samples are tested.

For instance, a field sample is taken at a random time point once within each hour. These twenty four field samples per day are mixed to form two composites. From each composite two sub-samples are taken and each subsample could also include two repeat samples.

4.5 Sample size, preservations and analysis

The sample size must be adequate. If a one milliliter water sample containing one ng/L of a pesticide is collected, the sample would contain such a small quantity of pesticide that it could not be detected by conventional analysis. Therefore, a larger volume of water must be collected, from which the pesticide can be concentrated.

Proper steps should be taken so that the pollutants are not lost or chemically altered during sample collection, preservation, and transport. Organic materials in water or soil samples, for instance, can be readily attacked and digested by bacteria present in the sample. A preservative to prevent bacterial action may be added as samples are collected, or the samples may be frozen or chilled to reduce these losses. Of course, the preservative must be carefully selected so that it does not interfere with the analyses to be done.

The most common environmental samples are water, soil, air, biological materials, and wastes (liquids, solids or sludges). Each matrix is sampled using different techniques, but the underlying concepts are the same in each case. As discussed in previous section, it is always good to know as much about the sampling site as possible, especially about the sources of the pollutants being investigated, and the mechanisms for their removal. For instance, before choosing a site for air sampling, pollution sources in the vicinity, such as industries and traffic should be surveyed. Knowledge of previous activities at a hazardous waste site may be helpful in finding the location of maximum contamination. Another important consideration is the physical environment. To predict the migration and distribution of pollutants in a contaminated site, for example, one should know about factors such as soil type, ground and surface water flows. Similarly, for river water sampling it is important to take into account factors such as pH, temperature, total dissolved solids (TDS), ionic conductivity and turbidity.

4.6 Sample preparation in the laboratory

When the sample, composited or not, reaches the laboratory, it may have to be reduced in size. A *reduced sample* is prepared by taking a representative portion of the original sample, usually by a mixing and dividing process. These processes depend strongly on the form of the sample and the analytes being sought. A loose solid, such as a soil sample, may be screened, ground, dumped into a pile and quartered, with opposite quarters being selected, and the other two-quarters discarded. This process can be repeated several times to reduce a large sample to a reasonably sized reduced sample.

Subsamples are portions of this sample, and after a reduced sample or a subsample is subjected to the laboratory processes needed to prepare it for analysis (grinding, dividing, mixing) it is referred to as a *test sample*. From the test sample, *test portions* are removed (kept separately) for the analysis. This must be of the proper size and concentration to be readily run on the instrument or to be analyzed by the chosen method. Often this test portion is dissolved, digested or extracted to obtain a *test solution*, and this is sometimes further treated with chemicals to derivatize or react some of its components. In that case, it becomes a *treated solution*. Sometimes the test solution is subdivided into equal portions, often to allow replication of the analytical method. These portions are termed *aliquots*, and this term almost always refers to a portion of a liquid. When a solution is made up in a 100 mL volumetric flask, and a 25 mL portion is taken out by pipet, that portion is a one-quarter aliquot of the original solution (see also in Chapter 3 for sample preparation).

4.7 Dealing with sample matrix

Within a sample there is likely a variety of substances present. The entire group of the substance that makes up a sample is called 'sample matrix'. The particular substance we are interested in measuring or studying in the sample is known as the analyte. In some cases, the analyte might be an atom, molecule, or ion while in others it might be a larger substance (e.g., organic pollutants, polymers or colloidal particles). The technique used for analyzing the analyte should produce a signal that is related to the presence of this analyte in the sample. Although we are not always interested to at other components in the sample, we still have to consider these components as we choose to use an analytical method because not all analysis methods are compatible with all types of samples. In addition, some sample components in the matrix may cause an error in the final result if they have not been properly dealt with before or during the analysis.

As much of the liquid sampling which is performed is concerned with drinking water or its precursors, most of the analytical tools will also be related to potable water, although the techniques can often be used for other liquids as well. When comes to potable water, strict control over its purity is both necessary and desirable.

Somewhere in excess of 700 organic compounds are found in water supplies, and different administrations (country, region, organization) take widely differing views about how they should be monitored and reduced. To overcome this difficulty, American Environmental Protection Agency (EPA) first listed the most dangerous chemicals and listed a priority pollutant list containing 129 pollutants with *maximum concentration levels (MCLs)* for each.

In Europe also similar system applies: there are red, gray and black lists where each substance has a *maximum allowable concentration (MAC)* or *prescribed concentration value (PCV)*.

Homogeneity of the sample: As discussed in the previous section, the analytical results should be as far as possible be representative of the whole and a true indication of the levels of any determined chemical contaminants. As even a small volume of water will contain a large number of particles, normally molecules or ions and as such, the sample taken will be amenable to statistical methods. But the problem remains that this sample must be representative of the whole. Although in the laboratory it is assumed that liquids are homogeneous, at least for a single phase, for large volumes the sample sites must be chosen carefully. It would not be sensible to assume that the whole Kaligandaki river or Phewa lake are homogeneous, with a constant concentration of analytes across its breadth and depth. But even for quite small volumes, positional variation can be important due to factors such as density variation, laminar flow at the surface, or lack of equilibrium in the bulk. Variability with time and meteorological conditions (rain, drought) impose further constrain upon the sampling method.

4.8 Dealing with variation in concentration range and sample stability

Since the range of analyte concentration can vary from the order of a few percent of the bulk, to traces at ppb level or even less, the volume of sample required can also vary considerably and often pre-concentration techniques are employed for obtaining measurable amounts of analyte. The stability of the collected sample can also cause problems since in many cases, without proper preservation or stabilization of the sample, changes will occur between collection and analysis. Finally it is important to be aware of the reason for doing the analysis.

Several approaches are used to obtain a gross liquid sample; they depend to a large extent upon the type of the system to be sampled:

1. Small static systems
2. Flowing contained volumes
3. Open flowing systems
4. Large static volumes

The sampling protocol also depends upon whether it is desirable to obtain a number of discrete samples which will be analyzed separately, or to bulk the individual samples as taken to give a composite sample. *Discrete sampling* is the more commonly employed method, although the problems of the need for a large number of analyses can make this a very expensive way of obtaining analytical results. If the average value of analyte concentration is required, it is often better to design and set up a composite sampling plan which will drastically reduce both the volume of sample required and the number of analyses needed. *Composite sampling* will not, however, be able to produce data that indicates variation in position or time, and often a combination of discrete and composite sampling is required. In either case make sure to observe cleanliness in all stages of the sampling protocol, this avoids contamination and false values.

4.9 Sampling procedure (water sample)
4.9.1 Water sample

Water sampling for chemical constituents including nutrients, anions, cations (such as NH_4^+ and metal cations) and other analytes are: Acidity, Alkalinity, Biological Oxygen Demand (BOD), Bromate, Chloride, Chlorite, Color, Conductivity, Fluoride, Foaming Agents, Nitrate, Nitrite, Odor, o-Phosphate, Residues, Silica, Sulfate, Surfactants, Total Dissolved Solids, Total Suspended Solids, Turbidity.

Sources of water and seasonal variation of chemical constituents in surface water: Water samples can come from many sources: ground water, precipitation (rain or snow), surface water (lakes, rivers, runoff, etc.), ice or glacial melt, saline water, waste water (domestic, landfill leachates, mine runoff, etc.), industrial process water and drinking water. Pollutants are distributed in the aqueous phase and in the particles suspended in the water. Solids and liquids with densities less than water (such as oils and grease) tend to float on the surface, while those with higher density sink to the bottom. The composition of stagnant water varies with the seasons and also with ambient temperatures. In rivers, lakes and oceans the concentration of pollutants varies with depth and may also depend on the distance from the shore.

Chemical constituents in precipitation water changes with meteorological conditions and atmospheric concentrations of the species of interest. The concentration of rainwater components may be higher when precipitation begins, and drop as the pollutants are washed out of the atmosphere. Concentration of water soluble gases such as H_2S, SO_2, NO_x are also higher in the early part of a precipitation event. Ground water shows seasonal variation and is especially affected by rain or snow. Many of these sources exhibit spatial and temporal variation and sampling devices should be chosen with these variations in mind.

Notes: In collecting from a lake or river, samples often are taken at different depths and different positions (edges, center) from demarcated area of certain dimension. Depending upon the type of analysis required, these samples can either be kept separate (for depth analysis or position dependence analysis) or can be mixed to give a "total" analysis. The necessary steps for water sampling, including choosing appropriate sample containers and method for preservation, are discussed below.

i. *Sampler (sampling bottle) preparation:* Many different types of manual and automatic samplers are commercially available. They are designed to collect grab samples or composite samples. Particular attention is given to the material of construction of the sampler. Water samples are typically collected as described below in screw-top containers. The simplest sampling device is a dipper (or a container) made of stainless steel or Teflon (see Fig. 4.3). Containers of these materials are preferred because of their inert nature.

ii. *Sampling Bottle Labeling:* Prior to taking a sample the following information should be provided on all sample bottles (Note: The specimen labels must be of a suitable material (use permanent marker) to prevent loss of identifying marks should they get wet or cold during the sampling or storage steps).
   * Sender reference number
   * Site code
   * Point of Collection
   * (Aquatic Facility Name and pool (i.e., toddler's pool)
   * Source (i.e. Pool outlet)

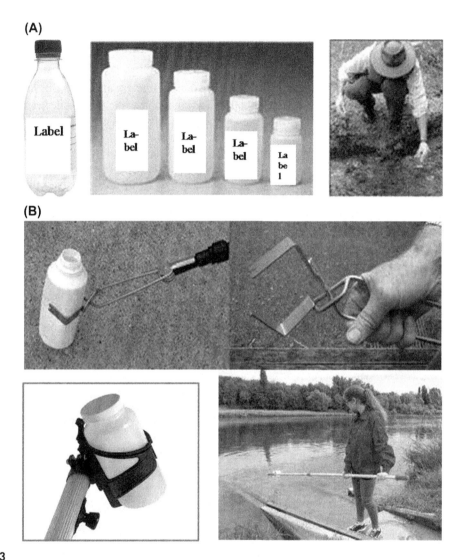

FIG. 4.3

(A) Left: 500 mL PET sterilized bottle clear round with screw cap and label. Middle: Various sized PET sterile bottles with screw pap and labeled. Right: An example of how water sample is collected. (B) Several different kinds of commercially available telescopic poles with a clamp for holding sample container.

- Date and time of collection
- Transport temperature (4 °C or ambient)
- Authority or Company Name

NOTES: Use a waterproof pen when marking sample bottles so the information will not rub off in the water.

iii. ***Water Sample collection (filling of sample: depth independent):*** Following procedure are recommended to follow during water sampling.
1. If wearing a long sleeve shirt, roll sleeves of shirt up past elbow.
2. Take a labeled sterile sample bottle of appropriate size (see below for labeling sample holder). Make sure you keep the lid on the bottle until you are ready to collect the sample.
3. As shown in Fig. 4.3A, hold the sterile bottle in one hand near the base, and then carefully remove and hold the screw cap with the other hand. Be careful not to touch the inside of the screw-cap when sampling.
4. For bacteriological (for monitoring microbial activity) water samples: Preferably use pre-cooled (either in refrigerator or in ice bath) bottles. Squat down on one hand and knee beside at the edge of the water body, and then plunge the bottle downwards into the water with continuous motion and direction away from the body but parallel with the edge of the water body, to a depth of approximately 30 cm below the water surface and then continue to move the bottle in a horizontal motion until finally removing the bottle from the water body with the same continued action and motion when full. Carefully replace the screw-cap immediately and tightly.
   NOTES: Tip enough of the water from the bottle to leave an air space of about 1–2 cm from the rim of the bottle. This air space is necessary to facilitate mixing of the sample in the laboratory during analysis.
5. Samples returned to the laboratory promptly after collection. Samples should be transported in an ice box with the aim of delivering the samples to the laboratory as soon as possible, or within 6 h of commencing sampling, while keeping the sample bottle temperatures at $4\,°C \pm 2\,°C$. Under exceptional circumstances (regional locations), the sampling and transport time may exceed 6 h but should never exceed 24 h.
6. For the analysis of tap water, it generally is recommended that the faucet be turned on and water allowed to flow for a couple of minutes before a water sample is collected. (This is to collect water that is representative of what is being delivered to the house, and does not indicate, for example, how much lead is picked up overnight while water sits in the pipe from the main to the house.)
7. Always collect microbiological samples before collecting other samples. Label the bottle before sampling. Discard damaged or contaminated bottles. If in doubt throw it out and take sample in a new bottle. Wash your hands thoroughly before and collecting samples. Also: If there is any reason to suspect that contamination has occurred during sampling, discard the sample and take another sample using a new sampling bottle.
   NOTES: In a flowing water stream, sampling should be carried out down stream before sampling upstream, because the disturbance caused by sampling may affect sample quality. Similarly, if water and sediment samples are to be collected at the same point, the water sample should be collected before the sediment is stirred up. However, sampling at depth in stratified sources (such as an outlet of drainage system from industrial area, agricultural land where pesticides contamination is likely to be high) can offer unique challenges. Prior to sampling, surface water drainage around the sampling site should be characterized.

Water Sample collection (filling of sample: depth dependent): Several different devices are commercially available for collecting samples at different depths (see, for example, in Fig. 4.3B). Most of them work on the general principle that a weighted bottle is lowered to the specified depth. At this point, a stopper or a cap is opened and the bottle is allowed to fill. Then the stopper is closed to prevent any water from flowing in or out and the bottle is pulled out.

Precautions: Following precautions must be taken to ensure a representative sample.

(i) The containers used must be well cleaned before the sample collection to prevent contamination of the sample.

(ii) Before sample collection, first, rinse the sample bottle with the water to be sampled. Sampling action should be one continuous motion in a direction away from the body, but parallel to/against the water body (i.e. in, along and out). Sample depth should be consistent.

(iii) Take extra care to avoid contaminating the sample container and water sample.

(iv) Keep hands away from the mouth of the bottle at all times.

(v) Do not contaminate the bottle and its rim by touching the inside of the bottle and inside of the rim.

(vi) Do not put the bottle lid on the ground while sampling.

(vii) Transport water samples to aquatic facility with care, maintaining temperature.

4.9.2 Soil sampling, processing and storage

Soil testing is an essential component of soil resource management. Each sample collected must be a true representative of the area being sampled. Utility of the results obtained from the laboratory analysis depends on the sampling precision. Hence, collection of large number of samples is advisable so that sample of desired size can be obtained by sub-sampling. In general, to gain a true representation of large area, sampling is done at the rate of one sample for every two hectare area. However, at-least one sample should be collected for a maximum area of five hectares.

i. **Tools required:** Following tools are required for soil sampling.
   1. Spade or auger (screw or tube or post hole type, see Fig. 4.4)
   2. Khurpi
   3. Core sampler (for sampling from deeper places)
   4. Sampling bags
   5. Plastic tray or bucket
   6. Wooden mallet
   7. Ceive (hole diam.: 2 mm and 0.2 mm)
ii. Points to be considered:
   1. Collect the soil sample during fallow period.
   2. In the standing crop, collect samples between rows.
   3. Sampling at several locations in a *zig-zag* pattern ensures homogeneity.
   4. Fields, which are similar in appearance, production and past-management practices, can be grouped into a single sampling unit.
   5. Collect separate samples from fields that differ in color, slope, drainage, past management practices like liming, gypsum application, fertilization, cropping system *etc.*
   6. Avoid sampling in dead furrows, wet spots, areas near main bund, trees, manure heaps and irrigation channels.

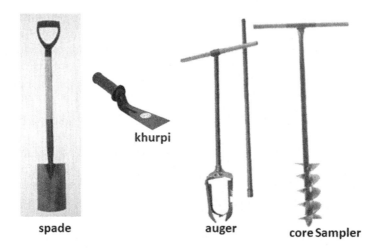

spade   khurpi   auger   core Sampler

FIG. 4.4

Sampling Augers, spade and khurpi for soil sampling.

7. For shallow rooted crops, collect samples up to 15 cm depth. For deep rooted crops, collect samples up to 30 cm depth. For tree crops, collect profile samples.
8. Always collect the soil sample in the presence of the farm owner who knows the farm better

iii. Procedure

1. Divide the field into different homogenous units based on the visual observation and farmer's experience.
2. Remove the surface litter at the sampling spot.
3. Drive the auger to a plow depth of 15 cm and draw the soil sample.
4. Collect at least 10 to 15 samples from each sampling unit and place in a bucket or tray.
5. If auger is not available, make a 'V' shaped cut to a depth of 15 cm in the sampling spot using spade.
6. Remove thick slices of soil from top to bottom of the exposed face of the 'V' shaped cut and place in a clean container.
7. Mix the samples thoroughly and remove foreign materials like roots, stones, pebbles and gravels. Reduce the bulk to about half to one kilogram by quartering or compartmentalization.
8. Quartering is done by dividing the thoroughly mixed sample into four equal parts. The two opposite quarters are discarded and the remaining two-quarters are remixed and the process repeated until the desired sample size is obtained.
9. Compartmentalization is done by uniformly spreading the soil over a clean hard surface and dividing into smaller compartments by drawing lines along and across the length and breadth. From each compartment a pinch of soil is collected. This process is repeated until the desired quantity of sample is obtained.
10. Collect the sample in a clean cloth or polythene bag.
11. Label the bag with information like name of the farmer, location of the farm, survey number, previous crop grown, present crop, crop to be grown in the next season, date of collection, name of the sampler *etc* (Fig. 4.5).

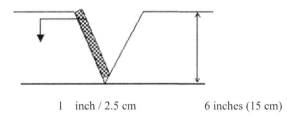

1   inch / 2.5 cm   6 inches (15 cm)

FIG. 4.5

V-shaped cut to a depth of 15 cm in the sampling spot.

iv. Laboratory processing and storage:
1. Assign the sample number and enter it in the laboratory soil sample register.
2. Dry the sample collected from the field in shade by spreading on a clean sheet of paper after breaking the large lumps, if present.
3. Spread the soil on a paper or polythene sheet on a hard surface and powder the sample by breaking the clods to its ultimate soil particle using a wooden mallet.
4. Sieve the soil material through 2 mm sieve.
5. Repeat powdering and sieving until only materials of >2 mm (no soil or clod) are left on the sieve.
6. Collect the material passing through the sieve and store in a clean glass or plastic container or polythene bag with proper labeling for laboratory analysis.
7. For the determination of organic matter it is desirable to grind a representative subsample and sieve it through 0.2 mm sieve.
8. If the samples are meant for the analysis of micronutrients at-most care is needed in handling the samples to avoid contamination of iron, zinc and copper. Brass sieves should be avoided and it is better to use stainless steel or polythene materials for collection, processing and storage of samples.
9. Air-drying of soils must be avoided if the samples are to be analyzed for NO_3–N and NH_4–N as well as for bacterial count.
10. Field moisture content must be estimated in un-dried sample or to be preserved in a sealed polythene bag immediately after collection.
11. Estimate the moisture content of sample before every analysis to express the results on dry weight basis.
v. Guidelines for sampling depth (Table 4.1):

4.9.3 Ore sampling, processing and storage

Ores and rock samples are notoriously inhomogeneous. Multiple samples are often required from either rock (drill) face or core samples so that a good average can be obtained. The process for sample preparation for ore is similar to soil sampling, except one additional step (Crushing of the rock sample, which is performed before sample preparation). Detailed processes for each step is discussed below.

4.9.3.1 Crushing of the rock sample

1. Rocks must be crushed (for example, using a jaw-crusher), or alternatively, the fine powder is produced by using a rotary air blast–reverse circulation drilling rig and is considered ideal for the task. If further crushing is required, this might have to be done in several steps using different

Table 4.1 Land coverage type and soil sampling depth.

| S.N. | Crop | Soil sampling depth | |
		Inches	cm
1	Grasses and grasslands	2	5
2	Rice, finger millet, groundnut, pearl millet, small millet *etc.*(shallow rooted crops)	6	15
3	Cotton, sugarcane, banana, tapioca, vegetables *etc.* (Deep rooted crops)	9	22
4	Perennial crops, plantations and orchard crops	Three soil samples at 12, 24 and 36 inches	Three soil samples at 30, 60 and 90 cm

With the permission from: http://agritech.tnau.ac.in/agriculture/agri_soil_sampling.html.

types of mechanical crushers to reduce the fragment size gradually. However, there is a risk of contamination from grinding surfaces, and this should be determined before using a particular grinder. Some rocks (especially silicate type) can be broken up more easily if they undergo a heating and cooling cycle first.

2. In all cases, the samples must be well identified, and any appropriate chain-of-custody documentation is used.

3. The sample containers must be cleaned so that the risk of contamination is minimized, and containers should not be reused unless they are cleaned thoroughly to prevent cross-contamination.

4.9.3.2 Sample preparation

The second step is the actual sample preparation, which is the process by which a rough sample is manipulated to produce a specimen (homogenization of the powder sample at the laboratory after following all the processes shown in Fig. 4.6 for soil sample) that can be presented to the spectrometer.

a. ***Homogenization:*** We usually homogenize materials by grinding them into a fine power or by dissolving an entire sample, described below.

Figs. 4.7 Shows two types of mortar and pestle: (Left) Agate and porcelain mortar and pestle. Agate is very hard and expensive. Less expensive porcelain mortars are widely used, but they are somewhat porous and easily scratched. These properties can lead to contamination of the sample by porcelain particles or by traces of previous samples embedded in the porcelain.

b. Sample Dissolution

Dissolution is required for atomic absorption spectrometry (AAS), inductively coupled plasma (ICP), photometry, and classical wet chemical methods such as gravimetry and titrimetry. The sample can be dissolved (or extract the element of interest) using an appropriate solvent. The solvent can be an acid or a combination of acids, alkali, or an organic solvent. This typically requires heating, either using a hot plate or microwave oven.

(A)   **Soil Sampling Locations in Kathmandu Valley**

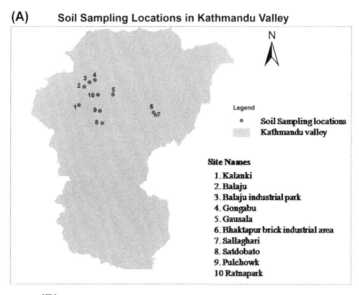

Legend

•   **Soil Sampling locations**
     Kathmandu valley

Site Names

1. **Kalanki**
2. **Balaju**
3. **Balaju industrial park**
4. **Gongabu**
5. **Gausala**
6. **Bhaktapur brick industrial area**
7. **Sallaghari**
8. **Satdobato**
9. **Pulchowk**
10 **Ratnapark**

(B)

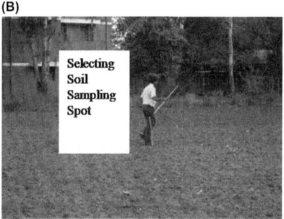

Selecting Soil Sampling Spot

FIG. 4.6

(A): Map of sampling spots. (B): Selecting sampling spots and removing the surface dirt. (C): Auger (spade or khurpi) is used to 'V' shaped cut to a depth of 15 cm in the sampling spot. (D): 'V' shaped cut to a depth of 15 cm in the sampling. (E): Mixing of the samples thoroughly. (F): Removal of the foreign materials like roots, stones, pebbles and gravel. (G): Quartering is done by dividing the thoroughly mixed sample into four equal parts. (H): Two opposite quarters are discarded and the remaining is mixed. (I): Further mixing of the separated sample. (J): Label preparation. (K): Collection of the sample in a clean cloth or polythene bag labeling with the required information.

—With the courtesy of Dr. Hemu Kafle.

(C)

(D)

FIG. 4.6

Cont'd

(E)

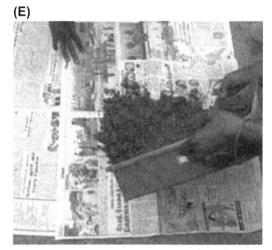

(F)

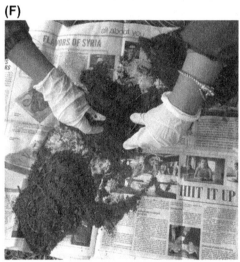

FIG. 4.6

Cont'd

(G)

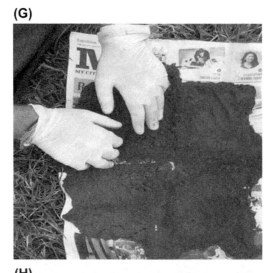

(H)

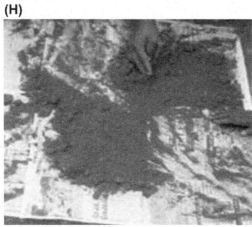

FIG. 4.6

Cont'd

(I)

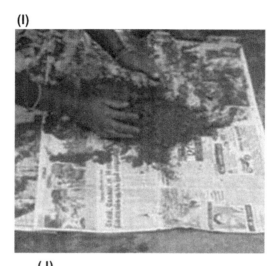

(J)

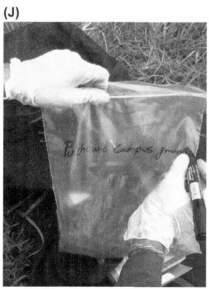

FIG. 4.6

Cont'd

(K)

FIG. 4.6

Cont'd

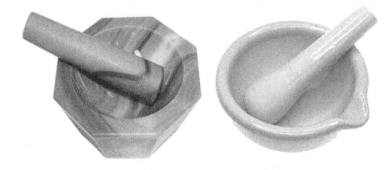

FIG. 4.7

Agate mortar and pestle. The mortar is the base and the pestle is the grinding tool.

Depending upon the chemical behavior of the elements to be analyzed as well as the other elements that are present in the sample, more than one dissolution techniques might be required for the analysis of all elements. The choice of appropriate solvent is critical because it must dissolve the element of interest reliably, but it might cause changes in viscosity in the solution, so care must be given in matching solvent strengths in samples and standards.

Caution: Some acids can attack the spectrometer sample introduction system, so appropriate materials must be selected. For example, hydrofluoric acid (HF) dissolves glass, so a sample introduction system of HF-resistant material must be used. Also, depending upon the type of spectrochemical analysis (mainly with AAS) other chemicals (such as suppressors and releasing agents) also can be added.

c. Dissolving Inorganic Materials with Strong Acids

The acids HCl, HBr, HF, H_3PO_4, and dilute H_2SO_4 dissolve most metals (M) with heating by the reaction,

$$M(s) + nH^+(aq) \rightarrow M^{n+}(aq) + (n/2)H_2(g)$$

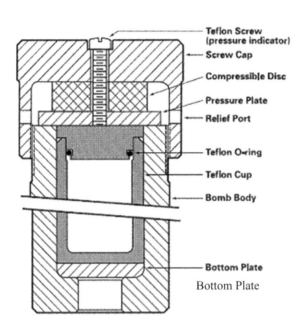

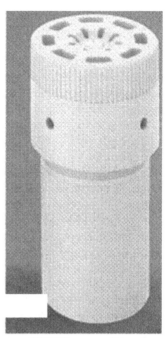

4782 Microwave Digestion

FIG. 4.8

Microwave digestion bombs lined with Teflon. Microwave digestion bomb lined with Teflon. A typical 23-mL vessel can be used to digest up to 1 g of inorganic material (or 0.1 g of organic material, which releases a great deal of gaseous CO_2) in up to 15 mL of concentrated acid. The outer container maintains its strength up to 150 °C, but rarely rises above 50 °C. If the internal pressure exceeds 80 atm., the cap deforms and releases the excess pressure.

Many other inorganic substances can also be dissolved. Some anions react with H^+ to form volatile products (species that evaporate easily), which are lost from hot solutions in open vessels. Examples include carbonate ($CO_3^{2+} + 2H^+$ (aq) → H_2CO_3 → $CO_2 + H_2O$) and sulfide ($S^{2-} + 2H^+$ (aq) → H_2S). Hot hydrofluoric acid (HF) dissolves silicates which is found in most rocks. HF also attacks glass, so Teflon is used for dissolving the sample powder, as Teflon is inert to attack by most chemicals and can be used up to 260 °C.

Substances that do not dissolve in the acids above mentioned acid reagents may dissolve as a result of oxidation by HNO_3 or concentrated H_2SO_4. Nitric acid dissolves most of the metals, except gold and platinum (For dissolution of these two metals 3:1 (vol/vol) mixture of HCl: HNO_3 called aqua regia is required).

Notes: Acid dissolution is carried out with a Teflon-lined sealed vessel (see Figs. 4.8) in a microwave oven, which heats the contents up to 200 °C in a minute. The vessel cannot be made of metal as it absorbs microwave and cannot reach up to sample. The vessel should be cooled prior to opening to prevent loss of volatile products.

d. Sample preservation

If not appropriate care is not taken, there may be both physical and chemical changes may occur which changes the composition and concentration of the sample after it is collected. Common physical processes which may degrade a sample are volatilization, diffusion, and absorption. Possible chemical changes include photochemical reaction, oxidation and microbial degradation. Below we discuss the method to prevent samples from each of these processes.

Volatilization: Analytes with high vapor pressures, such as volatile organic compounds and dissolved gases, such as HCN, SO_2, will readily escape from the sample by evaporation. Filling sample containers to the brim, so that they contain with no head space is the most common practice to minimize volatilization. The volatiles cannot not equilibrate between the water and the vapor phase above, if no air space is present at the top of the container. The samples are usually held at 4 °C, on ice, to lower the vapor pressure. Agitation during sample handling should also be avoided, to minimize air-sample interaction.

Choice of Proper Containers: The surface of the sample container may interact with the analyte. For example, metals can adsorb irreversibly on glass surfaces, so plastic containers are often chosen for water samples to be analyzed for their metal content. These samples are also acidified with HNO_3 to help keep the metal ions in solution.

Organic molecules may also diffuse in or out of the sample if the proper container is not used. Plasticizers, such as phthalate esters can diffuse from plastic containers into the sample. For organic analytes, it is best to collect samples in glass containers. Bottle caps should have Teflon liners to preclude contamination from the plastic caps.

Oily materials in water samples will adsorb strongly on plastic surfaces, and samples to be analyzed for such materials are usually collected in glass bottles. Oil which remains on the bottle walls should be removed by rinsing with a solvent and returned to the sample. Sometimes, oily samples are emulsified with a sonic probe to form a uniform suspension of the oil and then removed for analysis.

Absorption of Gases from the Atmosphere: Water samples can dissolve gases from the atmosphere as they are being poured into containers. Such components as O_2, CO_2 as well as volatile organic compounds may dissolve in the samples. Oxygen may oxidize species such as sulfite or sulfide to sulfate. Absorption of CO_2 may change conductance or pH measurements. This is one reason why pH measurements are always done in the field. Dissolution of organic compounds may lead to the detection of compounds that were actually absent. Field blanks should show if the samples have been contaminated with organic compounds which have been absorbed during sampling or transport.

Chemical Changes: A wide range of chemical changes in the sample are possible. For inorganic samples, controlling the pH can be useful in prevention of chemical reactions. For example, metal ions may react with oxygen to form insoluble oxides or hydroxides. The sample is usually acidified with HNO_3 to a pH less than 2, as most nitrates are soluble and excess nitrate ions will prevent precipitation. Other ions such as sulfide, or cyanide, are also preserved by pH control. Samples collected for NH_3 are acidified with sulfuric acid to stabilize the NH_3 as NH_4SO_4.

Organic species can also undergo changes due to chemical reactions. Photooxidation of poly-nuclear aromatic hydrocarbons, for example, is prevented by storing the sample in amber glass bottles. Organics can also react with free chlorine to form chlorinated organics. This type of problem is common for samples collected in treatment plants after the water has been chlorinated. Sodium thiosulfate, added to the sample, will remove chlorine.

Samples may also contain microorganisms which may biologically degrade the sample. High or low pH conditions, and chilling can minimize microbial degradation. The microbes can also be killed by addition of mercuric chloride or pentachlorophenol, if these preservatives will not interfere with the planned analyses.

Notes: Sample Preservation for Soil, Sludges, and Hazardous Wastes: Handling of water samples is better understood than solid and sludge samples, as these can be more varied in composition, but similar methods are used. Commonly encountered problems are biodegradation, oxidation-reduction and volatilization. Storing the sample at low temperature is always recommended to reduce volatilization, chemical reaction, and biodegradation.

A preservation temperature of 4 °C is most commonly used, because ice storage is convenient, and because a lower temperature may freeze the water, and separate the organic phase from the aqueous. Minimizing head space is also important for reducing volatilization losses. This also eliminates oxygen so that aerobic biodegradation or chemical oxidation are reduced. Samples to be analyzed for volatile organic compounds are sometimes collected directly into a known quantity of a solvent. In the laboratory, the analytes are either extracted or purged from the solvent. Methanol and polyethylene glycol have been used for this purpose.

4.10 Chemical analysis

The actual chemical analysis also consists of several steps: method creation, method validation, and sample analysis.

4.10.1 Method creation

In spectrochemical analysis, method creation is the process by which suitable spectral line(s) or regions are selected for: (i) analytical, (ii) background, (iii) interference, and (iv) reference measurements. The line selection process itself involves choosing suitable lines (wavelengths) that have the appropriate sensitivity for the concentration range required. The spectral line must not be so weak that the element cannot be detected at the concentration of interest. It is preferable to select spectral lines that are interference free; however, practically speaking, this is often difficult. Therefore, interfering elements must be identified and subsequently, the readings corrected.

Reference lines are spectral lines that will be used as internal standards.

The calibration curves can be derived in essentially two ways, empirically or theoretically. They can be generated empirically by running multiple standards (discussed in the Chapter 2 and 5). The calibration standards can be a series of single element standards (for example, single element solutions for AAS and UV–VIS spectroscopy).

This calibration curve is critical to the success of the analysis. Therefore, the analyst must have a good grasp of how the instrument is calibrated and what type of sample can be read against a particular calibration curve.

Precaution: Note with an example. Using an arc/spark spectrometer with 20 fixed lines (elements) in its optical system and then sparking a sample that contains an element not included will cause erroneous values. There might be spectral overlaps not considered in the calibration, and the typical "sum to 100%" algorithm used in this type of instrumentation will result in an incorrectly elevated reading for all the other elements.

Note: The "sum to 100%" algorithm mentioned here refers to the following rather elementary observation: The sum of all components of the sample must equal 100%. Typically, this is how the matrix element of the sample is calculated. We measure all analytes of interest (except for the matrix element), add these together, and subtract from 100 to obtain the matrix element concentration.

Also, even if the instrument is calibrated with the appropriate elements and the correct matrix, a particular sample might have a very high concentration of a particular element that is above the linear (or determined) portion of the calibration curve and will therefore still give an erroneous result.

4.10.2 Method validation

After an analytical method is created, it must be validated or checked for analytical integrity. This typically is performed by running certified reference materials or other well-characterized materials, preferably ones that were not used in the original calibration process. These standards are read against the calibration curve and the values with the associated error (standard deviation) recorded and checked against the certified values. If the standards are within expected ranges, the calibration curve is acceptable. If one or all of the standards falls outside of the allowable ranges, then the calibration curve should be revisited (recalibrate or check if the instrument's condition). Several check standards should be run during this method validation process.

The analysis of quality control (QC) samples is essential in testing out how well an instrument is performing. These samples should be taken through the identical process as the unknown samples. For example, reading a certified reference material that was dissolved last week and getting the "right" number only proves that the calibration on the spectrometer is reading appropriately. It does not tell you that a bad batch of acid or an incorrect dilution was made on the samples being prepared and analyzed today.

The instrument is now ready for the analysis of the unknown samples. It is advisable to run QC standards (blanks, reference standards) at the beginning and end of each analytical run, and depending upon how many samples are within a batch, maybe within the series, too. This is to verify the performance of the instrument through a run and to confirm the ongoing validity of the calibration curve.

4.10.3 Results

How do we evaluate the results obtained from our chemical analysis? The two most important criteria from the analytical point of view are *precision* and *accuracy.* Precision is defined as the reproducibility of repeat measurements. Specifically, it is a measure of the "scatter" or "dispersion" of individual measurements about the average or mean value. Accuracy, on the other hand, is defined as the relationship or correspondence of the measured value with the "true value." A quick review of some elementary statistics is required to better understand the difference.

Methods for *statistical analysis of the measured data* are discussed below:

i. *Average;* The average (or arithmetic mean) of a set of measurements is obtained by adding up the individual observations or measurements and dividing by the number of these observations. This is expressed mathematically in the following formula:

$$X_{mean} = \frac{\sum_1^i X_i}{N}.$$

(4.1)

where, X_{mean} is the average or arithmetic mean, x_i are the individual measurements, $X_1, X_2, 3_3, \ldots$ and N is the number of these measurements. The Greek symbol Σ indicates addition or summation.

ii. **Standard Deviation:** The standard deviation (σ) is a numerical measure of the variability of a set of results. It is a measure of the "spread" of measurements from the mean value. The standard deviation generally is calculated from a series of 10 or more independent measurements using the following formula:

$$\sigma = \sqrt{\frac{\sum |X_i - X_{mean}|^2}{N - 1}}, \qquad (4.2)$$

where σ is the lowercase Greek letter "sigma" used for the standard deviation, Xi, X_{Mean} and N are the values of the individual measurements, average and number of measurements, as defined previously.

What is the significance of the standard deviation of a set of measurements? As already noted, the standard deviation is a measure of the "spread" or "dispersion" of the measured values about the mean or average value. Statistically, we can say that with repeated measurements of the same sample, the following is true:

- 68% will fall within $\pm 1\sigma$ of the mean
- 95% will fall within $\pm 2\sigma$ of the mean
- 99.7 % will fall within $\pm 3\sigma$ of the mean

The way individual measurements are spread about a mean value is also shown in the familiar "bell curve," more technically called the "Gaussian Distribution Curve." The standard deviation can be estimated quickly by using a few handy approximations: Given only two measurements, take the difference and divide by the square root of 2.

Given a series of several measurements, take the difference of the maximum and minimum values (the range), and divide by the square root of the number of measurements.

It is important to note that the standard deviation is NOT a reference to accuracy, but rather an indication of the random errors encountered in the measurement process.

iii. **Relative Standard Deviation:** The relative standard deviation (*RSD*) is the standard deviation referenced to the mean measured value and expressed as a percentage:

$$RSD = \left(\frac{\sigma}{x_{mean}}\right) \times 100. \qquad (4.3)$$

The *RSD* is significant because, by reference to the mean value, it provides a measure of the dispersion of the observed values independent of their size or magnitude. It is sometimes called the "coefficient of variation" or *CV*.

Also, the RSD is a very useful tool for the analyst because it gives an immediate idea of the relationship of the measured values to the instrument detection limits. An RSD of 1% or less might be considered very good for routine measurements. This level of precision, however, typically is not found until the result is at least 500 times the detection limit.

In other words, for homogeneous samples and at concentrations well above the detection limit, the analyte RSD will be 1% or less. This holds regardless of the element, the matrix, and the instrumental technique used.

4.11 Problems

1. Discuss about systematic sampling?
2. Differentiate between systematic and random sampling? What is the advantage of this kind of sampling over systematic sampling?
3. A river, which flows through an agricultural land, falls in a lake. What is the appropriate sampling strategy to know if the lake is contaminated with pesticides?
4. Why preservation is necessary? Discuss the commonly used sample preservation techniques.
5. Government of Nepal made a strategy and located grant to access the water quality of all the rivers that flow through agricultural land and dense residential areas to know the feasibility for irrigation and for drinking. Make a sampling plan for a river water of your proximity.
6. Himalayan dwellers traditionally using iron wore prevailing in an iron mining. Describe the sampling plan for determining exact concentration of iron with spectroscopic measurements.

Further reading

[1] V. Thomsen, D. Schatzlein, Spectroscopy 21 (5) (2006) 44−48.
[2] K. Eckschlager, K. Danzer, Information Theory in Analytical Chemistry, John Wiley & Sons, New York, 1994.
[3] R.W. Gerlach, D.E. Dobb, G.A. Raab, J.M. Nocerino, J. Chemom. 16 (2002) 321−328.
[4] V.E. Burhke, R. Jenkins, D.K. Smith (Eds.), Practical Guide for the Preparation of Specimens for X-Ray Fluorescence and X-Ray Diffraction Analysis, John Wiley & Sons, New York, 2001.
[5] Environmental Health Directorate, Department of Health Western Australia, PO Box 8172, Perth Business Centre, Perth WA 6849.
[6] V. Thomsen, D. Schatzlein, D. Mercuro, Spectroscopy 21 (7) (2006) 32−40.
[7] V. Thomsen, Spectroscopy 17 (12) (2002) 117−120.
[8] V. Thomsen, Spectroscopy 22 (5) (2007) 46−50.
[9] ASTM, Method E882-82, Standard Guide for Accountability and Quality Control in the Chemical Analysis Laboratory, American Society for Testing and Materials, West Conshohocken, PA, 1982.
[10] V. Thomsen, D. Schatzlein, D. Mercuro, Spectroscopy 18 (12) (2003) 112−114.
[11] J. Workman, H. Mark, Spectroscopy 21 (9) (2006) 18−24.
[12] J. Workman, H. Mark, Spectroscopy 22 (2) (2007) 20−26.
[13] http://www.spectroscopyonline.com/chemical-analysis-process?id=&sk=&date=&pageID=2.
[14] D. Schatzlein, V. Thomsen, The Chemical Analysis Process 23 (2008).
[15] F.M. Schmalleger, Trial of the Century: People of the State of California vs. Orenthal James Simson, Prentice Hall, Englewood Cliffs, NJ, 1996.
[16] https://cirt.gcu.edu/research/developmentresources/research_ready/quantresearch/sample_meth (Accessed on 8/14/2018).
[17] F.F. Pitard, Pierre Gy, Sampling Theory and Sampling Practices, second ed., CRS Press, USA, 1993.
[18] J.W. Creswell, Research Design: Qualitative, Quantitative, and Mixed Methods Approaches, Sage Publications, Incorporated, 2013.
[19] M.D. Gall, W.R. Borg, J.P. Gall, Educational Research: An Introduction, seventh ed., Longman, White Plains, New York, 2003.

Application of UV–VIS spectrophotometry for chemical analysis

5.1 Types of chemical contamination in water/environment

Suitability of water for a particular use is identified by measuring/monitoring water quality parameters which includes measurement of physical (e.g., turbidity), chemical (cations and anions, organic pollutants), and biological (microbial). Public water supplies are tested frequently for contaminants, and suppliers are required to provide test results to their customers annually. As stated by US EPA, the *Safe Drinking Water Act* defines the term "contaminant" as meaning any physical, chemical, biological, or radiological substance or matter contained in water (That is, anything other than water molecules is defined as contaminant)[1]. Drinking water may reasonably be expected to contain at least small amounts of some contaminants. Some drinking water contaminants may be harmful if consumed at certain levels in drinking water while others may be harmless.

In general drinking water contaminants are classified into four categories:

I. *Physical contaminants* primarily impact the physical appearance or other physical properties of water. Examples of physical contaminants are sediment or organic material suspended in the water of lakes, rivers and streams from soil erosion.

II. *Chemical contaminants* are elements or compounds. These contaminants may be naturally occurring or man-made. Examples of chemical contaminants include nitrogen, bleach, salts, pesticides, metals, toxins produced by bacteria, and human or animal drugs.

III. *Biological contaminants* are organisms in water. They are also referred to as microbes or microbiological contaminants. Examples of biological or microbial contaminants include bacteria, viruses, protozoan, and parasites.

IV. *Radiological contaminants* are chemical elements with an unbalanced number of protons and neutrons resulting in unstable atoms that can emit ionizing radiation. Examples of radiological contaminants include cesium, plutonium and uranium.

Chemicals and other contaminants in groundwater are listed in Table 5.1. Here, we deal only with chemical contaminants.

Table 5.1 chemicals and contaminants in groundwater[1-6].

Inorganic contaminants found in groundwater		
Contaminant	**Sources to groundwater**	**Potential health and other effects**
Aluminum	Occurs naturally in some rocks and drainage from mines.	Can precipitate out of water after treatment, causing increased turbidity or discolored water.
Antimony	Enters environment from natural weathering, industrial production, municipal waste disposal, and manufacturing of flame retardants, ceramics, glass, batteries, fireworks, and explosives.	Decreases longevity, alters blood levels of glucose and cholesterol in laboratory animals exposed at high levels over their lifetime.
Arsenic	Enters environment from natural processes, industrial activities, pesticides, and industrial waste, smelting of copper, lead, and zinc ore.	Causes acute and chronic toxicity, liver and kidney damage; decreases blood hemoglobin. A carcinogen.
Barium	Occurs naturally in some limestones, sandstones, and soils in the eastern United States.	Can cause a variety of cardiac, gastrointestinal, and neuromuscular effects. Associated with hypertension and cardiotoxicity in animals.
Beryllium	Occurs naturally in soils, groundwater, and surface water. Often used in electrical industry equipment and components, nuclear power and space industry. Enters the environment from mining operations, processing plants, and improper waste disposal. Found in low concentrations in rocks, coal, and petroleum and enters the ground and	Causes acute and chronic toxicity; can cause damage to lungs and bones. Possible carcinogen.
Cadmium	Found in low concentrations in rocks, coal, and petroleum and enters the groundwater and surface water when dissolved by acidic waters. May enter the environment from industrial discharge, mining waste, metal plating, water pipes, batteries, paints and pigments, plastic stabilizers, and landfill leachate.	Replaces zinc biochemically in the body and causes high blood pressure, liver and kidney damage, and anemia. Destroys testicular tissue and red blood cells. Toxic to aquatic biota.
Chloride	May be associated with the presence of sodium in drinking water when present in high concentrations. Often from saltwater intrusion, mineral dissolution, industrial and domestic waste.	Deteriorates plumbing, water heaters, and municipal water-works equipment at high levels. Above secondary maximum contaminant level, taste becomes noticeable.
Chromium	Enters environment from old mining operations runoff and leaching into groundwater, fossil-fuel combustion, cement-plant emissions, mineral leaching, and waste incineration. Used in metal plating and as a cooling-tower water additive.	Chromium III is a nutritionally essential element. Chromium VI is much more toxic than chromium III and causes liver and kidney damage, internal hemorrhaging, respiratory damage, dermatitis, and ulcers on the skin at high concentrations.

Table 5.1 chemicals and contaminants in groundwater[1−6].—cont'd

	Inorganic contaminants found in groundwater	
Contaminant	**Sources to groundwater**	**Potential health and other effects**
Copper	Enters environment from metal plating, industrial and domestic waste, mining, and mineral leaching.	Can cause stomach and intestinal distress, liver and kidney damage, anemia in high doses. Imparts an adverse taste and significant staining to clothes and fixtures. Essential trace element but toxic to plants and algae at moderate levels.
Cyanide	Often used in electroplating, steel processing, plastics, synthetic fabrics, and fertilizer production; also from improper waste disposal.	Poisoning is the result of damage to spleen, brain, and liver.
Dissolved solids	Occur naturally but also enters environment from man-made sources such as landfill leachate, feedlots, or sewage. A measure of the dissolved "salts" or minerals in the water. May also include some dissolved organic compounds.	May have an influence on the acceptability of water in general. May be indicative of the presence of excess concentrations of specific substances not included in the safe water drinking Act, which would make water objectionable. High concentrations of dissolved solids shorten the life of hot water heaters.
Fluoride	Occurs naturally or as an additive to municipal water supplies; widely used in industry.	Decreases incidence of tooth decay but high levels can stain or mottle teeth. Causes crippling bone disorder (calcification of the bones and joints) at very high levels.
Hardness	Result of metallic ions dissolved in the water; reported as concentration of calcium carbonate. Calcium carbonate is derived from dissolved limestone or discharges from operating or abandoned mines.	Decreases the lather formation of soap and increases scale formation in hot-water heaters and low-pressure boilers at high levels.
Iron	Occurs naturally as a mineral from sediment and rocks or from mining, industrial waste, and corroding metal.	Imparts a bitter astringent taste to water and a brownish color to laundered clothing and plumbing fixtures.
Lead	Enters environment from industry, mining, plumbing, gasoline, coal, and as a water additive.	Affects red blood cell chemistry; delays normal physical and mental development in babies and young children. Causes slight deficits in attention span, hearing, and learning in children. Can cause slight increase in blood pressure in some adults. Probable carcinogen.
Manganese	Occurs naturally as a mineral from sediment and rocks or from mining and industrial waste.	Causes aesthetic and economic damage, and imparts brownish stains to laundry. Affects taste of water, and causes dark brown or black stains on plumbing fixtures. Relatively non-toxic to animals but toxic to plants at high levels.
Mercury	Occurs as an inorganic salt and as organic mercury compounds. Enters the environment from industrial waste, mining,	Causes acute and chronic toxicity. Targets the kidneys and can cause nervous system disorders.

Continued

Table 5.1 chemicals and contaminants in groundwater[1-6].—cont'd

Inorganic contaminants found in groundwater		
Contaminant	**Sources to groundwater**	**Potential health and other effects**
	pesticides, coal, electrical equipment (batteries, lamps, switches), smelting, and fossil-fuel combustion.	
Nickel	Occurs naturally in soils, groundwater, and surface water. Often used in electroplating, stainless steel and alloy products, mining, and refining.	Damages the heart and liver of laboratory animals exposed to large amounts over their lifetime.
Nitrate (as nitrogen)	Occurs naturally in mineral deposits, soils, seawater, freshwater systems, the atmosphere, and biota. More stable form of combined nitrogen in oxygenated water. Found in the highest levels in groundwater under extensively developed areas. Enters the environment from fertilizer, feedlots, and sewage.	Toxicity results from the body's natural breakdown of nitrate to nitrite. Causes "bluebaby disease," or methemoglobinemia, which threatens oxygen-carrying capacity of the blood.
Nitrite (combined nitrate/nitrite)	Enters environment from fertilizer, sewage, and human or farm-animal waste.	Toxicity results from the body's natural breakdown of nitrate to nitrite. Causes "bluebaby disease," or methemoglobinemia, which threatens oxygen-carrying capacity of the blood.
Selenium	Enters environment from naturally occurring geologic sources, sulfur, and coal.	Causes acute and chronic toxic effects in animals–"blind staggers" in cattle. Nutritionally essential element at low doses but toxic at high doses.
Silver	Enters environment from ore mining and processing, product fabrication, and disposal. Often used in photography, electric and electronic equipment, sterling and electroplating, alloy, and solder. Because of great economic value of silver, recovery practices are typically used to minimize loss.	Can cause argyria, a blue-gray coloration of the skin, mucous membranes, eyes, and organs in humans and animals with chronic exposure.
Sodium	Derived geologically from leaching of surface and underground deposits of salt and decomposition of various minerals. Human activities contribute through de-icing and washing products.	Can be a health risk factor for those individuals on a low-sodium diet.
Sulfate	Elevated concentrations may result from saltwater intrusion, mineral dissolution, and domestic or industrial waste.	Forms hard scales on boilers and heat exchangers; can change the taste of water, and has a laxative effect in high doses.
Thallium	Enters environment from soils; used in electronics, pharmaceuticals manufacturing, glass, and alloys.	Damages kidneys, liver, brain, and intestines in laboratory animals when given in high doses over their lifetime.
Zinc	Found naturally in water, most frequently in areas where it is mined. Enters environment from industrial waste, metal plating, and plumbing, and is a major component of sludge.	Aids in the healing of wounds. Causes no ill health effects except in very high doses. Imparts an undesirable taste to water. Toxic to plants at high levels.

Organic contaminants found in groundwater		
Contaminant	**Sources to groundwater**	**Potential health and other effects**
Volatile organic compounds	Enter environment when used to make plastics, dyes, rubbers, polishes, solvents, crude oil, insecticides, inks, varnishes, paints, disinfectants, gasoline products, pharmaceuticals, preservatives, spot removers, paint removers, degreasers, and many more.	Can cause cancer and liver damage, anemia, gastrointestinal disorder, skin irritation, blurred vision, exhaustion, weight loss, damage to the nervous system, and respiratory tract irritation.
Pesticides	Enter environment as herbicides, insecticides, fungicides, rodenticides, and algicides.	Cause poisoning, headaches, dizziness, gastrointestinal disturbance, numbness, weakness, and cancer. Destroys nervous system, thyroid, reproductive system, liver, and kidneys.
Plasticizers, chlorinated solvents, benzo[a]pyrene, and dioxin	Used as sealants, linings, solvents, pesticides, plasticizers, components of gasoline, disinfectant, and wood preservative. Enters the environment from improper waste disposal, leaching runoff, leaking storage tank, and industrial runoff.	Cause cancer. Damages nervous and reproductive systems, kidney, stomach, and liver.
Microbiological contaminants found in groundwater		
Contaminant	**Sources to groundwater**	**Potential health and other effects**
Coliform bacteria	Occur naturally in the environment from soils and plants and in the intestines of humans and other warm-blooded animals. Used as an indicator for the presence of pathogenic bacteria, viruses, and parasites from domestic sewage, animal waste, or plant or soil material.	Bacteria, viruses, and parasites can cause polio, cholera, typhoid fever, dysentery, and infectious hepatitis.
Physical characteristics of groundwater		
Contaminant	**Sources to groundwater**	**Potential health and other effects**
Turbidity	Caused by the presence of suspended matter such as clay, silt, and fine particles of organic and inorganic matter, plankton, and other microscopic organisms. A measure how much light can filter through the water sample.	Objectionable for aesthetic reasons. Indicative of clay or other inert suspended particles in drinking water. May not adversely affect health but may cause need for additional treatment. Following rainfall, variations in groundwater turbidity may be an indicator of surface contamination.
Color	Can be caused by decaying leaves, plants, organic matter, copper, iron, and manganese, which may be objectionable. Indicative of large amounts of organic chemicals, inadequate treatment, and high disinfection demand. Potential for production of excess amounts of disinfection byproducts.	Suggests that treatment is needed. No health concerns. Aesthetically unpleasing.

Continued

Table 5.1 chemicals and contaminants in groundwater[1−6].—cont'd		
Physical characteristics of groundwater		
Contaminant	**Sources to groundwater**	**Potential health and other effects**
pH	Indicates, by numerical expression, the degree to which water is alkaline or acidic. Represented on a scale of 0—14 where 0 is the most acidic, 14 is the most alkaline, and 7 is neutral.	High pH causes a bitter taste; water pipes and water-using appliances become encrusted; depresses the effectiveness of the disinfection of chlorine, thereby causing the need for additional chlorine when pH is high. Low-pH water will corrode or dissolve metals and other substances.
Odor	Certain odors may be indicative of organic or non-organic contaminants that originate from municipal or industrial waste discharges or from natural sources.	
Taste	Some substances such as certain organic salts produce a taste without an odor and can be evaluated by a taste test. Many other sensations ascribed to the sense of taste actually are odors, even though the sensation is not noticed until the material is taken into the mouth.	

5.2 Spectrophotometry: performance characteristics (basic terminology) and theory (recap)

5.2.1 Basic terminology and performance characteristics

5.2.1.1 Basic terminology

Calibration: The process that determines the sensitivity of an instrument to a particular analyte, or the relationship between the signal and amount of analyte in a sample.

External standard: A standard used in calibration that is prepared separately from the sample, usually consisting of the analyte itself and a blank, used when there are no matrix components to analyte solution.

Least-squares method: Linear regression, assumes there is a linear relationship between measured response y and standard analyte amount x, and that error originates from measurement only, not standards.

Standard-addition method: Calibration method that accounts for matrix affects by adding one or more increments of standard solution to sample aliquots containing identical volumes

Internal standard: Substance that is added in a constant amount to all samples, blanks, and calibration standards in analysis, calibration done by plotting ratio of analyte signal to this signal as function of analyte concentration, compensates for matrix effects and instrument method fluctuations.

Residual: Measure of how far the best straight line is from data points

Blank: Corrects baseline of raw analytic response, usually accounting for solvent/system without analyte.

5.2.1.2 Performance characteristics

Matrix effects: Effect of all the components in the analyte signal that are not the anayte, can enhance or suppress analyte signal.

Figures of merit: Numerical terms/expressions that permit us to narrow choice of instruments for a given analytical problem, examples are absolute standard deviation, relative systematic error, calibration sensitivity etc.

Precision: The degree of mutual agreement among data that have been obtained in the same fashion, reproducibility.

Bias: Measure of systematic error of an analytical method

Sensitivity: Measure of instrument's ability to discriminate between small differences in analyte concentration, connected to slope of calibration curve and precision of measuring device

Calibration sensitivity: Sensitivity figure of merit, the slope of the calibration curve, does not account for precision.

Analytical sensitivity: Sensitivity figure of merit, the slope of the calibration curve divided by standard deviation - accounts for precision but becomes concentration dependent

Detection limit: Figure of merit, minimum concentration or mass of analyte that can be detected at a known confidence level

Dynamic range: Range of concentrations between limit of quantification (LOQ) to limit of linearity (LOL) (See Fig. 5.1).

Limit of quantification (LOQ): It is the lowest analyte concentration that can be quantitatively detected with a stated accuracy and precision.

Limit of detection (LOD): It is the lowest quantity of a substance that can be distinguished from the absence of that substance (a blank value) with a stated confidence level (generally 99%).

Limit of Linearity (LOL): The point where a plot of concentration versus response goes non-linear (see Fig. 5.1).

Linearity or dynamic range: Linearity is a property that is between the limit of quantification (LOQ) and the point where a plot of concentration versus response goes non-linear.

*It may also defined as c*oncentration range over which a linearly changing instrumental response is observed.

Selectivity: Degree to which the method is free from interference by other species contained in the sample matrix

Selectivity coefficient: Figure of merit, gives the response of the method to species B relative to species A - will be negative for suppression and positive for enhancement.

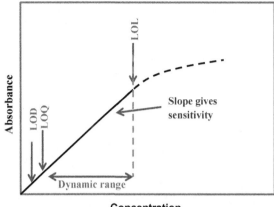

FIG. 5.1

A calibration curve plot showing limit of detection (LOD), limit of quantification (LOQ), dynamic range, and limit of linearity (LOL) [With permission from: https://en.wikipedia.org/wiki/Calibration_curve]. Notes: The terms; Precision, Bias, Sensitivity, Detection limit, Dynamic range and Selectivity are called Figure of Merits which reflects the Performance characteristics of an instrument.

5.2.2 Theory: Beer-Lambert's law (recap from Chapter 2)

As given in (Eq. 2.7 and 2.8), absorbance signal of an analyte and its concentration are related as,

$$A = \alpha c l$$

Here α represent the absorption coefficient. The absorption coefficient depends on wavelength. Therefore, its value is taken at the wavelength at which absorption of light is maximum, i.e. at λ_{max}, and it is represented as α_{max}, is the absorbance for unit path length.

α $(=A/c*l)$; is related with incoming radiation intensity $P = P_0 \times 10^{-\alpha l}$, where l represents the total path length (thickness) from which light passes in the sample container.

Molar absorption coefficient (or extinction coefficient) is the specific absorption coefficient for a concentration of 1 mol L^{-1} and a path length of 1 cm. That is, Eq. (2.7) rearranged as:

$$\alpha = \frac{A}{c \times l}$$

where $\varepsilon = \frac{\alpha}{2.303}$, is the molar absorption coefficient or molar absorptivity (which is called Beer-Lambert equation). The SI unit of absorption coefficient is the M^{-1}·cm^{-1} or the L·mol^{-1}·cm^{-1}, when the concentration is in molarity and the path length in centimeters.

5.2.2.1 Application of Beer's law

1. One of the most widely used applications of the Beer's law is to identify the concentration of an analyte by measuring absorbance (or transmittance). Specifically, by varying *concentration c and path length l*, the validity of Beer-Lambert law can be tested and the value of ε can be evaluated. When the value of ε is known, the concentration c_x of an unknown solution can be calculated from the relation, $c_x = \frac{log P_0/P_t}{\varepsilon l}$ (from Eq. (2.8) of Chapter 2).

Besides the wavelength of the incident radiation, the molar absorption coefficient ε depends upon the temperature and the solvent employed. For matched cells (i.e. l is constant), the Beer-Lambert law may be written as,

$$c \propto \frac{logP_0}{P} \quad (\text{i.e., } c \propto A)$$

Hence plotting A against concentration will produce a straight line, and this will pass through the point *concentration c = 0* and *A = 0 (T = 100%)*. This straight line of the form $y = mc_x + b$ may then be used to determine the concentrations c_x of analyte solutions (unknown solution): One need to find the slope of this linear line and employ it into *Eq.* $c_x = y/m$, in the case when the fitted linear line passes though origin *(c = 0* and *A = 0)*.

2. Also, Beer's law can be applied to a solution containing more than one kind of absorbing species, provided that the mixed substances do not interact. For more than one absorbing species, the absorbance, A is $\Sigma\varepsilon_i c_i l$ where ε_i and c_i are the molar absorption coefficient and concentration of ith species. Hence $A_{total} = A_1 + A_2 + \ldots \ldots + A_i, A_{total} = \varepsilon_1 c_1 l + \varepsilon_2 c_2 l + \ldots + \varepsilon_n c_n l$, [See also Eq. (2.9) in Chapter 2]

where the subscripts refer to absorbing components *1, 2, ..., n.*

Below we will learn the standard absorption spectrophotometric methods for determining both the inorganic and organic contaminants (mostly in the ionic form: anions, cations) prevailing in water sources.

5.3 Instrument calibration (by External Standard Method)

Among calibration methods, i) External Standard Calibration and ii) Standard Addition are frequently employed. The ist is appropriate if analyte is pure, whereas iind is suitable if the analyte is contaminated with other chemical contaminants that may influence the response from analyte (called matrix effect).

External standard Method: For an *external standard quantitation*, known data from a calibration standard and unknown data from the sample are combined to generate a quantitative report. It is called external standard because the standard or known material is separate or external to the unknown (in other world we call "external" because we prepare and analyze the standards separate from the samples. This is the most common method of standardization: It uses one or more external standards, each containing a known concentration of analyte.

1. A series of such external standards (usually, three or more) containing the analyte in known concentrations is prepared.
2. Then, the response signal (absorbance, peak height, peak area) as a function of the known concentration is measured.
3. A calibration curve is prepared by plotting the data and by fitting them to a suitable mathematical equation, such as the slope-intercept form ($y = m*c_x + b$, y-absorbance, m-slope, c_x-concentration of unknown analyte to be determined and b-intercept) used in the method of linear least squires.
4. The next step is the measuring the response signal (absorbance, peak height, peak area) of the sample of unknown concentration (c_x). Then estimate the concentration of unknown from the calibration curve or best-fit equation.

5. Here, y (absorbance) is the dependent variable and c_x-axis (concentration) is independent variable. Usually it is desirable that the plot approximates a straight line. However, because of the errors in the measurement process not all the data fall exactly on the line. Therefore, we need to draw a best straight line among the data points by using *regression analysis.* Regression analysis also provides the uncertainties associated with its subsequent use. The uncertainties are related to residual (residual are measure of how far away from the best straight line the data point lie). The least squire model is often applied to obtain an equation for the line on the basis of the following two a*ssumptions:*

I. The first is that there is actually a linear relationship between the measured response y and analyte's concentration (x-axis). The model that describes this assumption is called regression model, and represented as: $y = mc_x + b$.

II. Where m is the slope of the line and b is the intercept (gives the y value when x is equal to zero). Also, the second assumption is that any deviation of the data points is because of the error in the measurement. There is no error in the values in the x-axis.

Note, in cases where the data do not fit a linear model, nonlinear regression methods are applied. The slope m and intercept b are determined from the plot of y values obtained from the measurement for samples of known concentrations. For determining an unknown concentration c_x from the least squares line, the value of instrument response y is obtained for unknown concentration c_x, and the slope and intercept are used to calculate the unknown concentration, as,

$$c_x = (y - b)/m \tag{5.1}$$

The standard deviation (S_c) in c_x can be found from the standard error of the estimate S_y as,

$$S_c = \frac{S_y}{m} \sqrt{\frac{1}{M} + \frac{1}{N} + \frac{\left(\underline{y}_c - \underline{y}\right)^2}{m^2 S_{xx}}} \tag{5.2}$$

where M is the number of replicate results, N is the number of points in the calibration curve (number of standards), $\underline{y}_c$ is the mean response for the unknown, and y is the mean value of y for the calibration results. The quantity S_{xx} is the sum of the squares of the deviations of x values from the mean as given in Eq. (5.2).

5.4 Quantitative analysis of inorganic species (metals and nonmetals)
5.4.1 Anions
5.4.1.1 *Determination of nitrite* (NO_3^-) *present in water*
5.4.1.1.1 General discussion: *Occurrence and effects*
Nitrate is an inorganic compound that occurs under a variety of conditions in the environment, both naturally and synthetically. It occurs naturally in mineral deposits, soils, seawater, freshwater systems, the atmosphere, and biota. It is a more stable form of combined nitrogen in oxygenated water and found in the highest levels in groundwater under extensively developed areas. It enters the environment from fertilizer, feed yards, and sewage. Synthetically, nitrate in groundwater originates primarily from fertilizers, septic systems, and manure storage or spreading operations. For example, fertilizer nitrogen that is not taken up by plants, volatilized, or carried away by surface runoff leaches to the groundwater

in the form of nitrate. This not only makes the nitrogen unavailable to crops, but also can elevate the concentration in groundwater above the levels acceptable for drinking water quality.

Nitrate does not normally cause health problems unless it is reduced to nitrite. But, it is regulated in drinking water primarily because excess levels can cause methemoglobinemia, or "blue baby" disease. Although nitrate levels that affect infants do not pose a direct threat to older children and adults, they do indicate the possible presence of other more serious residential or agricultural contaminants, such as bacteria or pesticides.

The US standard for nitrate in drinking water is 10 mg per liter (10 mg/L) nitrate-N, or 45 mg/L nitrate-NO_3.

Determination of nitrate $(NO_3{}^-)$ is difficult due to the requirement of the relatively complex procedures. The main reasons for the complexity are the high probability of prevailing interfering constituents, and the limited concentration ranges. An ultraviolet visible (UV−Vis) spectroscopic technique that measures the absorbance of $NO_3{}^-NO_3^-$ at 220 nm is suitable for monitoring uncontaminated water. However it is recommended to use this technique only for samples that have low organic matter contents, i.e., uncontaminated natural waters and potable water supplies).

5.4.1.1.2 Principle for analysis
It is documented that the $NO_3{}^-$ calibration curve follows Beer's law up to 11 mg/L. Measurement of UV absorption at 220 nm enables rapid determination of $NO_3{}^-$. Because dissolved organic matter also may absorb at 220 nm and $NO_3{}^-$ does not absorb at 275 nm, a second measurement made at 275 nm has been used to correct the $NO_3{}^-$ value. The extent of this empirical correction is related to the nature and concentration of organic matter and may vary from one water to another. Correction factors for organic matter absorbance can be established by the *Standard Addition Method* in combination with analysis of the original $NO_3{}^-$ content by another method.

Sample filtration is required to remove possible interference from suspended particles. Acidification with 1 N HCl is considered to prevent interference from hydroxide or carbonate. Chloride has no effect on the determination.

5.4.1.1.3 Interference
$NO_3{}^-$ interferes with dissolved organic matter, surfactants, $NO_2{}^-$ and Cr^{6+}. Various inorganic ions not normally found in natural water, such as chlorite and chlorate, may also interfere. Inorganic substances can be compensated for by independent analysis of their concentrations and preparation of individual correction curves.

5.4.1.1.4 Storage of samples
Start $NO_3{}^-$ determinations promptly after sampling. If storage is necessary, store for up to 24 h at 4 °C. But for longer storage, preserve samples by adding 2 mL conc. H_2SO_4/L and storing at 4 °C.

Notes: When samples are preserved with acid, nitrite and nitrate cannot be determined as individual species. Measurements with Spectrophotometric methods require an optically clear sample. Filter turbid sample through 0.45 -μm-pore-diam membrane filter.

5.4.1.1.5 Apparatus
Spectrophotometer, for use at 220 nm and 275 nm with matched cells of 1-cm or longer light path, membrane filter (0.45 -μm-pore-diam).

5.4.1.1.6 Reagents

a. *Nitrate free water:* Use double distilled or distilled and deionized water of highest purity to prepare all solutions and dilutions.

b. *Stock nitrate solution:* Dry potassium nitrate (KNO_3: MW = 101.103 g/mol) in an oven at 105 °C for 24 h. Dissolve 0.7218 g in water and dilute to 1000 mL (1.0 mL = 700 µg NO_3^-). Preserve with 2 mL $CHCl_3$/L. This solution is stable for at least 6 months.

c. *Intermediate nitrate solution:* Dilute 100 mL stock nitrate solution to 1000 mL with water to give 70 mg/L (1.0 mL = 70.0 µg NO_3^-). Preserve with 2 mL $CHCl_3$/L. This solution is stable for 6 months.

d. *Hydrochloric acid solution:* Prepare 1 N HCl solution in distilled water.

Procedure:

a. *Treatment of sample:* To 50 mL clear sample, filtered if necessary, add 1 mL HCl solution of 1 N and mix thoroughly.

b. Preparation of standard curve (External Standard Method):

    I. *Preparation of NO_3^- standards:* Prepare NO_3^- standards (for calibration purpose) in the range 0–50 mg NO_3^-/L by diluting to 50 mL the following volumes of intermediate nitrate solution: 0.0, 1.0 mL (1.4 mg/L), 2.0 mL (2.8 mg/L), 4.0 mL (5.6 mg/L), 7.0 mL (9.8 mg/L), 10.0 mL (14 mg/L), 15 mL (21 mg/L), 20 mL (28 mg/L), 25 mL (35 mg/L) and 35 mL (49 mg/L) mL. Treat NO_3^- standards in the same manner as samples (that is, add 1 mL HCl solution in standard solutions as well). Note: if you suspect of contamination of organic matter follow the *Standard Addition Method* for calibration (see for example in sub-section 5.4.5).

    II. *Spectrophotometeric measurement:* Read absorbance or transmittance against redistilled water set at zero absorbance or 100% transmittance. Use a wavelength of 220 nm to obtain NO_3^- reading and a wavelength of 275 nm to determine interference due to dissolved organic matter.

    III. *Calibration curve:* Prepare the calibration curve by plotting measured absorbance versus concentration of standard reagent and make a least square fit (linear fit) to the measured data, as shown in Fig. 5.2.

5.4.1.2 Determination of nitrite (NO_2^-)

5.4.1.2.1 General discussion

Nitrate and nitrite ions are important indicators of pollution by organic materials as nitrogen from decomposing organic substances often ends up as nitrate or nitrite ions. Nitrite is commonly determined by a spectrophotometric procedure using the *Griess reaction*.

5.4.1.2.2 Principle

Nitrite is determined through formation of a reddish purple azo dye produced at pH 2.0 to 2.5 by coupling diazotized sulfanilamide with N-(1-naphthyl)-ethylenediamine dihydrochloride (NED dihydrochloride). In other words, the sample containing nitrite is reacted with Sulfanilimide and N-(1-naphthyl)-ethylenediamine to form a colored species that absorbs radiation at 550 nm. This photometric method is suitable for determining concentrations of 5–1000 µg NO_2 N/L in water.

Analytical calibration using a weighted linear curve fit, with error estimation

Calibration data

Weights	Concentration of standards	Instrument readings
	0.100	1.54
	0.200	2.03
	0.300	3.17
	0.400	3.67
	0.500	4.89
	0.600	6.73
	0.700	6.74
	0.800	7.87
	0.900	8.86
	1.000	10.35

Application to unknowns

Readings of the unknowns	Calculated concentration
1.54	0.1371
2.03	0.1871
3.17	0.3035
3.67	0.3545
4.89	0.4791
6.73	0.6669
6.74	0.6679
7.87	0.7832
8.86	0.8843
10.35	1.0354

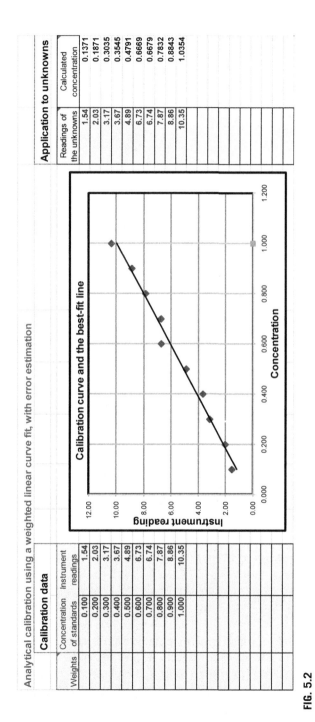

Calibration curve and the best-fit line

FIG. 5.2

Calibration curve prepared from measured concentration of an analyte versus instrument's response (e.g., absorbance, transmittance). Diamond shaped symbol represent measured data, whereas solid line is a model linear) fit (Ref. [12]).

With the permission from: Cengage Learning India Pvt. Ltd.

5.4.1.2.3 Interferences

The following ions interfere because of precipitation under test conditions and should be absent: Sb (III), Au (III), Bi (III), Fe (III), Pb (II), Hg (II), Ag (I), chloroplatinate $[PtCl_6^{2-}]$, and metavanadate $[VO_3^{2-}]$. Cupric ion may cause low results by catalyzing decomposition of the diazonium salt. Colored ions that alter the color system also should be absent. Co-existence nitrite, free chlorine, and nitrogen trichloride (NCl_3) is not possible due to chemical in compatibility.

5.4.1.2.4 Storage of samples

Start NO_2^- determinations promptly after sampling. Never use acid preservation for samples to be analyzed for nitrite. Make the determination promptly on fresh samples to prevent bacterial conversion of nitrite to nitrate or ammonia. For short-term preservation for 1–2 d, store at 4 °C. Also, as measurements with spectrophotometric methods require an optically clear sample, filter turbid sample through 0.45 -μm-pore-diam membrane filter.

5.4.1.2.5 Apparatus

Spectrophotometer for use at 550 nm, matched cells of 1-cm or longer light path, membrane filter (0.45-μm-pore-diam). Plastic containers (plastic bottles for sample collection) or glass containers (beakers, volumetric flasks for solution preparation, measuring cylinder, pipette).

5.4.1.2.6 Reagents

a. *Nitrite free water:* Use double distilled or distilled and deionized water of highest purity to prepare all solutions and dilutions. Sulfanilamide, N-(1-naphthyl)-ethylenediamine dihydrochloride (NED dihydrochloride), and nitrite.

Procedure (Standard addition method):

1. Pipette out 5 mL aliquots of the sample into five 50.0 mL volumetric flasks.
2. Prepare standard solution containing 10.0 μM nitrite.
3. Then add 0.0, 2.0, 4.0, 6.0, 8.0, mL of a standard solution of nitrite into each flask, and add the color-forming reagents.
4. After dilution to volume (volume makeup upto the mark of volumetric flask, in this case 50 mL), measure the absorbance for each of the five standard solutions at 550 nm.
5. Prepare the calibration curve by plotting measured absorbance versus volume of standard reagent added and make a least square fit (linear fit) to the measured data.

One of widely used approach, for evaluating the concentration of unknown analyte, following equation is used:

$$C_x = \frac{S_1 C_s V_s}{(S_2 - S_1)V_x}, \text{which is reduced to } C_x = \frac{C_s(V_s)_0}{V_x} \tag{5.3}$$

V_x is the volume of only the analyte of which we wish to analyze nitrite concentration (in this example, 5 mL), C_s is the concentration of the standard reagent, $(V_s)_0$ is the volume of the standard reagent equivalent to the amount of analyte in the unknown analyte. It is obtained from the difference between the volume of the standard added at the origin (zero mL, in this case) and the value of the volume at the intersection of the extrapolated portion of the fitted line with the x-axis.

Alternatively, following derivation can be used to illustrate how this method (standard edition) works. Only measurement of absorbance of unknown sample and of one standard sample is required. First, we need to assume that the signal we are measuring for either the unknown sample (of which we wish to know the analyte concentration) or the standard will be proportional to the concentration of the analyte in each. We can represent this for absorbance measurements by the following relationships,

$$\text{Absorbance of unknown sample (analyte)}: A_x = \varepsilon * b * C_x \tag{5.4}$$

$$\text{Absorbance of spiked sample (unknown sample + standard)}: A_{sp} = \varepsilon * b * C_{sp} \tag{5.5}$$

where C_x is the concentration of the analyte in the unknown sample, C_{sp} is the total concentration of analyte in the total volume of the sample that has been spiked with a known amount of the analyte. We are assuming that the absorption coefficient ε *and path length b* is the same for both the original and spiked samples, which should be true if they contain the same matrix and we are working in the linear range of Beer's law.

Along with the measured absorbance of the original and spiked samples, we also know:

(1) The original volume taken of the unknown sample (V_x),
(2) The volume of the standard solution (V_s) that we spiked (added) into this sample and
(3) The concentration of analyte in this standard solution (C_s).

This information can be combined to find the value of concentration C_x of analyte in the original sample, C_x, which is the goal of the standard addition method. To do this, we divide Eq. (5.5) by (5.4) to find the ratio of A_x/A_{sp} and substitute in the fact that $C_{sp} = (C_x V_x + C_s V_s)/(V_x + V_s)$.

$$\frac{A_x}{A_{sp}} = \frac{C_x (V_x + V_s)}{C_x V_x + C_s V_s} \tag{5.6}$$

With this combined equation we can use the measured ratio A_x/A_{sp} to calculate the value of C_x because we know the values of all other terms in this expression. The above equation shows how standard addition can be employed when using only one spiked sample. This process is illustrated in the following exercise.

Exercises: 5.1. A chemist performs a colorimetric assay that selectively measures iron. A 20.0 mL portion of the original sample gives an absorbance reading of 0.367 and a 20 mL portion of the same sample that has been spiked with 5.0 mL of a 2.00×10^{-2} M iron solution gives an absorbance of 0.538. What was the concentration of iron in the original sample?

We are given information of the measured absorbances ($A_x = 0.367$ and $A_{sp} = 0.538$), solution volumes, and concentration of the iron solution that was added into the sample. We simply have to place this information into Eq. (5.6) and rearrange this equation to solve for C_x.

$$\frac{0.367}{0.538} = \frac{C_x(0.02\ L + 0.005\ L)}{C_x(0.02\ L) + (0.005\ L) \times (2.0 \times 10^{-2})}$$

$$\text{Therefore,} \quad C_x = \frac{0.682 \times (1.0 \times 10^{-4}\ mol)}{0.025L - 0.02\ L} = 0.0136\ \text{M}$$

5.4.1.2.7 Alternative method (phosphomolybdenum blue complex method)

Also, nitrate and nitrite can be determined, using phosphomolybdenum blue complex, which is discussed below.

5.4.1.2.8 Principle

It is based on the reduction of phosphomolybdic acid to phosphomolybdenum blue complex by sodium sulfide. The obtained phosphomolybdenum blue complex is oxidized by the addition of nitrite causing a reduction in intensity of the blue color. The decrease in the absorbance of the blue color is directly proportional to the amount of nitrite added. The absorbance of the phosphomolybdenum blue complex is monitored spectrophotometrically at 814 nm and related to the concentration of nitrite. The main advantage of the proposed method over the method discussed in section 5.4.7 is related to the short analysis time and the low detection limit.

5.4.1.2.9 Reagents

Molybdenum (VI) solution 0.1 M is prepared by weighing accurately 1.44 g of MoO_3 and dissolving it in 40 mL of 1 M NaOH; the volume is completed to 100 mL with water.

Potassium dihydrogen phosphate solution 0.10 M is prepared by weighing accurately 1.33 g of KH_2PO_4 and dissolving it in water in a 100-mL volumetric flask.

Sodium sulfide solution 0.01% (w:v) is prepared by weighing accurately 1.0 g of Na_2S and dissolving it in water in a 100-mL volumetric flask.

Nitrite solution 0.10 M is prepared by dissolving exactly 0.69 g of $NaNO_2$ in water in a 100-mL volumetric flask; working solutions is prepared by diluting volumes of the stock solution to known volumes with water.

Nitrate standard solution 0.10 M is prepared by dissolving exactly 0.85 g of $NaNO_3$ in water in a 100-mL volumetric flask; working solutions are prepared by diluting volumes of the stock solution to known volumes with water. Modified Jones reductor was prepared as described in the AOAC (Association of Official Agricultural Chemists) official methods of analysis.

5.4.1.2.10 Apparatus

Spectrophotometer (with stable light source at 814 nm). Constant temperature cell holder.

The cells 1×1 cm glass cells. A pH meter (for pH measurements).

Procedures:

a. Preparation of phosphomolybdenum blue complex
  i. A 30-mL volume of 0.1 M molybdenum(VI) solution is transferred into a 100-mL volumetric flask, then 10 mL of 0.10 M potassium dihydrogen phosphate solution is added, followed by 10 mL of 0.01% (w:v) sodium sulfide solution and 13 mL of 11.2 M HCl, in that order.
  ii. The volume is completed with water.
  iii. The absorbance of the solution is measured after 30 min at 814 nm against water as a blank in a thermostatted bath at $25 \pm 0.2\,°C$.

b. Spectrophotometric determination of nitrite
  i. An aliquot of solution containing nitrite ions in the range 4.60–36.00 ppm is transferred into a 10-mL volumetric flask.
  ii. Then 3.0 mL of phosphomolybdenum blue complex is added and the volume is completed with water.

iii. A portion of the solution was placed in the cell and the absorbance versus time curve was recorded at 814 nm against water as a blank in a thermostatted bath at $25 \pm 0.2\ °C$.

iv. The concentration of nitrite can be calculated either by measuring the absorbance of the solution after exactly 30 min at 814 nm against water as a blank, or, in a different approach, by measuring the slope $dA{:}dt$ of the reaction curve at 5 min after initiating the reaction.

c. Spectrophotometric determination of nitrate

  i. A 5-mL portion of solution containing an amount of nitrate in the range $10-100$ ppm is transferred into the modified Jones reductor where the flow rate is adjusted to $3-5$ mL/min.

  ii. The reductor is then washed with 5 mL of water.

  iii. The nitrite solution obtained from the reductor is treated as described in the above procedure for spectrophotometric determination of nitrite.

d. Preparation of real samples for analysis

For water samples, an appropriate volume of water is treated using the above procedure for the determination of nitrite and nitrate.

Notes: The methods discussed in 5.4.1.1 and 5.4.1.2 (for $NO_3{}^-$ and for NO_2^- determination) are based on Ref. [8−16] and also listed below:

- APHA Method 4500-NO_3: Standard Methods for the examination of Water and Wastewater, 18th ed. 1992.
- M. McCasland, N. M. Trautmann, K. S. Porter and R. J. Wagenet, Nitrate: Health Effects in Drinking Water, Pesticide Safety Education Program (PSEP), Kornell University: < http://psep. cce.cornell.edu/facts-slides-self/facts/nit-heef-grw85.aspx>.
- F. A. J. Armstrong, Determination of Nitrate in Water Ultraviolet Spectrophotometry, Anal. Chem. 35 [9] (1963) 1292−1294.
- J. Patton and J. R. Kryskalla, Colorimetric Determination of Nitrate Plus Nitrite in Water by Enzymatic Reduction, Automated Discrete Analyzer Methods, Book Chapter 8 Section B, Methods of the National Water Quality Laboratory, Laboratory Analysis. Techniques and Methods 5−B8. USGS. Virginia, 2011.
- D. A. Skoog, F. J. Holler and S. R. Crouch, Instrumental Analysis, Cengage Learning, New Delhi (Reprint), 2007, pp. 420.
- D. S. Hage and J. D. Carr, Analytical Chemistry and Quantitative Analysis, Pearson Publication (Int. Ed.), 2011, pg.442.
- N. A. Zatar, M. A. Abu-Eid, and Abdullah F. Eid, Spectrophotometric determination of nitrite and nitrate using phosphomolybdenum blue complex, Talanta 50 (1999) 819−826.
- A. Shrivastava, V. B. Gupta, Methods for the determination of limit of detection and limit of quantitation of the analytical methods, Chronicles of Young Scientists, **2** (2011).

5.4.1.3 (A) determination of phosphorus (P) in the phosphate form in water (ammonium molybdate method)

5.4.1.3.1 General discussion

Occurrence and effect: Both phosphorus and nitrogen are essential nutrients for the plants and animals that make up the aquatic food web. Since phosphorus is the nutrient in short supply in most fresh waters, even a modest increase in phosphorus can, under the right conditions, set off a whole chain of undesirable events in a stream including accelerated plant growth, algae blooms, low dissolved oxygen, and the death of certain fish, invertebrates, and other aquatic animals.

Table 5.2 Range of phosphorous concentration and suitable wavelength for absorption measurement.

P range [mg/L]	Wavelength [nm]
1.0–5.0	400
2.0–10	420
4.0–18	470

There are many sources of phosphorus, both natural and human. These include soil and rocks, wastewater treatment plants, runoff from fertilized lawns and cropland, runoff from animal manure storage areas, disturbed land areas, drained wetlands, water treatment, and commercial cleaning preparations.

Phosphorus occurs in natural water and waste waters almost solely as phosphate (PO_4) and pure, "elemental" phosphorus (P) is rare. Phosphorus in aquatic systems occurs as organic phosphate and inorganic phosphate. It can occur in dissolved and undissolved forms. These are classified as ortho-phosphates, condensed phosphates (pyro-, metal- and other polyphosphates) and organically bound phosphates. They occur in solution, in particles or in the aquatic organisms. Phosphorus can be determined accurately with the UV–Vis spectrophotometric method in above sources in the range from 0.01 mg/L to 0.5 mg/L.

The phosphorus has been divided into three chemical types: reactive, acid hydrolysable, and organic phosphorus.

Reactive phosphorus: Phosphates that respond to colorimetric tests without preliminary hydrolysis or oxidative digestion of the sample are termed as "reactive phosphorus." Reactive phosphorus, which occurs in both dissolved and suspended forms, is largely a measure of orthophosphate.

Acid-hydrolyzable phosphorus: Acid hydrolysis at boiling-water temperature converts dissolved and particulate condensed phosphates to dissolved orthophosphate. This is called "acid-hydrolyzable phosphorus".

Organically bound phosphorus: The phosphate fractions that are converted to orthophosphate only by oxidation destruction of the organic matter are called "organic" or "organically bound phosphorus".

Principle: in a dilute orthophosphate solution, ammonium molybdate reacts under acid conditions to form a heteropoly acid, molybdophosphoric **acid**. In the presence of vanadium, yellow vanado-molybdophosphoric acid is formed. The intensity of the yellow color is proportional to phosphate concentration (Table 5.2).

Interferences: The wavelength at which absorbance is measured depends on sensitivity desired, because sensitivity varies tenfold with wavelengths 400–490 nm. Ferric iron causes interference at low wavelengths, particularly at 400 nm. A wavelength of 470 nm usually is used. Concentration ranges for different wavelengths are given in Table 5.1:

5.4.1.3.2 Sample storage
Minimum detectable concentration: The minimum detectable concentration is 200 µg/L in 1-cm spectrophotometer cells.

5.4.1.3.3 Apparatus

I. Spectrophotometer, provided with a blue or violet filter exhibiting maximum transmittance between 400 and 490 nm.

II. Cuvettes (a pair)

III. 1 mL variable micropipette

IV. Autoclave or pressure cooker, capable of operating at 98−137 kPa.

V. *Acid-washed glassware:* Because of its absorption on glass surfaces, use acid-washed glassware for determining low concentration of phosphorus in the form of phosphate contamination is common. Avoid using commercial detergents containing phosphate. Clean all glassware with hot dilute HCl and rinse well with distilled water. Preferably, reserve the glassware only for phosphate determination and after use, wash and keep filled with water until needed. If this is done, acid treatment is required only occasionally.

5.4.1.3.4 Reagents

a. Phenolphthalein indicator aqueous solution.

b. *Hydrochloric acid, 1+1HCl∗.* The acid concentration in the determination is not critical but a final sample concentration of 0.5 N is recommended. H_2SO_4, $HClO_4$, or HNO_3 may also be substituted for HCl.

c. *Vanadate-Molybdate reagent*: Prepare commercially available *Vanadate-Molybdate reagent* or prepare it by the following method:

d. *Solution A:* Dissolve 2.5 g of ammonium molybdate $((NH_4)_6Mo_7O_{24},4H_2O)$ in 300 mL distilled water.

e. *Solution B:* Dissolve 1.25 g ammonium metavanadate (NH_4VO_3) by heating to boiling in 300 mL distilled water, cool and add 330 mL conc. HCl.

f. Cool *Solution B* to room temperature, pour *Solution A* into *Solution B*, mix, and dilute to 1 L.

g. *Standard phosphate solution:* Dissolve in distilled water 219.5 mg anhydrous KH_2PO_4 and dilute to 1000 mL [1.0 mL = 50 µg PO_4^{3-} -P].

h. *∗1 + 1 HCl:* It means take one volume of concentrated hydrochloride acid and mix with one volume of water, doesn't matter what the volume is, it is the same as 50% dilution. For example 10 mL of acid mixed with 10 mL of water is a 1:1 dilution or a 1 + 1 dilution or a 50% dilution.

5.4.1.3.5 Method

This experiment makes use of a series of phosphorus standard solutions with known concentrations and their resulting absorbances (measured by a spectrophotometer) to create an equation for a line of concentration versus absorbance. This equation is then used to calculate the concentration of an unknown phosphorus solution. The overall purpose of the calibration curve is to determine the concentration of unknown analytes based on what is known.

a. Sample preparation and measurement process

    I. Transfer 1 mL of sample water into a clean cuvette.

    II. Add 0.25 mL of vanadate-molybdate reagent and mix by pipetting up and down several times.

    III. Wait at least 10 min for color to develop

IV. Place the cuvette with your sample into the spectrophotometer (or colorimeter) and start measurement. The program will return the absorbance that is proportional to the phosphate concentration in standard analyte sample.

V. In a dilute orthophosphate solution, ammonium molybdate reacts under acid conditions to form a heteropoly acid, molybdophosphoric acid. In the presence of vanadium, yellow vanadomolybdophosphoric acid is formed. The intensity of the yellow color is proportional to the phosphate concentration.

b. *Sample* pH *adjustment:* If sample pH is greater than 10, add 0.05 mL (1 drop) phenolphthalein indicator to 50.0 mL sample and discharge the color red with 1 + 1 HCl before diluted to 100 mL.

c. *Color removal from sample:* Remove excessive color in sample by shaking about 50 mL with 200 mg activated carbon in an erlenmeyer flask for 5 min and filter to remove carbon. Check each batch of carbon for phosphate because some batches produce high reagent blanks.

d. *Color development in sample:* Place 35 mL or less of sample (containing 0.05–1.0 mg P) in a 50-mL volumetric flask. Add 10 mL vanadate-molybdate reagent and dilute to the mark with distilled water.

e. Prepare a blank in which 35 mL distilled water is substituted for the sample.

f. After 10 min or more, measure absorbance of sample vs. a blank at a wavelength of 400–490 nm, depending on the sensitivity desired. The color is stable for days and its intensity is unaffected by variation in room temperature.

g. *Preparation of calibration curve:* Prepare a calibration curve by using suitable volumes of standard phosphate solution and proceeding with the same procedure from c. When ferric ion is low enough not to interfere, plot a calibration curves of one series of standard solutions for various wavelengths. This permits wide latitude of concentrations in one series of determinations.

h. Analyze at least one standard with each set of samples.

i. Calculation:

$$mgP/L = \frac{\text{mg } P(\text{in 50 mL final volume}) \times 1000}{\text{sample volume (mL)}}$$

5.4.1.4 (B) determination of phosphorus (all forms of phosphorus-molybdate/ascorbic acid method with single reagent)

Notes: This is an alternative method for determination of total phosphorus.

5.4.1.4.1 Principle of the method

Ammonium molybdate and antimony potassium tartrate react in an acidic medium with dilute solutions of phosphorus to form an antimonyphospho-molybdate complex. This complex is reduced to an intensely blue colored complex by ascorbic acid. The color is proportional to the phosphorus concentration.

Only orthophosphate forms a blue color in this test. Polyphosphate (and some organic phosphorus compounds) can be converted to the orthophosphate form by sulfuric acid hydrolysis. Organic phosphorus compounds may be converted to the orthophosphate form by persulfate digestion.

Interferences: Positive interference is caused by silica and arsenate only if the sample is heated. Blue color is caused by ferrous iron but this does not affect results if ferrous iron concentration is less than 100 mg/L. Sulphide interference can be removed by oxidation with bromine water. No interference is caused in concentrations up to 1000 mg/L by Al^{3+}, Fe^{2+}, Mg^{2+}, Ca^{2+}, Ba^{2+}, Sr^{2+}, Li^{+}, Na^{+}, K^{+}, NH^{4+}, Cd^{+}, Mn^{2+}, Pb^{2+}, Hg^{+}, Hg^{2+}, Sn^{2+}, Cu^{2+}, Ni^{2+}, Ag^{+}, AsO_3^{-}, Br^{-}, CO_3^{-}, ClO_4^{-}, CN^{-}, IO_3^{-}, SiO_4^{-}, NO_3^{-}, NO_2^{-}, SO_4^{2-}. If HNO_3 is used in the test, Cl^{-} interferes at 75 mg/L.

The range of detectable label of phosphorus concentration is 0.01−1.2 mg in 10-mm spectrophotometric cells.

5.4.1.4.2 Sample handling and preservation

If benthic (bottom of river, lake or other water source) deposits are present in the area being sampled, special care should be taken not to include these deposits.

Sample containers may be of plastic material, such as cubitainers, or of Pyrex glass.

If the analysis cannot be performed the day of collection, the samples should be preserved by the addition of 2 mL conc. H_2SO_4 per liter and refrigeration at 4 °C.

5.4.1.4.3 Apparatus and reagents

The absorbance measurements are performed using UV/Vis spectrophotometer. Operation parameters for these measurements are: wavelength 880 nm and cell 10-mm.

Apparatus: Volumetric flasks, volume 50 mL, 100 mL, 250 mL, 500 mL, Erlenmeyer flasks, volume 250 mL, 100 mL, Polyethylene bottle, pH meter, Water bath (95 °C).

Sulfuric acid solution, 2.5 mol/L, Antimony potassium tartrate solution, Ammonium molybdate solution, Ascorbic acid, 0.1 mol/L, Combined reagent, Phosphorus stock solution (50 mg/L P), Phosphorus working solution (0.5 mg/L P).

Acid washed glassware: All glassware used should be washed with hot 1:1 HCl and rinsed with distilled water and treated with all the reagents to remove the last traces of phosphorus that might be absorbed on the glassware. Preferably, this glassware should be used only for the determination of phosphorus and after use it should be washed with distilled water and kept covered until needed again. If this is done, the treatment with 1:1 HCl and reagents is only required occasionally. Commercial detergents should never be used.

5.4.1.4.4 Reagents

  i. *Ammonium molybdate-antimony potassium tartrate solution:* Dissolve 8 g of ammonium molybdate and 0.2 g antimony potassium tartrate in 800 mL of distilled water and dilute to 1 L.
 ii. *Ascorbic acid $C_6H_8O_6$ solution:* Dissolve 60 g of ascorbic acid in 800 mL of distilled water and dilute to 1 L. Add 2 mL of acetone. This solution is stable for two weeks.
iii. *Sulfuric acid, 11 N:* Slowly add 310 mL of conc. H_2SO_4 to approximately 600 mL distilled water. Cool and dilute to 1000 mL.
 iv. *Sodium bisulfite (NaHSO₃) solution:* Dissolve 5.2 g of $NaHSO_3$ in 100 mL of 1.0 N H_2SO_4.
  v. Ammonium persulfate.
 vi. *Stock phosphorus solution:* Dissolve 0.4393 g of pre-dried (105 °C for 1 h) KH_2PO_4 in distilled water and dilute to 1000 mL (1.0 mL = 0.1 mg P or 100 mg/L).
vii. *Standard phosphorus solution:* Dilute 100 mL of stock phosphorus solution to 1000 mL with distilled water. 1.0 mL = 0.01 mg P (10 mg/L). Prepare an appropriate series of standards by diluting suitable volumes of standard or stock solutions to 100 mL with distilled water.

Procedure:

a. Determination of total phosphorus

   i. Transfer 50 mL of sample or an aliquot diluted to 50 mL into a 125 mL Erlenmeyer flask and add 1 mL of 11 N sulfuric acid (reagent: iii).

   ii. Add 0.4 g ammonium persulfate (reagent: v), mix and boil gently for approximately 30–40 min or until a final volume of about 10 mL is reached. Alternatively heat for 30 min in an autoclave at 121 °C (with pressure: 15–20 psi). Cool, dilute to approximately 40 mL and filter.

   iii. For samples containing arsenic or high levels of iron, add 5 mL of sodium bisulfite (reagent: 4), mix and place in a 95 °C water bath for 30 min (20 min after the temperature of the sample reaches 95 °C). Cool and dilute to 50 mL.

   iv. Determine phosphorus as outlined in (5.3).

b. Hydrolyzable Phosphorus

   i. Add 1 mL of H_2SO_4 solution (reagent: 3) to a 50 mL sample in a 125 mL Erlenmeyer flask.

   ii. Boil gently on a pre-heated hot plate for 30–40 min or until a final volume of about 10 mL is reached. Do not allow sample to go to dryness. Alternatively, heat for 30 min in an autoclave at 121 °C (15–20 psi). Cool, dilute to approximately 40 mL and filter.

   iii. Treat the samples as in procedure mentioned in 5.1.3.

   iv. Determine phosphorus as outlined in (5.3) orthophosphate.

c. Orthophosphate

   i. To 50 mL of sample and/or standards, add 1 mL of 11 N sulfuric acid (reagent:3) and 4 mL of ammonium molybdate-antimony potassium tartrate (reagent: 1) and mix. NOTE: If sample has been digested for total or hydrolyzable phosphorus do not add acid.

   ii. Add 2 mL of ascorbic acid solution (reagent:2) and mix.

   iii. After 5 min, measure the absorbance at 650 nm with a spectrophotometer and determine the phosphorus concentration from the standard curve. The color is stable for at least 1 h. For concentrations in the range of 0.01–0.3 mg P/L, a 5 cm cell should be used. A one cm cell should be used for concentrations in the range of 0.3–1.2 mg P/L.

Notes: The methods discussed in 5.4.1.3 (for phosphorus with two different methods) are based on Ref. [16–20] and also listed below:

S. Lim (2011). Determination of phosphorus concentration in hydroponics solution.
EPA *Method 365.1 (1993), Revision 2.0: Determination of Phosphorus by Semi-Automated Colorimetry, US ENVIRONMENTAL PROTECTION AGENCY (EPA (Ed. by: J. W. O'Dell) Water.epa.gov.. Environmental Protection Agency*, 1978. Visited on: 16 Sept. 2015.
Method 4500-P, APHA "Standard Methods for the Examination of Water and Wastewater", 1992.
https://archive.epa.gov/water/archive/web/html/vms56.html

5.4.1.5 *Sulfate* $SO_4{}^{2-}$ *(turbimetric method)*
5.4.1.5.1 General discussion
Salts, acid derivatives, and peroxides of sulfate are widely used in industry. Sulfates occur widely in everyday life. Sulfates are salts of sulfuric acid and many are prepared from that acid Many of them are highly soluble in water. Some of the sulfate do not dissolve, however. For e.g., calcium sulfate,

strontium sulfate, lead(II) sulfate, and barium sulfate, which are poorly soluble. Although sulfate is not particularly toxic if the concentration is too high it may cause other problems, such as build up in high-pressure boilers, the taste of the water being bitter, and at concentrations above 1000 ppm, the water may cause diarrhea. Therefore, regular monitoring of this ion is necessary to maintain the water quality standard.

When testing the water quality, we need to keep in mind that an acceptable concentration for drinking water is 250 ppm sulfate.

5.4.1.5.1.1 Principle. The turbidimetric method to determine sulfate concentration is based on the fact that light is scattered by particulate matter in aqueous solution. For example, when barium and sulfate react in water, they make the solution turbid, which means the concentration of the sulfate can be measured by using a spectrophotometer (or colorimeter). The equation for the reaction of barium and sulfate is shown below:

$$SO_4^{2-}(aq) + Ba_2^+(aq) \rightarrow BaSO_4(s)$$

Sulfate ion is converted to a barium sulfate suspension under controlled conditions. The resulting turbidity is determined by a spectrophotometer and compared with a curve prepared from standard sulfate solution.

5.4.1.5.2 Scope and application

This method is applicable to groundwater, drinking and surface waters, and domestic and industrial wastes and is suitable for all concentration ranges of sulfate (SO_4^-); however, in order to obtain reliable readings, use a sample aliquot, containing not more than 40 mg/L of SO_4^-. The minimum detectable limit is approximately 1 mg/L of SO_4^{2-}.

5.4.1.5.3 Interferences

Color and turbidity due to the sample matrix can cause positive interferences which must be accounted for by use of blanks. Silica in concentrations over 500 mg/L will interfere.

5.4.1.5.4 Apparatus

1. *Magnetic stirrer:* Use identical shapes and sizes of magnetic stirring bars for preparing solution homogeneous. Maintain constant speed (just below splashing).
2. *Spectrophotometer:* For measurement use at 420 nm wavelength and sample cell of light path of 4–5 cm.

5.4.1.5.5 Reagents

1. *Double distilled water* (analytical grade- Type II water is the best).
2. *Conditioning reagent:* Slowly add 30 mL concentrated HCl to 300 mL distilled water, 100 mL 95% ethanol or isopropanol, and 75 g NaCl in solution in a container. Add 50 mL glycerol and mix.
3. *Barium chloride (BaCl₂):* Crystals with particle size range from 20 to 30 mesh.
4. *Sodium carbonate solution:* (approximately 0.05 N): Dry 3–5 g primary standard Na_2CO_3 at 250 °C for 4 h and cool in a desiccator. Weigh 2.5 g, transfer to a 1-L volumetric flask, and fill to the mark with distilled water.

5. *Proprietary reagents:* Such as Hach Sulfaver or equivalent are acceptable.
6. *Standard sulfate solution of 0.1 N (1.00 mL = 100 μg SO₄⁻):* Prepare standard sulfate solution from H_2SO_4 as follows.

Standard sulfuric acid, 0.1 N: Dilute 3.0 mL concentrated H_2SO_4 to 1 L with distilled water. Standardize against 40.0 mL of 0.05 N Na_2CO_3 solution with about 60 mL water by titration potentiometrically to a pH of about 5. Lift electrodes and rinse into beaker. Boil gently for 3–5 min under a watch glass cover. Cool to room temperature. Rinse cover glass into beaker. Continue titration to the pH inflection point. Calculate the normality of H_2SO_4 using:

$$N = \frac{A \times B}{53.0 \times C} \tag{5.8}$$

where, A = weight (gm) of Na_2CO_3 weighed into 1 L flask (see above); B = volume (mL) of Na_2CO_3 solution used in the standardization; C = volume (mL) acid used in titration.

Standard acid, 0.02 N: Dilute appropriate amount of standardized H_2SO_4 acid to 0.02 N (for, e.g., if standard acid solution is of exactly 0.1 N, use 200.00 mL to make 1 L stock solution). Check again by standardization against 15 mL of 0.05 N Na_2CO_3 solutions.

Standard acid, 0.002 N: Place 10 mL standard sulfuric acid, 0.02 N in a 100-mL volumetric flask and dilute to the mark.

Standard sulfate solution from Na₂SO₄, 0.1 N: Also, standard sulfate solution of 0.1 N can be prepared from sodium sulfate as follows: Dissolve 142.04 mg anhydrous Na_2SO_4 in distilled water in a 1-L volumetric flask and dilute to the mark with distilled water.

5.4.1.5.6 Sample preservations and handling

1. Preserve by refrigerating at 4 °C.

5.4.1.5.7 Measurement procedure

1. Formation of barium sulfate turbidity:
    i. Place a 100-mL sample (or a suitable portion diluted to 100 mL) into a 250-mL Erlenmeyer flask.
    ii. Add exactly 5.0 mL conditioning reagent (See section:5.7.5).
    iii. Mix in the stirring apparatus.
    iv. While the solution is being stirred, add a measured spoonful of $BaCl_2$ crystals (See point 3 in section: 5.7.5) and begin timing immediately.
    v. Stir exactly 1.0 min at constant speed.
2. Measurement of barium sulfate turbidity:
    2.1 Immediately after the stirring period has ended, pour solution into absorbance cell.
    2.2 Measure turbidity at 30-sec intervals for 4 min.
    2.3 Record the maximum reading obtained in the 4-min period.
3. Preparation of calibration curve:
    3.1 Prepare calibration curve using standard sulfate solution (see point 6 of "*Reagent*" section).
    3.2 Maintain increment of standards at 5-mg/L in the 0–40 mg/L sulfate range (Above 50 mg/L the accuracy decreases and the suspensions lose stability).
    3.3 Check reliability of calibration curve by running a standard with every three or four samples.

4. *Calculations*: Estimate amount of SO_4^{2-} from linear calibration curve, using the following relation:

$$\text{mg } SO_4{}^{2-} - /L = \frac{SO_4^{2-}(\text{mg}) \times 1000}{\text{sample (mL)}}$$

5.4.1.5.8 Quality control

1. All quality control data should be maintained and available for easy reference or inspection.
2. Calibration curves must be composed of a minimum of a blank and three standards.
3. Dilute samples if they are more concentrated than the highest standard or if they fall on the plateau of a calibration curve.
4. Verify calibration with an independently prepared standard solutions and also check every 15 samples during measurements.
5. Run one spike duplicate sample for every 10 samples (A spike duplicate sample is a sample brought through the whole sample preparation and analytical process).

5.4.1.6 Chlorine determination using N,N-diethyl-P-phenylenediamine (Dpd) reagent
5.4.1.6.1 General discussion: forms of chlorine and effects of water chlorination
Chemicals used in chlorination are elemental Chlorine, Cl_2 and hypochlorite (Calcium Hypochlorite, $Ca(OCl)_2$ and Sodium Hypochlorite, $NaOCl$). Chlorine is added to water as chlorine gas or as sodium or calcium hypochlorite. Chlorine gas and sodium hypochlorite react with water as:

$$Cl_2 + H_2O \rightarrow HOCl + H^+ + Cl^-$$

$$NaOCl + H_2O \rightarrow Na^+ + HOCl + Na^+ + OH^-$$

They form hypochlorous acid and hydrochloric when added to water.

Forms of chlorine and reactions: Cl exists in water as free, total, and combined form. The free chlorine is defined as the Cl existing in water as hypochlorous acid or hypochlorite ion and is known as chlorine residual. Whereas, combined Cl includes chloramines (such as monochloramines, dichloramines), and nitrogen trichloride. The total is defined as the sum of free and combined forms of chlorine. Chlorine applied to water in its molecular or hypochlorite form initially undergoes hydrolysis to form free chlorine consisting of aqueous molecular chlorine, hypochlorous acid, and hypochlorite ion. As mentioned above, the chlorine reacts with ammonia present in water to form the combined chlorines; chloramines (dichloramine, and nitrogen trichloride). Chlorinated wastewater effluents, as well as certain chlorinated industrial effluents, normally contain only combined chlorine. The relative proportion of these combined forms of chlorine and free chlorine is pH and temperature dependent. At the pH of most waters, hypochlorous acid and hypochlorite ion will predominate.

We need to conduct chlorine residual analysis to ensure proper concentration of residual chlorine as:

1. Required by Groundwater Rule (recommended minimum of 0.2 mg/L).
2. Determine chlorine level and potential generation of disinfection by-products,
3. Determine residual disinfectant in water and determine requirements for dechlorination.

Effects of Chlorination: Positive effects: chlorination of household water supplies and polluted waters serves primarily to destroy or deactivate disease-producing microorganisms. A secondary benefit, particularly in treating drinking water, is the overall improvement in water quality resulting from the reaction of chlorine with ammonia, iron, manganese, sulfide, and some organic substances.

Negative effects: Chlorination may also produce adverse effects. For example, taste and odor characteristics of phenols and other organic compounds present in a water supply may be intensified. Potentially carcinogenic chloro-organic compounds such as chloroform may be formed. Combined chlorine formed on chlorination of ammonia- or amine-bearing waters adversely affects some aquatic life. To fulfill the primary purpose of chlorination and to minimize any adverse effects, it is essential that proper testing procedures be used with a foreknowledge of the limitations of the analytical determination.

Chlorine (both the total and free Cl) can be determined by gravimetric, titrimetric, amperometric and Spectrophotometric method (or colorimetric method) methods. Here, we will discuss only the spectrophotometric method. This method is operationally simpler for determining chlorine in water than the amperometric titration and titrimetric method (titration with standard ferrous ammonium sulfate, FAS, solution).

Procedures are given for estimating the separate mono- and dichloramine and combined fractions with spectrophotometric technique.

5.4.1.6.2 Principle

The DPD (N, N-diethyl-p-phenylenediamine) colorimetric method for residual chlorine was first introduced by Palin in 1957 (Ref. [21]: Palin, A. T.; Jour. Am. Water Works Assoc.; 49,873 (1957)). Over the years, it has become the most widely used method for determining free and total chlorine in water and wastewater.

Hach Company introduced its first chlorine test kit based on the DPD chemistry in 1973. The DPD amine is oxidized by chlorine to two oxidation products. At a near neutral pH, the primary oxidation product is a semi-quinoid cationic compound known as a *Würster dye.* This relatively stable free radical species accounts for the magenta color in the DPD colorimetric test. When DPD reacts with small amounts of chlorine at a near neutral pH, the *Würster dye* (magenta color) is the principal oxidation product. Note that at higher oxidant levels, the formation of the unstable colorless imine is favored — resulting in apparent "fading" of the colored solution.

The DPD *Würster dye* color has been measured photometrically at wavelengths ranging from 490 nm to 555 nm. The absorption spectrum indicates a doublet peak with maxima at 512 and 553 nm. For maximum sensitivity, absorption measurements can be made between 510 and 515 nm. *Hach Company* (see Ref. [22]) has selected 530 nm as the measuring wavelength for most of its DPD systems. This "saddle" between the peaks minimizes any variation in wavelength accuracy between instruments and extends the working range of the test on some instruments.

Two "standard" DPD colorimetric methods generally are recognized in the international community. These are the *Standard Methods* 4500-Cl G (EPA: United States Environmental Protection Agency) and the International Organization for Standardization (ISO) Method 7393/2). The ISO method has been adopted by most of the members of the European Union (Germany's DIN Standard 38 408 G4 for free and total chlorine is modeled after ISO 7393/2).

5.4.1.6.3 Application

This method determines free (in the form of Cl_2) and total chlorine residual levels on drinking water, industrial and municipal wastes, swimming pools, natural and treated waters. Minimum detectable concentration, i.e. the detection limit of this method has been established at 0.05 mg/L and the Minimum Level (mL) for reporting results is 0.15 mg/L (150 µg/L). This method is capable of measuring Cl_2 in the range of 0.15−5.0 mg/L.

5.4.1.6.4 Interference

Additional halogens and halogenating agents produce positive interferences. Oxidized manganese and ozone produce positive interferences in spectrophotometric or colorimetric measurements.

Color and suspended matter may interfere with the photometric measurement. Counter this potential positive interference; use the sample without any addition of reagent as the sample blank.

Chlorine greater than 500 mg/L will oxidize the DPD to a colorless amine, which can be interpreted as a low chlorine value. Analysts should take note of sample odor, which may indicate high levels of chlorine present.

High pH levels cause dissolved oxygen to react with the reagents. Very low pH causes a positive free chlorine residual when mono-chloramines are present. The test should be conducted between 20 and 25 °C. Compensate for color and turbidity by using sample to zero photometer. Minimize chromate interference by using the thioacetamide blank correction.

5.4.1.6.5 Apparatus

a. *Photometric equipment:* Spectrophotometer, for use at a wavelength of 515 nm and providing a light path of 1 cm or longer and cuvettes with light path length of 1 cm or longer can also be used.
b. *Glassware:* Use separate glassware, including separate spectrophotometer cells, for free and combined (dichloramine) measurements, to avoid iodide contamination in free chlorine measurement.

Reagents.

(i) *Phosphate buffer solution:* Dissolve 24 g anhydrous Na_2HPO_4 and 46 g anhydrous KH_2PO_4 in distilled water. Combine above solution with 100 mL distilled water in which 800 mg disodium ethylenediamine tetraacetate dihydrate (EDTA) have been dissolved. Dilute to 1 L with distilled water and optionally add either 20 mg $HgCl_2$ or 2 drops toluene to prevent mold growth. Interference from trace amounts of iodide in the reagents can be negated by optional addition of 20 mg $HgCl_2$ to the solution. (CAUTION: *$HgCl_2$ is toxic—take care to avoid ingestion.*)
(ii) N,N-*Diethyl-p-phenylenediamine (DPD) indicator solution:* Dissolve 1 g DPD oxalate, or 1.5 g DPD sulfate pentahydrate, or 1.1 g anhydrous DPD sulfate in chlorine-free distilled water containing 8 mL H_2SO_4 (25%) and 200 mg disodium EDTA. Make up to 1 L, store in a brown glass-stoppered bottle in the dark, and discard when discolored. Periodically check solution blank for absorbance and discard when absorbance at 515 nm exceeds 0.002/cm. (The buffer and indicator sulfate are available commercially as a combined reagent in stable powder form)

CAUTION: *The oxalate is toxic-take care to avoid ingestion.*

c. *Standard ferrous ammonium sulfate (FAS) titrant:* Dissolve 1.106 g $Fe(NH_4)_2(SO_4)_2.6H_2O$ in distilled water containing 1 mL H_2SO_4 (25%) and make up to 1 L with freshly boiled and cooled distilled water.

d. *Potassium iodide,* KI, crystals.

e. *Potassium iodide solution:* Dissolve 500 mg KI and diluted to 100 mL, using freshly boiled and cooled distilled water. Store in a brown glass-stoppered bottle, preferably in a refrigerator. Discard when solution becomes yellow.

f. *Sodium arsenite solution:* Dissolve 5.0 g $NaAsO_2$ in distilled water and dilute to 1 L. (CAUTION: Toxic-take care to avoid ingestion.)

g. *Thioacetamide solution:* Dissolve 250 mg CH_3CSNH_2 in 100 mL distilled water. (CAUTION: *Cancer suspect agent. Take care to avoid skin contact or ingestion.*)

h. *Chlorine-demand-free water:* Prepare chlorine-demand-free∗ water from good-quality distilled or deionized water by adding sufficient chlorine to give 5 mg/L free chlorine. After standing 2 days this solution should contain at least 2 mg/L free chlorine; if not, discard and obtain better-quality water. Remove remaining free chlorine by placing container in sunlight or irradiating with an ultraviolet lamp. After several hours take sample, add KI, and measure total chlorine with a spectrophotometric method using a quartz cuvettes. Does not use before last trace of free and combined chlorine has been removed. Distilled water commonly contains ammonia and also may contain reducing agents. Collect good-quality distilled or deionized water in a sealed container from which water can be drawn by gravity. To the air inlet of the container add an H_2SO_4 trap consisting of a large test tube half filled with H_2SO_4 (50%) connected in series with a similar but empty test tube. Fit both test tubes with stoppers and inlet tubes terminating near the bottom of the tubes and outlet tubes terminating near the top of the tubes. Connect outlet tube of trap containing H_2SO_4 to the distilled water container; connect inlet tube to outlet of empty test tube. The empty test tube will prevent discharge to the atmosphere of H_2SO_4 due to temperature-induced pressure changes. Stored in such a container, chlorine-demand-free water is stable for several weeks unless bacterial growth occurs.

∗Notes: Chlorine demand is the difference between the amount of chlorine added to water or wastewater and the amount of residual chlorine remaining after a given contact time.

Procedure:

a. *Calibration of photometric equipment:* Calibrate instrument with chlorine or potassium permanganate solutions.

(1) Chlorine solutions-Prepare chlorine standards in the range of 0.05–4 mg/L from about 100 mg/L chlorine water standardized as follows: Place 2 mL acetic acid and 10–25 mL chlorine-demand-free water in a flask. Add about 1 g KI. Measure into the flask a suitable volume of chlorine solution. In choosing a convenient volume, note that 1 mL 0.025 N $Na_2S_2O_3$ titrant is equivalent to about 0.9 mg chlorine. Titrate with standardized 0.025 N $Na_2S_2O_3$ titrant until the yellow iodine color almost disappears. Add 1–2 mL starch indicator solution and continue titrating to disappearance of blue color. Determine the blank by adding identical quantities of acid, KI, and starch indicator to a volume of chlorine-demand-free water corresponding to the sample used for titration. Perform blank titration A or B, whichever applies, according to B.3*d.* of APHA 4500-Cl G (1992) and evaluate the chlorine as:

$$\text{Chlorine as } Cl_2/mL = \frac{(A+B) \times N \times 35,45}{\text{Sample Volume (mL)}}$$

where: N-normality of $Na_2S_2O_3$. A-Volume of titrant (in mL) for sample. B− Volume of titrant (in mL) for blank (to be added or subtracted according to required blank titration.

Use chlorine-demand-free water and glassware to prepare these standards. Develop color by first placing 5 mL phosphate buffer solution and 5 mL DPD indicator reagent in flask and then adding 100 mL chlorine standard with thorough mixing as described in '*b*' and '*c*' below. Fill photometer cell from flask and read color at 515 nm. Return cell contents to flask and titrate with standard FAS titrant as a cross check on chlorine concentration.

(2) *Potassium permanganate solutions*: Prepare a stock solution containing 891 mg $KMnO_4$/L. Dilute 10.00 mL stock solution to 100 mL with distilled water in a volumetric flask. When 1 mL of this solution is diluted to 100 mL with distilled water, a chlorine equivalent of 1.00 mg/L will be produced in the DPD reaction. Prepare a series of $KMnO_4$ standards covering the chlorine equivalent range of 0.05−4 mg/L. Develop color by first placing 5 mL phosphate buffer and 5 mL DPD indicator reagent in flask and adding 100 mL standard with thorough mixing as described in *b* and *c* below. Fill photometer cell from flask and read color at 515 nm. Return cell contents to flask and titrate with FAS titrant as a check on any absorption of permanganate by distilled water. Obtain all readings by comparison to color standards or the standard curve before use in calculation.

b. *Volume of sample:* Use a sample volume appropriate to the photometer or colorimeter. The following procedure is based on using 10-mL volumes; adjust reagent quantities proportionately for other sample volumes. Dilute sample with chlorine-demand free water when total chlorine exceeds 4 mg/L.

c. *Free chlorine:* Place 0.5 mL each of buffer reagent and DPD indicator reagent in a test tube or photometer cell. Add 10 mL sample and mix. Read color immediately (Reading A).

d. *Monochloramine:* Continue by adding one very small crystal of KI (about 0.1 mg) and mix. If dichloramine concentration is expected to be high, instead of small crystal add 0.1 mL (2 drops) freshly prepared KI solution (0.1 g/100 mL). Read color immediately (Reading B).

e. *Dichloramine:* Continue by adding several crystals of KI (about 0.1 g) and mix to dissolve. Let stand about 2 min and read color (Reading C).

f. *Nitrogen trichloride:* Place a very small crystal of KI (about 0.1 mg) in a clean test tube or photometer cell. Add 10 mL sample and mix. To a second tube or cell add 0.5 mL each of buffer and indicator reagents; mix. Add contents to first tube or cell and mix. Read color immediately (Reading N).

g. *Chromate correction using thioacetamide:* Add 0.5 mL thioacetamide solution (F.2*i*) to 100 mL sample. After mixing, add buffer and DPD reagent. Read color immediately. Add several crystals of KI (about 0.1 g) and mix to dissolve. Let stand about 2 min and read color. Subtract the first reading from Reading A and the second reading from Reading C and use in calculations.

h. *Simplified procedure for total chlorine:* Omit Step *d* above to obtain monochloramine and dichloramine together as combined chlorine. To obtain total chlorine in one reading, add the full amount of KI at the start, with the specified amounts of buffer reagent and DPD indicator. Read color after 2 min.

5.4.1.7 Chlorine residual (alternative procedure for chlorine residual measurement)
5.4.1.7.1 Principle
Free Chlorine oxidizes DPD indicator at a pH of 6.3−6.6 to form a magenta-colored compound.

$$\text{Free } Cl_2 + DPD \rightarrow \text{magenta-colored compound}$$

Free chlorine and potassium iodide is added to the DPD reagents. Then the chloramines oxidizes iodide to iodine which, along with free chlorine, oxidizes DPD to form pink color.

Free Chlorine + Chloramines + KI + DPD → magenta-colored compound

With this procedure the ranges of Chlorine Residual that can be analyzed are as follows: Low Range (0.02–2.0 mg/L), High Range (0.1 mg/L to 8.0 mg/L).

Drinking Water Standard: Maximum contaminant level Goal (MCLG): 4.0 mg/L and maximum permitted contaminant level (MCL): 4.0 mg/L.

i. Fill a 10 mL cell with sample (the blank). Cap it. Note: Samples much be analyzed immediately and cannot be preserved for later analysis.
ii. Press the power key to turn the spectrophotometer meter on.
iii. Open the meter cap. Place the blank in the cell holder. Wipe excess liquid and finger prints off sample cells.
iv. Set to auto zero by pressing AUTO ZERO button.
v. Fill a second 10 mL cell to the 10 mL line with sample. Note: Do not use the sample cells for free and total chlorine analysis without thoroughly rinsing the cells with sample between free and total tests.
vi. Add the contents of one DPD free chlorine powder packet or one DPD total chlorine powder packet to the sample cell (the prepared sample).
vii. Cap and shake gently for 20 s. Note: Shaking dissipates bubbles that may form in samples with dissolved gases. Also note that a pink color will develop if chlorine is present.
viii. For free chlorine, wipe excess liquid and fingerprints from the sample cell with a clean and soft tissue paper. Put the prepared sample cell in the cell holder, then cover the cell with the instrument cap. Proceed to step 10 within 1 min after the DPD free packet. Accuracy is not affected by undissolved powder.
ix. For total chlorine, wait 3–6 min after adding the DPD total packet. After the reaction time, wipe excess liquid and fingerprints from the sample cell. Put the prepared sample in the cell holder and cover the cell with the instrument cap.
x. Press START button. The instrument will show the absorbance reading, which can be compared with calibration curve to estimate the chlorine concentration.
xi. Prepare calibration curve with standard addition method using the ampule breaker to snap the neck off a high range chlorine standard solution Ampule, 50–70 mg/L Cl_2.
xii. Use pipette of appropriate size to add 0.1, 0.2 and 0.3 and 0.4 mL of standard to three 5 mL samples. Swirl gently to mix.
xiii. Analyze sample as described from 1 to 10. Each 0.1 mL of standard will cause an incremental increase in chlorine. The exact value depends on the concentration of the ampule standard. Check the certificate enclosed with ampoules for calculation of the expected increase in the chlorine concentration.

Notes: The methods discussed in 5.4.1.5 (for Chlorine determination) are based on Ref. [21–23] and also listed below:

• Palin, A. T.; Jour. Am. Water Works Assoc.; 49 (1957) 873.

- Danial L. Harp, Current Technology of Chlorine Analysis for Water and Wastewater, Technical Information Series of Hach Company — Booklet No.17, Lit. no. 7019, L21.5, Printed in US, 2002.
- APHA 4500-Cl G (Chlorine residual), Standard Methods for the examination of water and wastewater, Inorganic, nonmetals (4000), 18th Ed., 1992.

5.4.1.8 Chlorine determination (alternative method: with mercury (II) thiocyanate method)

Notes: Chlorine can also be determined with Mercury (II) thiocyanate Method as prescribed in Ref. [24] (Vogel's Quantitative Analysis, (fifth Ed), 1994)).

5.4.1.8.1 Principle

This second method for the determination of trace amounts of chloride ion depends upon the displacement of thiocyanate ion from mercury (II) thiocyanate by chloride ion; in the presence of iron (III) ion a highly colored iron (III) thiocyanate complex is formed, and the intensity of its color is proportional to the original chloride ion concentration;

$$2Cl^- + Hg(SCN)_2 + 2Fe^{3+} = HgCl_2 + 2[Fe(SCN)]^{2+}$$

This method is applicable to the range 0.5−100 μg of chloride ion.
Procedure:

i. Place a 20 mL aliquot of the chloride solution in a 25 mL graduated flask.
ii. Add 2.0 mL of 0.25 M ammonium iron (III) sulfate 2[Fe(NH$_4$) (SO$_4$)$_2$.12H$_2$O] prepared by dissolving in 9 M nitric acid, followed by 2.0 mL of a saturated solution of mercury (II) thiocyanate in ethanol.
iii. After 10 min measure the absorbance of the sample solution and the absorbance of the blank; use 5 cm cells in a spectrophotometer at 460 nm and put water in the reference cell.
iv. The amount of chloride ion in the sample corresponds to the difference between the two absorbances (of sample and blank) and is obtained from a calibration curve. Construct a calibration curve using a standard sodium chloride solution containing 10 μg/mL Cl$^-$; cover the range 0−50 μg as above. Plot absorbance against micrograms of chloride ion.

5.4.2 Cations

5.4.2.1 Determination of total iron (sum of ferrous (Fe^{++}) and ferric (Fe^{+++}) in the water sample)

5.4.2.1.1 Discussion

Occurrence and effects: Iron can occur in all kinds of waters in the form of ferrous, Fe(ll), or ferric, Fe(lll), in different concentrations. It prevails in true solution, colloidal solution, in organic compounds or as a complex compound. Iron exists also in soils and minerals mainly as insoluble ferric oxide and iron sulphide (pyrite) and in some areas, also as ferrous carbonate (siderite), which is very slightly soluble.

The United States Environmental Protection Agency (EPA) recommends that the concentration of iron in drinking water should not exceed 0.30 mg/L. Since the daily nutritional requirement of iron is

1–2 mg, the standard is for aesthetic reasons rather than toxicity. It should be pointed out, however, that iron concentrations of above 1.0 mg/L are harmful to many freshwater fish, especially trout.

5.4.2.1.2 Principle

The *phenanthroline method* is the preferred standard procedure for the measurement of iron in water except when phosphate or heavy metal interferences are present. Iron is brought into solution, reduced to the ferrous state by boiling with acid and hydroxylamine, and treated with 1,10-phenanthroline ($C_{12}H_8N_2$) at pH 3.2 to 3.3. Three molecules of phenanthroline chelate each atom of ferrous iron to form an orange-red complex. The equation for the reduction of the iron(III) ion to the iron(II) ion is:

$$4Fe^{3+} + 2NH_3O \rightarrow 2N_2O + 4Fe^{2+} + 6H^+$$

The colored solution obeys Beer's law; its intensity is independent of pH from 3 to 9. By measuring the intensities of transmitted and incident light through a colored solution and knowing its absorbance, we can prepare a calibration curve and subsequent concentration can be read. A pH between 2.9 and 3.5 insures rapid color development in the presence of an excess of phenanthroline.

Therefore, the reaction is buffered. Color standards are stable for at least 6 months. The iron present in the form of iron(III) must be reduced to the iron(II) state using hydroxylamine hydrochloride.

This method is suitable for determining Iron in drinking and surface waters, domestic and industrial wastes in the range from 0.5 mg/L to 5 mg/L Fe.

5.4.2.1.3 Interferences

Strong oxidizing agents, cyanide, nitrite and phosphates (polyphosphates more than orthophosphates) interfere. To overcome this problem, the sample is boiled initially with acid which converts poly-phosphate to orthophosphate and removes cyanide and nitrite that otherwise would interfere. Also, adding excess hydroxylamine eliminates errors caused by excessive concentrations of strong oxidizing reagents. Chromium and zinc in concentrations exceeding 10 times that of iron, cobalt and copper in excess of 5 mg/L, nickel in excess of 2 mg/L cause interference.

In the presence of interfering metal ions, use a larger excess of phenanthroline to replace that complexed by the interfering metals. Where excessive concentrations of interfering metal ions are present, extraction by isopropyl ether can remove the interference. If noticeable amounts of color or organic matter are present, it may be necessary to evaporate the sample, gently ash the residue and redissolve in acid. The ashing may be carried out in silica, porcelain or platinum crucibles that have been boiled for several hours in HCl (50%). The presence of excessive amounts of organic matter may necessitate digestion by nitric acid/sulfuric acid or nitric acid/perchloric acid before use of the extraction procedure.

5.4.2.1.4 Apparatus and reagents

Apparatus: (1) Spectrophotometer with Wavelength 508 nm, with measurement mode transmittance or absorbance. (2) Curvets with path length 10 mm, (3) pH meter, (4) Volumetric flasks (volume 100 mL and 250 mL), (5) Graduated cylinder, (6) Polyethylene bottle, (7) Burette 50 mL.

Reagents: Concentrated sulfuric acid (H_2SO_4), 1.0 mL, 1,10-phenanthroline, $C_{12}H_8N_2.H_2O$, 0.1 g, Acetic acid (glacial) CH_3CO_2H, 9 mL, Hydroxylamine hydrochloride, $H_2NOH.HCl$, 0.5 g, Iron(II) ammonium sulfate hexahydrate, $Fe(NH_4)_2(SO_4)_2.6H_2O$, 35.1 mg, 100-mL, Sodium acetate ($CH_3CO_2Na.3H_2O$) 10 g.

Safety Precautions: The sodium acetate buffer, hydroxylamine hydrochloride solution, glacial acetic acid, and 1,10-phenanthroline are toxic by ingestion. Sulfuric acid and glacial acetic acid are severely corrosive to the eyes, skin, and other tissue. Wear chemical splash goggles, chemical-resistant gloves, and a chemical-resistant apron. All solutions may be disposed of down the drain with excess water. Please consult standard method (e.g. Scientific Catalog/Reference Manual) for general guidelines and specific procedures, state and local regulations that may apply, before proceeding.

Procedure:

a. Preparation of standard iron solutions and calibration curve

   i. Prepare a stock iron solution by dissolving 35.1 mg of iron(II) ammonium sulfate hexahydrate in 100 mL of distilled water in a 250-mL volumetric flask. Add 1.0 mL of concentrated sulfuric acid, H_2SO_4, and fill to the mark with distilled water. Unknown solutions can be made by diluting the stock iron solution. Make sure to keep track of the concentration.

   ii. Prepare a sodium acetate buffer by dissolving 10 g of sodium acetate in 50 mL of distilled water in a 100-mL volumetric flask. Slowly add 9 mL of glacial acetic acid and diluted to 100 mL with distilled water.

   iii. Prepare a hydroxylamine hydrochloride solution by dissolving 0.5 g of hydroxylamine hydrochloride in 50 mL of distilled water.

   iv. Prepare a 1,10-phenanthroline solution by dissolving 0.1 g of 1,10-phenanthroline in 30 mL of distilled water with the aid of stirring and gentle heating. Cool and dilute to 50 mL. Store in a dark place.

   v. Using a burette, transfer 2.00, 4.00, 8.00, 10.00 and 12.00 mL of the stock iron solution into separate 100-mL volumetric flasks. Fill each flask about half full with deionized or distilled water. Using a graduated cylinder, add 10.0 mL of the sodium acetate buffer to each flask.

   vi. Using a pipette or a graduated cylinder, measure and add 2.0 mL of hydroxylamine hydrochloride solution to each flask and mix.

   vii. Finally, add 5.0 mL of the 1,10-phenanthroline solution to each flask. Fill each flask to the 100-mL mark with distilled or deionized water and mix. Calculate the concentration of iron in mg/100 mL.

   viii. To prepare a blank, add 10.0 mL of buffer, 2.0 mL of hydroxylamine hydrochloride solution, and 5.0 mL of 1,10- phenanthroline solution to a 100-mL flask. Dilute to the mark with distilled or deionized water and mix. This is your blank solution.

   ix. You now have five standard iron solutions and one blank. Transfer portions of the solutions to the spectrophotometer test tubes. Using the blank, zero the instrument at a wavelength of 508 nm. Take readings for each of the standard solutions.

   x. Plot a calibration curve of Absorbance versus Concentration of iron

(b) Analysis of Unknown

   i. Obtain 50.0 mL of an unknown iron solution or a sample of tap water. Place in a 100 mL volumetric flask and treat as above (steps 2 through 5). Transfer a portion of the solution to a spectrophotometer test tube.

   ii. Read the absorbance.

   iii. Locate the absorbance of the unknown solution on the calibration curve and determine the concentration of iron in that solution.

   iv. Also, prepare several replicates (about 10) of each sample, for good statistics: Find the standard deviation from the measured data of the replicate samples.

Notes: The methods discussed in 5.4.2.1 (for total iron) are based on Ref. [25–27] and also listed below:

- Flinn Scientific Spectrophotometer Laboratory Manual; Flinn Scientific: Batavia, IL, 1994; pp 55–60.
- APHA, 3500-Fe B. Phenanthroline Method, "Standard Methods for the Examination of Water and Wastewater".
- CFR Part 136 Appendix B – Definition and Procedure for the Determination of the Method Detection Limit.

5.4.2.2 Chromium (cr)

5.4.2.2.1 Discussion

Occurrence and effects: Chromium can occur as Cr(III) or Cr(VI) in dissolved form in different concentrations in all kinds of waters. Cr(VI) is more toxic than Cr(III) and is of specific interest.

Applications: This method is suitable for determining hexavalent chromium in drinking and surface waters, domestic and industrial wastes in the range from 0.1 mg/L to 1.0 mg/L Cr.

5.4.2.2.2 Principle of the method

This procedure measures only hexavalent chromium. To determine total chromium, prior oxidation with potassium permanganate is necessary: For this, the sample is digested with a sulfuric-nitric acid mixture and then oxidized with potassium permanganate before reaction with diphenylcarbazide. Cr(VI) containing chromates combine with diphenylcarbazide in strong acid solution to produce a red-violet complex.

5.4.2.2.3 Interferences

Heavy metals such as molybdenum, vanadium and mercury interfere. Iron in concentrations greater than 1 mg/L may produce a yellow color but the ferric iron color is not strong and is not normally encountered, when the absorbance is measured photometrically at the appropriate wavelength. Interfering amounts of molybdenum, vanadium, iron and copper can be removed by extraction of the cupferrates form of these metals into chloroform. But do not use this unless necessary because residual cupferron and chloroform in the aqueous solution complicate the later oxidation. Therefore, follow the extraction by additional treatment with acid fuming to decompose these compounds. Nitrite interferes in concentrations higher than 20 mg/L. Cr(III) and other interfering metals are precipitated from a phosphate buffered solution by the addition of aluminum sulfate prior to determination. The precipitate is filtered off and the solution retained.

5.4.2.2.4 Apparatus and reagents

Apparatus: The absorbance measurements are performed using UV/Vis spectrophotometer at wavelength 510 nm with cuvette of path length of 10 mm. Measurement mode can be either the transmittance or absorbance.

Other required apparatus are: volumetric flasks, volume 50 mL, 100 mL, conical flask, Eppendorf® micropipettes, Polyethylene bottle, Erlenmeyer flask 250 mL, pH meter, Beaker, Nessler tubes (matched, 100 mL, tall form).

Reagents: Sulfuric acid 0.2 N, Diphenylcarbazide solution, Potassium dichromate, Phosphoric acid.

i. Sulfuric acid 0.2 N
ii. *Potassium dichromate stock solution (50 mg/L):* Dissolve 14.14 mg of dried potassium dichromate in distilled water and diluted up to the mark with distilled water in 100-mL volumetric flask.
iii. *Diphenylcarbazide solution:* Dissolve 250 mg 1,5 Diphenylcarbazide in 50 mL acetone in a dark brown colored bottle.

5.4.2.2.5 Procedure for sample preparation and measurements

i. Prepare a series of reference solutions, as shown in Table 5.3, by pipetting suitable volumes of chromium stock solution into 100-mL volumetric flasks.
ii. Make volume up to 95 mL with distilled water.
iii. Adjust the pH of the solution to 2.0 ± 0.5 with 0.25 mL conc. phosphoric acid and 0.2 N sulfuric acid, mix well and diluted up to the mark with distilled water.
iv. Add 2.0 mL of diphenylcarbazide solution to the solutions mix well and allow the solutions to stand for 10 min for color development. (The procedure was followed by assuming 100 mL of sample volume.).
v. Perform background correction with the blank solution and measure the absorbance of the solutions at 510 nm using a 10-mm quartz cuvette. Prepare a table similar to Table 5.3. Tabulate the recorded data and prepare a calibration curve followed by plot of the absorbance of the standard.
vi. Analyte Sample preparation and measurement:
   1. Tape water sample may be used as received but wastewater sample may need to be filtered through a 0.45 μm filter.
   2. Calibration linearity curve is generated using above eight different levels of calibration standards in the range from 0.1 mg/L to 1.0 mg/L, including the blank as the first level. Also determine the correlation coefficient of the fitted linear fit.
   3. Sample of drinking mineral water and wastewater need to be prepared in duplicate and analyze as per the procedure mentioned above.

Table 5.3 A spreadsheet table for concentration versus absorbance measurements.

	Amount of taken chromium stock solution in 100 mL	Concentration mg/L	Recorded Absorbance
Blank	0	0	
Reference 1	0.2	0.2	
Reference 2	0.4	0.2	
Reference 3	0.8	0.4	
Reference 4	1.0	0.5	
Reference 5	1.2	0.6	
Reference 6	1.6	0.8	
Reference 7	2.0	1.0	

5.4.2.3 AMMONIA (NH₃)

5.4.2.3.1 Discussion

Occurrence and effect: Ammonia can occur in all kinds of surface waters, in some ground waters and wastewaters from industry and households. Whether ammonia nitrogen is found as NH_4^+ ion or as NH_4OH or NH_3 depends on the pH value of the water.

Applications: This method is suitable for determining ammonia - nitrogen in drinking, surface and saline waters, domestic and industrial wastes in the range from 0 00.01 mg/L - 2 00.0 mg/L NH_3 as N(nitrogen present in the form of NH_3). Higher concentrations can be determined by sample dilution.

5.4.2.3.2 Principle of the method

The sample is buffered at a pH of 9.5 with a borate buffer to decrease hydrolysis of cyanates and organic nitrogen compounds and is then distilled into a solution of boric acid. Alkaline phenol and hypochlorite react with ammonia to form indophenol blue that is proportional to the ammonia concentration.

5.4.2.3.3 Interferences

i.   Cyanate, which may be encountered in certain industrial effluents, will hydrolyze to some extent even at a pH of 9.5 at which distillation is carried out.
ii. Residual chlorine must be removed by pretreatment of the sample with sodium thiosulfate or other reagents before distillation.
iii. Interferences may be caused by contaminants in the reagent water, reagents, glassware, and other sample processing apparatus that bias analyte response.
iv. Calcium and Magnesium: Ions of these metals can cause errors by precipitating during analyses. Use EDTA solution (for river and industrial waters) or potassium tartrate solution (for sea water) to reduce this interference.
v.   Turbidity: Sample turbidity should be removed by filtration.
vi. Color: Color absorbing in the range of the indophenol blue complex can interfere.

5.4.2.3.4 Safety

Sulfuric acid, Phenol and Sodium nitroprusside have the potential to be highly toxic or hazardous, and therefore, safe handling of these chemicals with appropriate safety percussions is required. A reference file of Material Safety Data Sheets (MSDS) should be made available to all personnel involved in the chemical analysis.

5.4.2.3.5 Apparatus and reagents

The absorbance measurements are performed using UV/Vis spectrophotometer. Operation parameters for these measurements are: wavelength 640 nm and cell 10-mm.

a.   *Apparatus:* Volumetric flasks, volume 50 mL, 100 mL, 250 mL, Erlenmeyer flasks, volume 250 mL, 100 mL, Polyethylene bottle, pH meter, Water bath (95 °C).

Acid washed glassware: All glassware used should be washed with hot 1:1 HCl and rinsed with distilled water and treated with all the reagents to remove the last traces of phosphorus that might be absorbed on the glassware. Preferably, this glassware should be used only for the determination of phosphorus and after use it should be washed with distilled water and kept covered until needed again. If this is done, the treatment with 1:1 HCl and reagents is only required occasionally.

b. Reagents and Standards:

   i. *Reagent water - Ammonia free:* All solutions must be made with ammonia-free water.

   ii. *Boric acid solution (20 g/L):* Dissolve 20 g H_3BO_3 (analytical grade) in reagent water and dilute to 1 L.

   iii. *Borate buffer:* Add 88 mL of 0.1 N NaOH solution to 500 mL of 0.025 M sodium tetraborate solution (5.0 g anhydrous $Na_2B_4O_7$ or 9.5 g $Na_2B_4O_7 \cdot 10H_2O$ per L) and dilute to 1 L with reagent water.

   iv. *Sodium hydroxide, 1 N:* Dissolve 40 g NaOH in reagent water and dilute to 1 L.

   v. *Dechlorinating reagents:* A number of dechlorinating reagents may be used to remove residual chlorine prior to distillation. These include:

     • *Sodium thiosulfate:* Dissolve 3.5 g $Na_2S_2O_3 \cdot 5H_2O$ in reagent water and dilute to 1 L. One mL of this solution will remove 1 mg/L of residual chlorine in 500 mL of sample.

     • *Sodium sulfite:* Dissolve 0.9 g Na_2SO_3 in reagent water and dilute to 1 L. One mL removes 1 mg/L Cl per 500 mL of sample.

   vi. *Sulfuric acid 5 N:* Air scrubber solution. Carefully add 139 mL of conc. sulfuric acid to approximately 500 mL of reagent water. Cool to room temperature and diluted to 1 L with reagent water.

   vii. *Sodium phenolate:* Using a 1-L Erlenmeyer flask, dissolve 83 g phenol in 500 mL of distilled water. In small increments, cautiously add with agitation, 32 g of NaOH. Periodically cool flask under water faucet. When cool, dilute to 1 L with reagent water.

   viii. *Sodium hypochlorite solution:* Dilute 250 mL of bleach solution containing 5.25% NaOCl to 500 mL with reagent water. Due to the instability of this product, storage over an extended period should be avoided.

   ix. *Disodium ethylenediamine-tetraacetate (EDTA) (5%):* Dissolve 50 g of EDTA (disodium salt) and approximately six pellets of NaOH in 1 L of reagent water.

   x. *Sodium nitroprusside (0.05%):* Dissolve 0.5 g of sodium nitroprusside in 1 L of reagent water.

   xi. *Stock solution:* Dissolve 3.819 g of anhydrous ammonium chloride, NH_4Cl, dried at 105 °C, in reagent water, and dilute to 1 L. 1.0 mL = 1.0 mg NH_3–N.

   xii. *Standard Solution A:* Dilute 10.0 mL of stock solution of NH_4Cl) (above) to 1 L with reagent water. 1.0 mL = 0.01 mg NH_3 -N.

   xiii. *Standard Solution B*: Dilute 10.0 mL of *standard solution A* to 100.0 mL with reagent water. 1.0 mL = 0.001 mg NH_3 -N.

5.4.2.3.6 Sample collection, preservation and storage

   i. Samples should be collected in plastic or glass bottles. All bottles must be thoroughly cleaned and rinsed with reagent water. Volume collected should be sufficient to insure a representative sample, allow for replicate analysis (if required), and minimize waste disposal.

   ii. Samples must be preserved with H_2SO_4 to a pH < 2 and cooled to 4 °C at the time of collection.

   iii. Samples should be analyzed as soon as possible after collection. If storage is required, preserved samples are maintained at 4 °C and may be held for up to 28 days.

5.4.2.3.7 Calibration, standardization and measurements

i. Prepare a series of at least three standards, covering the desired range, and a blank by diluting suitable volumes of *standard solutions A and B* (section: 5.11.5 (b): xii and xiii) to 100 mL with reagent water.

ii. Process standards and blanks as described in Procedure (section: 5.11.7).

iii. Place appropriate standards in the sampler (if spectrophotometer has multiple sample holding compartments) in the order of decreasing concentration and perform analysis.

iv. Prepare standard curve by plotting instrument response against concentration values. A calibration curve may be fitted to the calibration solutions concentration/response data using computer or calculator based regression curve fitting techniques. Acceptance or control limits should be established using the difference between the measured value of the calibration solution and the "true value" concentration.

v. After the calibration has been established, it must be verified by the analysis of a suitable quality control sample (QCS). The QCS is the solution of known standard, is obtained from a source external to the laboratory and different from the source of calibration. If measurements exceed $\pm10\%$ of the established QCS value, the analysis should be terminated and the instrument recalibrated. The new calibration must be verified before continuing analysis. Periodic reanalysis of the QCS is recommended as a continuing calibration check.

Procedure.

i. *Preparation of equipment:* Add 500 mL of reagent water to an 800 mL Kjeldahl flask (A round-bottomed flask that has a very long neck used for distillation process). The addition of boiling chips that have been previously treated with dilute NaOH will prevent bumping. Steam out the distillation apparatus until the distillate shows no trace of ammonia.

ii. *Sample preparation:* Remove the residual chlorine in the sample by adding dechlorinating agent (section: 5.11.5 (b): v) equivalent to the chlorine residual. To 400 mL of sample add 1 N NaOH (5.11.5 (b): iv), until the pH is 9.5, check the pH during addition with a pH meter or by use of a short range pH paper.

iii. *Distillation:* Transfer the sample, the pH of which has been adjusted to 9.5, to an 800 mL Kjeldahl flask and add 25 mL of borate buffer (5.11.5 (b): iii). Distill 300 mL at the rate of 6—10 mL/min into 50 mL of 2% boric acid (5.11.5 (b): ii) contained in a 500 mL Erlenmeyer flask.

Note: The condenser tip or an extension of the condenser tip must extend below the level of the boric acid solution.

iv. Since the intensity of the color used to quantify the concentration is pH dependent, the acid concentration of the wash water and the standard ammonia solutions should approximate that of the samples.

v. Allow analysis system to warm up as required. Feed wash water through sample line.

vi. Arrange ammonia standards in sampler in order of decreasing concentration of nitrogen. Complete loading of sampler tray with unknown samples.

vii. Measure the absorbance of the standard samples with UV—Vis spectroscopy at 640 nm and prepare the calibration curve. Then analyze the water sample.

Notes: The method discussed in 5.4.2.4 (for ammonia determination) is based on Ref. [28–30] and also listed below:

- World Health Organization (WHO), Ammonia. Geneva, 1986 (Environmental Health Criteria, No. 54).
- International Organization for Standardization. Water quality—determination of ammonium. Geneva, 1984, 1986 (ISO5664:1984; ISO6778:1984; ISO7150−1:1984; ISO7150−2:1986).
- APHA, Standard methods for the examination of water and wastewater, 17th ed. Washington, DC, American Public Health Association/American Water Works Association/Water Pollution Control Federation, 1989.

5.4.2.4 Manganese (mn)
5.4.2.4.1 Discussion
Occurrence and Effects: Manganese exists in the soil principally as manganese dioxide, which is very insoluble in water containing carbon dioxide. Under anaerobic conditions, the manganese in the dioxide form is reduced from an oxidation state of IV to II and solution occurs. The manganese in groundwater is of divalent form. It may also be present in trivalent or quadrivalent form.

Consumption a small amount of manganese each day is important in maintaining our health. The amount of manganese in a normal diet (about 2−9 mg/day) seems to be enough to meet our daily need, and no cases of illness from eating too little manganese have been reported in humans. In animals, eating too little manganese can interfere with normal growth, bone formation, and reproduction. Too much manganese, however, can cause serious illness. Although there are some differences between different kinds of manganese, most manganese compounds seem to cause the same effects. Manganese miners or steel workers exposed to high levels of manganese dust in air may have mental and emotional disturbances, and their body movements may become slow and clumsy. This combination of symptoms is a disease called manganism. Workers usually do not develop symptoms of manganism unless they have been exposed for many months or years. Manganism occurs because too much manganese injures a part of the brain that helps control body movements. Some of the symptoms of manganism can be reduced by medical treatment, but the brain injury is permanent. It is not certain whether eating or drinking too much manganese can cause manganism or not. In another report, people who drank water with above average levels of manganese seemed to have a slightly higher frequency of symptoms such as weakness, stiff muscles, and trembling of the hands. However, these symptoms are not specific for manganese, and might have been caused by other factors. Studies in animals have shown that very high levels of manganese in food or water can cause changes in the brain. This information suggests that high levels of manganese in food or water might cause brain injury (but it does not appear that this is of concern to people exposed to the normal amounts of manganese in food, water, or air). The chances of harm from exposure near a waste site can only be evaluated on a site-by-site basis.

5.4.2.4.2 Principle
The concentration of manganese in potable water seldom exceeds a few mg/L. So spectrophotometric or colorimetric methods are applicable. The methods are [1] persulphate method, and [2] periodate method. Both methods depend upon oxidation of manganese from its lower oxidation state to VII where it forms the highly colored permanganate ion. The color produced is directly proportional to the

concentration of manganese present over a considerable range of concentration in accordance with Beer's law. So it can be measured by colorimetric photometric means, besides atomic absorption spectroscopy. Persulphate method is suitable because pretreatment of samples is not required. Persulphate oxidizes manganese to permanganate in the presence of silver nitrate as catalyst. Provisions must be made to overcome the influence of chlorides: Its (chloride) concentration is reduced by using mercuric sulfate. The color intensity is observed at a wavelength of **525** nm in a spectrophotometer.

Also, *periodate method* is used when concentrations are below 0.1 mg/L. To obtain complete oxidation of small amounts of manganese, silver nitrate is added and the heating time is increased.

5.4.2.4.3 Apparatus and reagents

 i. Spectrophotometer, for use at 525 nm, providing a light path of 1 cm or longer.
 ii. Nessler tubes, matcheds, 100 mL tall form.
 iii. Glassware like conical flasks, measuring cylinder and pipette.
5.4.2.4.3.1 Persulphate method

 i. *Special reagent:* Dissolve 75 g mercuric sulfate ($HgSO_4$) in 400 mL concentrated nitric acid (HNO_3) and 200 mL distilled water. Add 200 mL 85% phosphoric acid and 35 mg silver nitrate to the above solution. Dilute the cooled solution to 1 L.
 ii. *Ammonium persulphate:* $(NH_4)_2S_2O_3$ solid.
 iii *Standard manganese solution:* Prepare a 0.1 N potassium permanganate ($KMnO_4$) solution by dissolving 3.2 g of $KMnO_4$ in distilled water and making it up to 1 L. Age for several days in sunlight or heat for several hours near the boiling point and then filter through fritted glass filter crucible and standardize against sodium oxalate. Calculate the volume of this solution necessary to prepare 1 L solution of such strength that 1 mL = 50 µg Mn as follows:

$$V_{KMNO4(ml)} = \frac{4.55}{Normality \ of \ KMnO_4}$$

where V_{KMNO4} is the volume (in mL) of $KMnO_4$.

 iv. To this solution add 2–3 mL concentrated H_2SO_4 and sodium bisulphite solution (10 g $NaHSO_3$ + 100 mL distilled water). Boil to remove excess SO_2, cool and dilute to 1000 mL with distilled water.

 Procedure:

 i. Take 50 mL of the sample in a conical flask. Add 50 mL distilled water to it.
 ii. Pipette 1, 2, 3, 4, and 8 mL of standard manganese solution to different flasks, and dilute each to 100 mL using distilled water.
 iii. Add 5 mL special reagents to all the flasks.
 iv. Concentrate the solutions in all the flasks to about 90 mL boiling.
 v. Add 1g ammonium persulphate to all the flasks, bring to boil and boil for 1 min.
 vi. Remove all the flasks from the heat source and let stand for 1 min.
 vii. Then cool the flasks under the tap water.
 viii. Prepare one distilled water blank along with the color standards.
 ix. Measure the absorbance of each solution in a 1 cm cell at 525 nm against the reference blank prepared by treating distilled water.

x. Prepare the calibration graph taking meter reading along y-axis and concentration of manganese (in µg) in color standards on x-axis.

xi. Keep the sample in the spectrophotometer and note down the meter reading.

xii. Read off from the graph, the corresponding concentration of manganese in µg.

5.4.2.4.3.2 Periodate method. *Preparing known standards and unknown solution for analysis:* As mentioned in the previous section, the aqueous solutions that contains manganese ions $Mn^{2+}(aq)$ is an almost colorless solution. The manganese ions are easily oxidized in acidic solution to form permanganate ions, $MnO_4^-(aq)$, an intensely purple species. The very intense color means that the analysis can be very sensitive because the light absorption will be relatively large, even with small amounts of manganese in the sample.

5.4.2.4.4 Principle

In this method of analysis, potassium periodate, $KIO_4(s)$, will be used to oxidize $Mn^{2+}(aq)$ to the purple MnO_4^- (aq) ion, according to the following balanced chemical equation (all species aqueous):

$$2Mn^{2+} + 5KIO_4 + 3H_2O \rightarrow 2MnO_4^- + 5KIO_3 + 6H^+$$

$$(purple)$$

5.4.2.4.5 Apparatus and materials

The following are required equipment:

(1) 100.0-mL volumetric flask 1 [2], 25.00-mL volumetric flasks 3 [3], cuvettes for the spectrophotometer [4], 100 mL volumetric pipette 1 [5], 5.00 mL volumetric pipette 1 [6], 10.00 mL volumetric pipette 1 [7], Rubber pipette bulb 1 and [8] Beaker tongue 1.

Procedure.

i. Clean pipette with 1:1 HCl solution followed by distilled water and finally with the stock solution before using. Pipette 5.00 mL of $Mn^{2+}(aq)$ stock solution into a clean 400-mL beaker.

ii. Using a graduated cylinder to measure volumes, add 30 mL of deionized water to the 5.00-mL stock solution. Then add 10 mL of 9 M phosphoric acid, $H_3PO_4(aq)$. (It is important to add the acid before adding the $KIO_4(s)$ in step (v) below).

iii. Based on the concentration and volume of the known solution and the balanced reaction between Mn^{2+} and KIO_4, calculate the minimum mass of $KIO_4(s)$ required to oxidize all of the $Mn^{2+}(aq)$ in solution to $MnO_4^-(aq)$. Note: you will use double this amount to be certain that all of the $Mn^{2+}(aq)$ is actually oxidized.

iv. A watch glass, weigh out double the amount of solid potassium periodate, $KIO_4(s)$, you just calculated. Do not use paper for weighing as KIO_4 reacts with paper.

v. Add the $KIO_4(s)$ to your solution in the hood and stir to dissolve using a glass stirring rod. Rinse any $KIO_4(s)$ that sticks to the watch glass into the beaker using deionized water and your squirt bottle. Leave the stirring rod in the beaker to help control bubble formation when boiling.

vi. Boil the solution gently. A purple color should appear. If it does not add more $KIO_4(s)$ (add an additional 25% of your calculated value) and continue heating. Avoid splattering and boiling over! If you lose any material due to splattering you must start over repeating steps 1−4.

Continue boiling gently for one to 2 min after the color changes to deep purple to be certain that the color change is complete.

vii. Cool the solution and transfer the entire contents of the beaker to a 100.0-mL volumetric flask using a clean funnel. Rinse the inside of the beaker and the funnel with deionized water from your wash bottle and then transfer the rinse water into the volumetric flask as well to be sure you have transferred all of the manganese solution.

viii. Dilute the solution in the volumetric flask with deionized water to exactly the index mark and swirl to mix thoroughly.

ix. Calculate the concentration of MnO_4^- (aq) in this standard solution. Label this solution "standard." You may want to transfer some of this solution to a separate container in case of "accidents."

x. For preparing standards of various concentrations, clean and dry 25.00-mL volumetric flasks: 1, 2, and 3. Pipette accurately 1.00, 5.00, and 10.00-mL portions of your standard solution into these flasks. Be sure to rinse the pipette with deionized water and standard solution first. Fill each flask with deionized water exactly to the index mark and mix well. Calculate the concentration (molarity, M) of MnO_4^- (aq) ions in each of these flasks based on the molarity of your standard solution, the volume of this that you pipetted, and the total volume of your new diluted solution (25.00 mL). Rinse two cuvettes using deionized water and dry the outside of each cuvette using a Kim Wipe.

xi. Use one cuvette for the blank and fill it ¾ full with deionized water. You will use the other to measure the absorbance of each of the three solutions in the 25.00-mL volumetric flasks and the pure standard solution in the 100.0-mL volumetric flask. Be sure to rinse the second cuvette well between trials. You should first rinse this cuvette using a small amount of deionized water and then a second time using a small amount of the sample to be added.

xii. Before measuring the absorbance of your solutions you will need to zero your spectrometer and choose a wavelength (525 nm).

xiii. Once you have zeroed the instrument and selected a wavelength, measure the absorbance of each of the three solutions in the 25.00-mL volumetric flasks and the pure standard solution in the 100.0-mL volumetric flask at this wavelength. You will have five measured data points including the deionized water blank (which should have a zero absorbance if you properly zeroed the spectrometer).

xiv. Make a calibration curve plotting the concentration (M) of the five solutions on the x-axis and the absorbance of each on the y-axis. Draw a best-fit-line through your points and determine the molar absorptivity, E, the slope of your line. (Be sure to measure the slope of the best-fit-line; not specific data points). Your intercept should be at the origin, since your blank solution counts as one of the points: you may use a computer program such as Excel to make the best fit. Analyzing your unknown sample:

xv. Prepare your unknown sample in exactly the same way you prepared your standard solution following steps 1–8 above using 5.00 mL of your unknown solution instead of the known stock solution in step 1. You do not need to repeat steps (ix − xv) with the unknown.

xvi. Measure the absorbance of the 100.0-mL purple solution prepared from your unknown at the same wavelength that you used to construct your standard curve. You can now use the equation of your best-fit-line to determine the concentration of MnO_4^- (aq) ions in this solution. Because this is a diluted solution, you will now need to calculate backwards from this concentration to determine the actual concentration of manganese in your original unknown sample.

5.4.2.5 Lead (Pb)

5.4.2.5.1 Discussion
Occurrence and effects: Lead (Pb) is a serious cumulative body poison. Natural waters seldom contain more than 5 μg/L, although much higher values have been reported. Lead in a water supply may come from industrial, mine, and smelter discharges or from the dissolution of old lead plumbing. Tap waters that are not suitably treated may contain lead resulting from an attack on lead service pipes or solder pipe joints. The most common problem is with fixtures with lead solder, from which significant amounts of lead can enter into the water, especially hot water.

5.4.2.5.2 Principle
An acidified sample containing micrograms quantities of lead is mixed with ammonical citrate-cyanide reducing solution and extracted with dithizone in chloroform ($CHCl_3$) to form a cherry-red lead dithizonate. The color of the mixed color solution is measured spectrophotometrically.

Note: Minimum detectable concentration: 1.0 μg Pb/10 mL dithizone solution.

5.4.2.5.3 Interference
In a weak ammoniacal cyanide solution (pH 8.5 to 9.5), dithizone forms colored complexes with bismuth, stannous tin, and monovalent thallium. In strongly ammoniacal citrate-cyanide solution (pH 10 to 11.5) the dithizonates of these ions are unstable and are extracted only partially. This method uses a high pH, mixed color, single dithizone extraction.

5.4.2.5.4 Preliminary sample treatment
At time of collection acidify with conc. HNO_3 to pH < 2 but avoid excess HNO_3. Add 5 mL 0.1 N iodine solution to avoid losses of volatile organo-lead compounds during handling and digesting of samples. Prepare a blank of lead-free distilled water and carry through the procedure.

Digestion of samples: Unless digestion is shown to be unnecessary, digest all samples for dissolved or total lead.

5.4.2.5.5 Apparatus and reagents
Apparatus: (a) Spectrophotometer for use at 510 nm, providing a light path of 1 cm or longer (b) pH meter (c) Separatory funnels: 250-mL Squibb type. Clean all glassware, including sample bottles, with $1 + 1$ HNO_3. Rinse thoroughly with distilled or deionized water (d) Automatic dispensing burettes: Use for all reagents to minimize indeterminate contamination errors.

Reagents: Prepare all reagents in lead-free distilled water.

a. *Stock lead solution:* Dissolve 0.16 g lead nitrate. $Pb(NO_3)_2$ (minimum purity 99.5%), in approximately 200 mL water. Add 10 mL conc HNO_3 and dilute to 1000 mL with water.
b. *Working lead solution:* Dilute 2.0 mL stock solution to 100 mL with water; 1 mL = 2.00 μg Pb.
c. *Ammonia-cyanide—sulphite reducing solution:* Prepare by diluting 35 mL of concentrated ammonia solution and 3.0 mL of 10% potassium cyanide solution (**Caution poisonous**) to 100 mL and adding 0.15 g of sodium sulfite.
d. *Stock Diphenylthiocarbazone (dithizone), $C_6H_5.N=N.CS.NH.NH.C_6H_5$ solution:* The compound is insoluble in water and dilute mineral acids, but is readily soluble in dilute aqueous ammonia or in chloroform. It is an important selective reagent for quantitative determination of metals by colorimetric or spectrophotometric method.
e. *Stock dithizone solution:* Dissolve 5 gm of the solid **dithizone** in 100 mL of chloroform.

Procedure:

i. Place 10.0 mL of the working lead solution in a 250 mL separatory funnel.

ii. Add 75 mL of the Ammonia-cyanide –sulphite solution and then by carefully addition of dilute hydrochloric acid to adjust the pH of the solution to 9.5. Note that this process should be carried out slowly so that the pH of the solution does not fall below 9.5 (if the pH of the solution falls even temporarily below 9.5, HCN may be liberated (Caution, for this reason this operation must be carried out in fuming hood).

iii. Now add 7.5 mL of dithizone reagent to the separatory funnel, followed by the 17.5 mL of chloroform.

iv. Shake for 1 min, allow the layers to separate, then remove the chloroform layer.

v. Measure the absorbance of this against a blank solution, using a 1 cm cell and a wavelength of 510 nm.

vi. Repeat the procedure with 5.0 mL, 7.5 mL and 15.0 mL portions of the working lead solution and then with 10 mL of the test solution.

5.4.2.6 Arsenic (As)

5.4.2.6.1 Discussion

Occurrence and effects: Arsenic (As), a naturally occurring element with atomic number 33 and metalloid properties, is hazardous to both the human health and ecosystem. International Agency for Research on Cancer (IARC) had identified it as one element causing cancer. It often presents in environmental samples (for e.g., soil, sediment, water, aerosol, rain, aquatics, vegetation, milk, etc) (See Refs. [31–33]). In nature, as a heavy metal, it exists in the form of both organic and inorganic compounds in four oxidation states: −III (arsine), 0 (arsenic), +III (arsenite) and +V (arsenate). Among its inorganic forms, trivalent arsenic (As-III) mainly presented in water as arsenite (H_3AsO_3 or $HAsO_2$) is recognized as the most toxic and highly mobile. Due to being readily soluble in water at pH 8 and above, it enters into the human and animal body mainly by consumption of contaminated water and also food. The mobility and toxicity of arsenic are determined by its oxidation state, thus the behavior of arsenic species will change depending on the biotic and abiotic conditions in water. In groundwater, arsenic is predominantly present as As (III) and As (V), with a minor amount of methyl and dimethyl arsenic compounds being detected.

Arsenic mobilization in the water residing in aquifers is understood to occur by over withdrawal of groundwater during cultivation. Generally, arsenic is available in much higher concentration in groundwater compared to surface water. The widespread use of arsenic contaminated groundwater in irrigation for a prolonged period of time could elevate its concentrations in surface soil and eventually into vegetations for example, rice plants and rice grains.

The World Health Organization (WHO) has set 0.01 mg per liter (10 parts per billion) as the limit for the safety of water for drinking purposes. Nepal, including India and Bangladesh, has adopted the limit of 50 ppb. Lack of expertize and knowledge of the implementation, economic consideration and technical ability to measure arsenic concentration below 50 ppb in the field are the main reasons behind the national standard. The lowest standard currently set for acceptable arsenic concentration in drinking water is implemented by Australia, which has a national standard of 7 ppb (See Ref. [33]). The drinking water standard for arsenic is likely to be lowered in the coming year(s) because of a threat to human health. The major issue of arsenic contaminated water (and also food grains) is to find out the level of contamination of arsenic and it is not so easy because of no color, no odor and no taste even in the highly contaminated water.

World's highest concentration of arsenic is presently observed to be located in Bangladesh and West Bengal State of India. It has also been a measure problem in Terai region of Nepal (See Ref. [34,35]).

Therefore, it is very important to determine these inorganic arsenic species (total As, As (III) and As (V) accurately with a low cost and easier method.

5.4.2.6.2 Principle

The molybdenum blue method is well-established for the determination of phosphate (H_3PO_4) and inorganic As in solution[7]. Further optimization (modification) has been made in this method to achieve fast and more accurate measurements of H_3AsO_4 and H_3PO_4 in solution at concentrations below 10 μmol/L (see Ref. [38] for more details).

5.4.2.6.3 Apparatus and reagents

  i. *Apparatus:* Spectrophotometer, for use at 525 nm, providing a light path of 1 cm or longer.
 ii. Matched cells with path length of 10-mm.
iii. Glassware like conical flasks, measuring cylinder and pipette.

5.4.3 Reagents

Sulfuric acid (H_2SO_4), L-ascorbic acid ($C_6H_8O_6$), Ammonium molybdate tetrahydrate (($NH_4)_6Mo_7O_{24} \cdot 4H_2O$), Antimony Potassium Tartrate hydrate (($K(SbO)C_4H_4O_6)_2 \cdot H_2O$), *Disodium hydrogen arsenate heptahydrate* ($Na_2HAsO_4 \cdot 7H_2O$), potassium dihydrogenphosphate (KH_2PO_4), Sodium dithionite ($Na_2S_2O_4$) are used in the preparation of solutions. All solutions are prepared at room temperature using deionized water.

Stock solutions of arsenate and phosphate (100 μmol/L) are prepared by dissolving of 10 μmol of arsenate/phosphate salts in 100 mL of water and are used for sample preparation. The masses of all solids are measured using an analytical balance.

5.4.4 Preparation of sample solutions of arsenate

Sample solutions of arsenate are prepared by adding an accurately measured aliquot of stock solution into 40 mL of water placed in a 50 mL volumetric flask. Prepared solutions are color developed and diluted to the final volume with deionized water. This procedure is used to prepare solutions with different concentrations of arsenate. For accurate measurement of solutions with the same concentration of arsenate, an aliquot of stock solution is placed into a 250 or 500 mL of solution (final volume), which subsequently are divided into multiple samples (45 mL) in 50 mL volumetric flasks. These solutions were diluted to 50 mL by the color development reagents (3 mL), sulfuric acid (0.5 mL) and deionized water (1.5 mL).

The volume of aliquots required for sample preparation is to be varied from 5 μL to 5 mL. Reduction of *Arsenate is carried out* by sodium dithionite dissolving in the aqueous sample and heating the solution at 80 °C. Heated solutions are cooled to room temperature and then color is developed as described below. The concentration of reductant (sodium dithionite) in solution is varied from 0.4 to 0.8 mmol/L.

Sample solutions were color developed according to the procedure described below and are measured with a spectrometer in a 1 cm optical path cell at $\lambda_{max} = 880$ nm.

The photometer can be used in the determination of low concentrations of H_3AsO_4 in solution.

Color development procedure is carried out using two solutions [1]: ascorbic acid and [2] molybdate. The procedure includes the addition of 2 mL of molybdate solution to the sample with subsequent addition of 1 mL of ascorbic acid solution and bringing sample solution to a final volume of 50 mL.

Aqueous solutions of ascorbic acid is prepared maintaining the concentration of ascorbic acid (=0.57 mol/L, 10.8 g in 100 mL) following Ref. [8].

Stock Solutions of molybdate is prepared by adding 5.2 g solid ammonium molybdate (AM) and potassium antimony tartrate (PAT) salts to in 30 mL 9 mol/L solution of sulfuric acid and diluted to a final volume of 50 mL in a volumetric flask:

5.4.4.1 Measurement of the analyte solution

i. 45 mL of analyte solution (V_{sample}) is measured out in a 50 mL volumetric flask (V_{flask}).
ii. For acidification of *Solutions,* add 0.5 mL of 98% H_2SO_4 directly to the aqueous sample prior to color development and, shake.
iii. Wait 45 s and add 2 mL of molybdate *stock solution*, shake
iv. Wait 45 s and add 1 mL of ascorbic acid *stock solution*, shake;
v. Top up solution to a final volume of 50 mL with deionized water, shake;
vi. Wait ~10 min and measure absorbance at 880 nm (C_{meas}).
vii. The analyte concentration is calculated using Eq. (1):

$$C_{sample} = \frac{V_{flask} C_{meas}}{V_{sample}} \qquad \text{Eq. (1)}$$

Notes: The method discussed in 5.4.2.5–5.4.2.8 (for As, Cr, Pb and Mn) are based on Ref. [31–38] and also listed below:

- P. J. Craig, *Organometallic Compounds in the Environment.* New York: John Wiley & Sons, Inc. 1986.
- WHO *Arsenic in Drinking Water*, Geneva, 210, 1999.
- J. C. Ng. *Environmental contamination of arsenic and its toxicological impact on humans.* Environmental Chemistry, **2**(3) (2005) 146–160.
- J. K. Thakur, K. R. Thakur, A. Ramanathan, M. Kumar, and S.K. Singh (2011). *Arsenic contamination of Groundwater in Nepal- An overview.* Water **3**: 1–20.
- WHO, Arsenic contamination in Groundwater Affecting Some Countries in the South-East Asia Region. WHO: Washington, DC, USA (2001).
- FAO, Arsenic threat and irrigation management in Nepal, Rome (2004).
- G.H. Jeffery, J. Bassett, J. Mendham, R.C. Denney, Vogel's Textbook of Quantitative Chemical Analysis sixth ed., Longman Scientific & Technical, England. Longman Group UK Ltd.,1989.
- Lenoble, V. Deluchat, B. Serpaud, J.C. Bollinger, Arsenite oxidation and arsenate determination by the molybdene blue method, Talanta 61 (2003) 267.

5.4.5 Probing organic compounds with UV–VIS spectrophotometry

5.4.5.1 Important terminologies used in interpretation of nature of absorption spectra

Changes in chemical structure or the environment lead to changes in the absorption spectrum of molecules and materials. There are several terms which are commonly used to describe these shifts, are given below.

i. **Chromophore:** A covalently unsaturated group responsible for absorption due to electronic transitions (e.g., C=C, C=O, and NO_2).

ii. **Auxochrome:** A saturated group with nonbonded electrons which, when attached to a chromophore, alters both the wavelength and the intensity of the absorption (e.g., OH, NH_2 and Cl) (Fig. 5.3).

iii. **Bathochromic Shift** (from Greek: *bathys*, "deep"; and *chrōma*, "color"): The shift of absorption to a longer wavelength (lower energy side) due to substitution of a functional group or solvent effect. The cause for such a phenomenon is understood as follows: If a group of (or compound of interest) is more polar in the excited state, the non bonding electrons in the excited state are relatively stabilized by interaction with a polar solvent. Thus the absorption is shifted to longer wavelength with increasing solvent polarity. This is commonly called *Bathochromic Shift* (see in Fig. 5.4 right). Because the red color in the visible spectrum has a longer wavelength than most other colors, the effect is also commonly called a **red shift**.

FIG. 5.3

A simplistic picture of shift in an excited state energy level during *redshift* and *blueshift*. The energy levels are changed, as shown in Figure 5.3 to ΔE_b for blue shift (energy level shown on left) and ΔE_r for red shift (energy level shown on right).

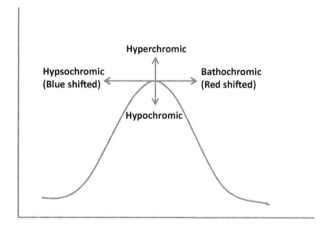

FIG. 5.4

Summarization of important terminologies: Bathochromic, Hypsochromic, Hyperchromic, Hypochromic shifts.

iv. ***Blue shift or Hypsochromic shift*** *(from ancient Greek (upsos) "height"; and chrōma, "color":* The shift of absorption to a shorter wavelength (higher energy side) due to substitution or solvent effect (see in Fig. 5.4 left). Opposite to red shift, on the other hand, if the group (or compound of interest) is more polar in the ground state than in the excited state, the non-bonding electrons in the ground state of the compound are relatively stabilized by hydrogen bonding or due to electrostatic interaction with the polar solvent. As a result, the absorption is shifted to shorter wavelength with increasing solvent polarity. This is commonly called ***Hypsochromic Shift.*** Because the blue color in the visible spectrum has a shorter wavelength than most other colors, this effect is also commonly called a ***blue shift.***

v. ***Hyperchromic Effect***: As demonstrated in Fig. 5.4, an increase in absorption intensity is termed as hyperchromic effect.

vi. ***Hypochromic Effect:*** A decrease in absorption intensity is known as hyperchromic effect.

5.4.5.2 Application of UV—Vis spectroscopy for qualitative information: understanding molecular structure from absorption spectra

(a) ***Types of transitions undergoing in an absorbing species:*** Determining true molecular structure of chemical compounds (organic, inorganic and metallo-organic complex compounds) employing various spectroscopic techniques (such as NMR-, Mass-, UV VIS−, FTIR-spectroscopy) is a routine work in analytical chemistry and organic chemistry. In particular, spectroscopic investigations confirm the molecular structures of new reaction products or new chemical species (isolated from natural sources), and size of nanomaterials and their properties. Usually increased molecular size results in complexity in NMR, IR and Mass spectra. However, this is not the case for UV spectra, therefore, spectrophotometry serve as a major tool to confirm the molecular structure and features associated with the chemical species or material under investigation.

Depending upon the energy of an absorbed photon, the energy gained by a molecule may bring about increased vibration or rotation of the atoms, or may raise electrons to higher energy levels (called vibrational, rotational or electronic transition, respectively). The changes occur in vibrations or rotations or electronic states (by the gain of the energy) that are permitted to a molecule of that structure. Therefore, the absorption spectrum of a compound of interest has long been regarded as closely connected with its molecular structure and therefore depends on the electronic transitions that can occur and the effect of the atomic environment of the transitions.

Fundamentally, absorption of a photon of ultraviolet and visible radiation comes about as the result of transitions of valence shell electrons between different energy levels. A summary of the electronic structures and transitions that are involved in UV absorption is presented in Fig. 5.5.

Consider electrons in covalent bonds and lone pairs of electrons on atoms in saturated organic compounds (i.e., compounds containing single bonds only, e.g., ethane and diethyl ether) (Fig. 5.6).

These electrons can be excited by UV radiation however they are held so tightly that the λ_{max} values fall well below 200 nm where it is technically very difficult to measure and, as a consequence, is uninformative for spectroscopic analysis.

However, still the molecular structural information of the compounds that have double or triple bonds (π) or compounds with lone pair electrons (n) as part of a saturated structure such as in the

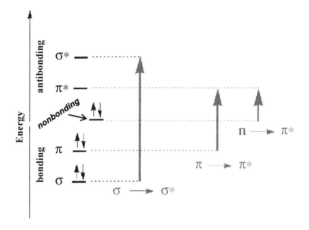

FIG. 5.5

Schematic energy level diagram of a compound with a lone pair of electrons (non-bonding- n orbital).

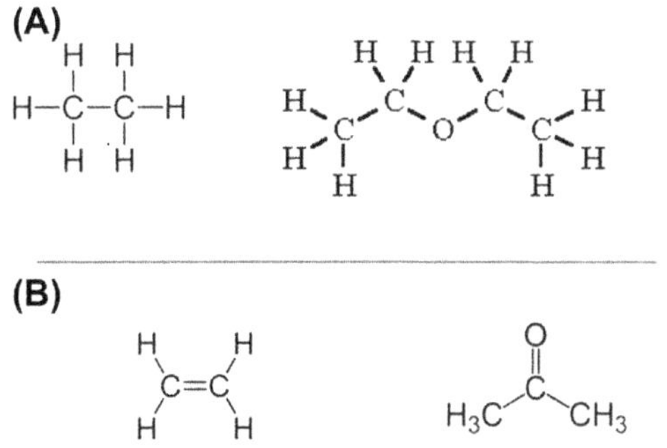

(A)

(B)

FIG. 5.6

(A): Molecular structures of ethane (Upper: Left) and diethyl ether (Upper: Right). Here, ethane possesses all covalently bonded atoms (constituents), whereas, the later possesses an oxygen atom with unshared pairs of electrons. (B): Molecular structural representation of ethene (lower: left), possesses only σ and π bonding, and acetone (lower: right) possesses σ, π and n-bondings.

carbonyl ($C{=}O$) functional group can be extracted by employing UV−Vis spectroscopy. These species undergo mainly two kinds electronic transitions,

$$\pi \rightarrow \pi^* \text{ and } n \rightarrow \pi^*$$

which are initiated by absorption of radiation above 200 nm (note $\pi*$ is the excited state and is termed an 'antibonding π orbital'). On the other hand, because the electrons involved in $n \rightarrow \pi^*$ and $\pi \rightarrow \pi^*$ transitions are bound less tightly within the molecular structure they require less energy to excite and therefore have λ_{max} values at longer wavelengths. This expectation from molecular picture (structure) is clearly observed in measured UV−Vis spectra.

Shift of peak position of absorption spectrum to longer λ_{max} values is even more pronounced when conjugation is present i.e. where there is a sequence of unsaturated bonds separated by a single bond. There are many organic compounds that have conjugated double bond systems (hereafter referred to as "conjugated systems"), in which every other bond is a double bond. These conjugated systems have a large influence on peak wavelengths and absorption intensities. An isolated carbonyl group or an isolated double bond do not have a strong maximum above 200 nm. As indicated above, this circumstance changes with conjugation: The longer a conjugated system, the longer will be the wavelength of the absorption. In other words, conjugation pushes λ_{max} to longer wavelengths (absorption intensity, ε_{max}, normally increases as well).

Fig. 5.7 shows the structures of benzene, naphthalene, and anthracene and their absorption spectra obtained by dissolving these compounds in ethanol and analyzing the resulting solutions. In this measurement, the concentrations were adjusted so that the absorption intensities of the components were roughly the same. As demonstrated in Fig. 5.7, it can be seen that peak wavelengths tend to be shifted toward the long wavelength region as the conjugated system gets larger.

The molar absorption coefficient is a measurement of how strongly a substance absorbs light. The larger its value, the greater the absorption. With larger conjugated systems, besides the absorption peak wavelengths tend to be shifted toward the longer wavelength region, the absorption peaks tend to be larger.

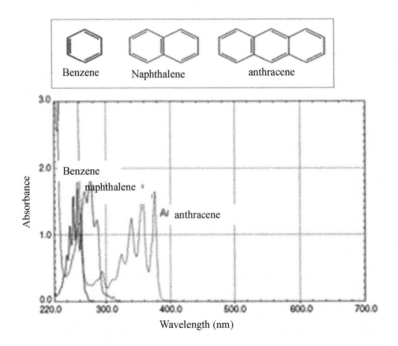

FIG. 5.7

Structure of Benzene, Naphthalene and Anthracene (up) and their absorption spectra (down).

Types of electronic transitions: Several notation systems have been used to designate UV absorption bands. Using the simplest one assigned by Burawoy, the electronic transitions is explained below.

R bands: Selection rule from quantum mechanics dictates us that $n \rightarrow \pi^*$ transitions in a single chromophoric groups such as carbonyl and nitro groups and therefore the corresponding bands are characterized by low molar absorptivities, ε_{max} generally less than 100. These weak bands are known as ***R bands:*** These species also show blue shift (hypsochromic shift) with an increase in solvent polarity index.

Molecules with conjugated π system such as butadiene, mesityl oxide or cholest-4-en-3-one show absorption bands due to $\pi \rightarrow \pi^*$ transitions. Fig. 5.8 shows UV spectra (log ε vs. λ) to represent such transition. Note that mesityl oxide, a simple compound, very closely resembles the spectral feature of

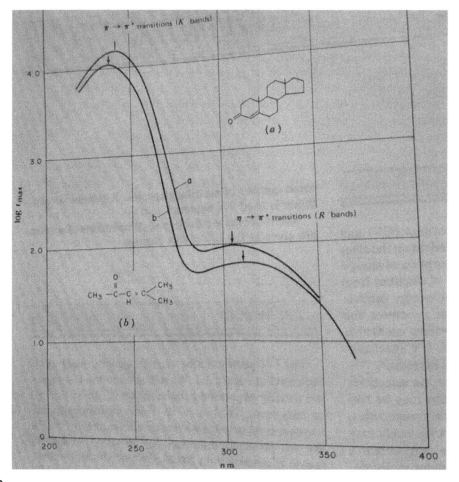

FIG. 5.8

Ultraviolet spectra of (A) cholest-4-en-3-one and (B) mesityl oxide.

more complex molecule of steroid. In Fig. 5.8, it is apparently seen that an n → π^* requires less energy than a $\pi \rightarrow \pi^*$ or $\sigma \rightarrow \sigma*$transition. These transitions involve compounds that have double or triple bonds or compounds with lone pair electrons as part of a saturated structure such as in the carbonyl (C= O) functional group. Moreover, aromatic molecules possessing chromophoric substitution such as styrene, benzaldehyde or acetophenone show high absorptivities ꞌ 10,000 L.mole-1cm-1.

The $\pi \rightarrow \pi^*$ transition of conjugated di- or polyene systems (e.g., propylene) can be distinguished from those of enone (e.g., ketone) systems by observing the effect of changing solvent polarity: The di- or polyene systems do not response to solvent polarity and also the hydro carbon double bonds, are nonpolar. The corresponding absorptions of enones, however, undergo a bathochromic shift (red shift), frequently accompanied by increasing intensity, as the polarity of the solvent is increased. The red shift assumed to result from a reduction in the energy level of the excited state accompanying dipole-dipole interaction and hydrogen bonding.

For example, the effect of solvent has been measured for the n → π^* of acetone: With maximum at 279 nm in hexane, and decreases to 272 and 264.5 nm for the solvents ethanol and water, respectively (see in Fig. 5.9).

B band (also called Benzenoid band): UV–Vis spectra of aromatic or heteroaromatic molecules possess *B bands* (*benzenoid bands*), as their finger print. For example, benzene possesses broad peak from 230 nm–270, with λ_{max} = 255 nm. When chromophoric group is attached to aromatic ring of a molecule like benzene the B band is observed at longer wavelength than its main peak due to $\pi \rightarrow \pi^*$ transitions. For example, styrene molecule which possesses conjugated π electrons, has $\pi \rightarrow \pi^*$ transitions at λ_{max} = 255 nm (εmax = 12,000), and B band at λ_{max} = 282 nm (εmax = 450).

Moreover, when n → $\pi*$ transitions appear in the aromatic compound including (B band) the n → π^* band shifts to longer wavelength. The fine structure is usually destroyed, if polar solvents are used.

E bands (Ethylenic bands): Like the B bands, E bands are also the characteristic of aromatic structures. The E1 and E2 bands of benzene are observed near 180 nm and 200 nm, respectively. When a heteroatom with lone pair of electrons is substituted in the aromatic ring of a molecule, lone pair of

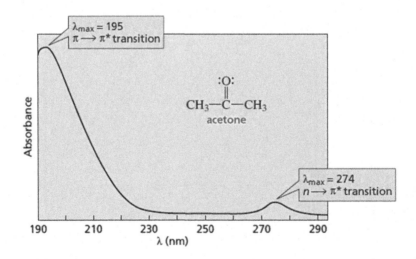

FIG. 5.9

Ultraviolet spectra of acetone in ethanol.

Table 5.4 Types of chromophores and their nature of transitions.

Chromophore	Example	Excitation	λ_{max}, nm	E	Solvent
C=C	Ethene	$\pi \rightarrow \pi*$	171	15,000	Hexane
C≡C	1-Hexyne	$\pi \rightarrow \pi*$	180	10,000	Hexane
C=O	Ethanal	$n \rightarrow \pi*$	290	15	Hexane
		$\pi \rightarrow \pi*$	180	10,000	Hexane
N=O	Nitromethane	$n \rightarrow \pi*$	275	17	Ethanol
		$\pi \rightarrow \pi*$	200	5000	Ethanol
C-X X = Br	Methyl bromide	$\pi \rightarrow \sigma*n \rightarrow \sigma*$	205	200	Hexane
X = I	Methyl iodide		255	360	Hexane

electrons of heteroatom share with π electron system of the ring, facilitating the $\pi \rightarrow \pi*$ transitions and thus causing bathochromic shift of the *E bands*. The molar absorptivity of *E bands* generally varies between 2000 and 14,000 (Table 5.4).

(b) Probing organic compounds and functional groups

From the chart above it should be clear that the only *chromophores* (which are part or functional groups of a molecule) likely to absorb light in the 200–800 nm region are pi-electron functions and hetero atoms having non-bonding valence-shell electron pairs. A list of some simple chromophores and their light absorption characteristics is provided on Table 5.3. The oxygen non-bonding electrons in alcohols and ethers do not give rise to absorption above 160 nm. Consequently, pure alcohol and ether solvents may be used for spectroscopic studies.

The presence of chromophores in a molecule is best documented by UV–Visible spectroscopy, but the failure of most instruments to provide absorption data for wavelengths below 200 nm makes the detection of isolated chromophores problematic. Fortunately, as demonstrated in Table 5.3, conjugation generally moves the absorption maxima to longer wavelengths, as in the case of isoprene, so conjugation becomes the major structural feature identified by this technique.

Molar absorptivities may be very large for strongly absorbing chromophores (>10,000) and very small if absorption is weak (10–100). The magnitude of absorption coefficient reflects both the size of the chromophore and the probability that light of a given wavelength will be absorbed when it strikes the chromophore.

(c) *Understanding the relationship between UV–VIS absorption, color and structure of organic compounds:* There are many colored organic compounds, such as dyes and pigments. How is it that these colors come about? There is a close relationship between the color of an organic compound and its structure. Below we will know about this relationship using absorption spectra of organic compounds obtained with UV–VIS spectrophotometer. To understand the phenomena, below we discuss with some practical examples.

Absorption spectra of food dyes with large conjugated systems and their colors: Fig. 5.9 shows the structures of food dyes New Coccine (Red No. 102) and Brilliant Blue FCF (Blue No. 1) and their absorption spectra. Food dyes tend to have large conjugated systems, like those shown in, and Fig. 5.9, therefore, their peak wavelengths tend to be shifted toward the long wavelength region, with peaks appearing in the visible region (400–700 nm). This is why they are recognized as colors (Table 5.5).

Table 5.5 Absorption peaks and molar absorption coefficients of various organic substances.

Substance	Absorption peak	Molar absorption coefficient
Ethylene(CH_2=CH_2)	180 nm	10,000
1.3-butadiene	217 nm	21,000
Vitamin A	328 nm	51,000
β-carotene	450 nm	140,000
Benzene	255 nm	180
Naphthalene	286 nm	360
Anthracene	375 nm	7100
Naphthacene	477 nm	110,000

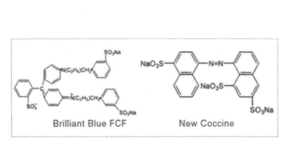 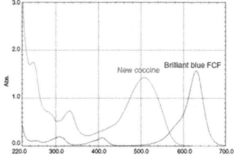

FIG. 5.10

Structures of New Coccine and Brillinant Blue FCF food dyes (left) and their absorption spectra (absorbance vs. wavelength (nm)).

Incidentally, the color that we see is the color that is not absorbed by the substance (which is called the "complementary color"). Apparently, as shown in absorption spectrum, New Coccine absorbs blue and green light in the range 450–550 nm, and so the complementary color, red, is seen by the human eye. Brilliant Blue FCF absorbs yellow light in the range 560–650 nm and so blue is seen by the human eye (Fig. 5.10).

Absorption peaks are also influenced by functional groups. Fig. 5.11 shows the absorption spectra of benzene, phenol, which consists of a hydroxyl group bonded to a benzene ring, and pnitrophenol, which consists of a hydroxyl group and a nitro group bonded to a benzene ring. The functional groups influence the conjugated systems, causing the absorption peaks to appear at longer wavelengths than the peak wavelength of benzene, although they do not go beyond 400 nm and enter the visible region. The color of organic compounds, then, is influenced more strongly by the size of the conjugated system.

- A. Skoog, F. J. Holler and S. R. Crouch, Instrumental Analysis, Cengage Learning, New Delhi (Reprint), 2007.
- <https://www.shimadzu.com/an/uv/support/uv/ap/apl.htmlwww.shimadzu.com/an/uv/support/uv/ap/apl.html>
- Masayoshi Nakahara: "The Science of Color", Baifukan (2002), p. 108.
- <https://www2.chemistry.msu.edu/faculty/reusch/VirtTxtJml/Spectrpy/UV-Vis/spectrum.htm>.

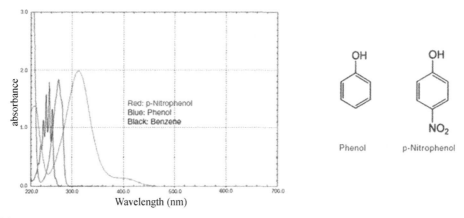

FIG. 5.11

Absorption spectra of benzene, phenol and p-nitrophenol (left) and their molecular structures (right). Notes: section (5.17): Probing organic compounds with UV—Vis spectroscopy was discussed on the basis of the Ref. ([12,39—41]). following references.

5.4.5.3 Application for quantitative analysis of organic compounds (or functional groups)

Quantitative determination concentration of organic species or functional groups present in organic compounds is possible. Below in the consecutive section, we will learn concentration determination of, for example, amines, carbonyl group, anionic detergent, phenol, and also in enzymatic catalysis, cholesterol in mayonnaize and aromatic hydrocarbons and binary mixtures.

5.4.5.3.1 Primary amines

5.4.5.3.1.1 General discussion. In organic chemistry, amines are compounds and functional groups that contain a basic nitrogen atom with a lone pair. Amines are formally derivatives of ammonia, wherein one or more hydrogen atoms have been replaced by a substituent such as an alkyl or aryl group. Important amines include amino acids, biogenic amines, trimethylamine, and aniline. The substituent $-NH_2$ is called an amino group. Inorganic derivatives of ammonia are also called amines, such as chloramine ($NClH_2$).

Classification of amines: Amines can be classified according to the nature and number of substituents on nitrogen. Aliphatic amines contain only H and alkyl substituents. Aromatic amines have the nitrogen atom connected to an aromatic ring. Amines, alkyl and aryl alike, are organized into three subcategories based on the number of carbon atoms adjacent to the nitrogen:

Primary (1°) amines—Primary amines arise when one of three hydrogen atoms in ammonia is replaced by an alkyl or aromatic group. Important primary alkyl amines include, methylamine, most amino acids, and the buffering agent tris, while primary aromatic amines include aniline.

Secondary (2°) amines—Secondary amines have two organic substituents (alkyl, aryl or both) bound to the nitrogen together with one hydrogen. Important representatives include dimethylamine, while an example of an aromatic amine would be diphenylamine.

Tertiary (3°) amines—In tertiary amines, nitrogen has three organic substituents. Examples include trimethylamine, a distinctively fishy smell, and EDTA.

Solubility in water: The small amines of all types are very soluble in water. In fact, the ones that would normally be found as gases at room temperature are normally sold as solutions in water - in much the same way that ammonia is usually supplied as ammonia solution.

All of the amines can form hydrogen bonds with water - even the tertiary ones.

5.4.5.3.1.2 Principle. For determining aromatic primary amines present in a given sample on the micro scale the diazotization method is employed. Whereas, the naphthaquinone method can be applied to both aliphatic and aromatic primary amines.

A. Diazotization method: In this procedure the amine is diazotized and then coupled with N-(1-naphyl)enthlenediamine. This causes to develop the colored product whose concentration can be determined with a spectrophotometer.

5.4.5.3.1.3 Reagents. *N-(1-naphyl)enthlenediamine dihydrochloride:* Dissolve 0.3 g of the solid in 100 mL of 1% v/v HCl (*solution A*).

Sodium nitrite: Dissolve 0.7 g sodium nitrite in 100 mL distilled water (*solution B*).

1 M HCl: 90 % Ethanol (rectified spirit).

Procedure:

i. Weigh out 10–15 mg of the desired amine sample and dissolved in 1 M HCl solution in 50 mL graduated flask.
ii. Clamp a small conical flask in a 500 mL beaker filled with tap water. Place 2.0 mL of the amine solution prepared in (i) and then add 1 mL of the *solution B*; allow to stand for 5 min.
iii. Now add 5 mL of 90% ethanol and after waiting a further 3 min, add 2 mL of *solution A*. A red coloration develops quickly.
iv. Then the absorbance of the above solution is measured against blank solution containing all the reagents except the amine sample at 550 nm (this value may vary slightly with the nature of the amine).
v. A calibration curve can be prepared using a series of solutions of the pure amine of appropriate concentrations, which are treated following the steps [1–4].

5.4.5.3.2 Primary amines (alternative method: naphthaquinone method)

5.4.5.3.2.1 Principle. Many primary amines develop a blue color when treated with ortho-quinones. The preferred reagent is the sodium salt of 1,2-naphthaquinone-4-sulphonic acid.

5.4.5.3.2.2 Reagents

(I) *1,2-naphthaquinone-4-sulphonic acid sodium salt.* Dissolve 0.4 gm of this sodium salt in 100 mL distilled water (solution C).
(II) *Buffer solution:* Dissolve 4.5 g of disodium hydrogen phosphate (Na_2HPO_4) in 1 L of distilled water and carefully add 0.1 M sodium hydroxide (NaOH) solution to give a pH of 10.2–10.4.

Procedure:

i. Mix 25 mL of an aqueous solution of the amine, 10 mL of solution C and 1 mL of buffer solution in a 100 mL conical flask. Note that amine should be less than 10 µg.
ii. After 1 min add 10 mL chloroform and stir on a magnetic stirrer for 15–20 min.
iii. Transfer to a separatory funnel, and after the phases have separated, run off the chloroform layer.
iv. Measure the absorbance at a wavelength of 450 nm against a chloroform blank.
v. A calibration curve can be prepared using a series of solutions of the pure amine of appropriate concentrations, which are treated following the steps (i) to (iv).

5.4.5.3.3 Carbonyl group

5.4.5.3.3.1 General discussion. A carbonyl group is an organic functional group composed of a carbon atom double-bonded to an oxygen atom and represented as $(C=O)$. The simplest carbonyl groups are aldehydes and ketones usually attached to another carbon compound. These structures can be found in many aromatic compounds contributing to smell and taste.

5.4.5.3.3.2 Principle. Aldehydes and ketones can be determined with gravimetric technique (converting them to the 2,4-dinitrophenylhydrazone complex), when present in large quantities. But when present in micro scale (10^{-3} M or less), spectrophotometric or colorimetric method is thought to be appropriate.

5.4.5.3.3.3 Reagents. *Methanol (analytical grade).* Be sure that methanol does not contain aldehyde and ketone.

   2,4-dinitrophenylhydrazine. Prepare a saturated solution in 20 mL of the purified methanol (*Solution A*). The solution should be discarded after one week.

   Potassium hydroxide (KOH): Dissolve 10 mg of solid (KOH) in 20 mL distilled water and make up to 100 mL with pure methanol.

   Procedure.

   i. Dissolve 0.1 g of the sample in 10 mL purified methanol and transfer 1.0 mL of this solution to a stoppered test-tube.

   ii. Add 1.0 mL of *solution A* and 1 mL of conc. HCl then place the stopped test tube in a beaker of boiling water for 5 min. Cool and then add 5.0 mL of the KOH solution.

   iii. Measure the absorbance of the solution at 480 nm (Note: use methanol as a blank and set to zero for background correction, before measurement for the standard or analyte samples).

5.4.5.3.4 Anionic detergents

5.4.5.3.4.1 General discussion. Anionic surfactants are currently the types most used, being incorporated in the majority of detergent and cleaning-product formulas in daily use. Linear-chain alkyl benzenesulfonate types, are the most popularly used synthetic anionic surfactants. In the majority of the detergent and cleaning-products, anionic surfactants have been extensively used for over 40 years. These surfactants pass into sewage-treatment plants, where they are partially aerobically degraded and partially adsorbed to sewage sludge that is applied to land. Finally, they are dumped into the waterways and onto soil, where they constitute some of the main factors affecting the natural ecosystem. Therefore, it is important to determine the concentration of anionic surfactants with accuracy and have quick and simple procedures to monitor their biodegradation over time.

5.4.5.3.4.2 Principle. Anionic surfactants are usually determined by spectrophotometric methods using methylene blue, this standard method being used to determine the surface agents in tap-water samples (ISO 7875-1, 1996). The method involves treatment of an aqueous solution of the detergent that contains anionic surfactants (AS), with methylene blue (MB), in the presence of chloroform according to the reaction,

$$(MB^+)Cl^- + RSO_3Na \rightarrow (MB^+)(RSO_3) + NaCl,$$

where MB^+ indicates the cation of methylene blue. The reaction product can be extracted by chloroform, at the same time as the original dye is insoluble in this medium, and the absorbance recorded (or intensity of the color in the chloroform) is proportional to the concentration of the detergent.

However, this official method is not only long and tedious but also requires large quantities of chloroform and sample.

A simplified method of the spectrophotometric methylene blue method that can be useful for determining anionic surfactants in relatively clean aqueous samples has already been developed. We will discuss that method.

5.4.5.3.4.3 Apparatus. Spectrophotometer (required 650 nm).

5.4.5.3.4.4 Reagents. *Stock linear alkylbenzenesulfonate (LAS) solution of 1 g/L:* 1 g of the commercial product dodecylbenzene sulfonate acid is dissolved in 750 mL of distilled water, adjusted to pH 7.0 by the addition of NaOH solution, and level to 1L.

Standard LAS solution of 10 mg/L: is prepared by dissolving 1 mL of the above stock solution into 100 mL distilled water.

Methylene blue reagent is established at a slightly acid pH *of g/L (3.13 mM):* 0.1 g of methylene blue is dissolved in 100 mL of tetraborate buffer solution 10 mM. The p^H of this solution should be 5.0–6.0. This solution is kept in a topaz-coloured flask.

Sodium tetraborate buffer, 50 mM at pH 10.5: 19 g of sodium decahydrated tetraborate $(Na_2B_4O_7.10H_2O)$ is dissolved in 850 mL of distilled water. The pH is adjusted to 10.5 and leveled to 1 L.

Phenolphthalein indicator: 1 g of phenolphthalein is dissolved in 50 mL of ethanol $(C_2H_5OH, 95\%$ v/v) and, under constant stirring; 50 mL of water is added. Any precipitate is eliminated by filtration.

5.4.5.3.4.5 Analytical procedure

i. Prepare a calibration curve of the detergent sample of which we wish to analyze in the unknown solution, preparing its standard solution and measuring its absorbance.

ii. Next, in a quartz cuvette or test tube, 5 mL of the sample are added and alkalinized by the addition of 50 mM $Na_2B_4O_7.10H_2O$ at pH 10.5 to the color change of phenolphthalein (pH > 8.3). 200 μL of $Na_2B_4O_7.10H_2O$ generally should be sufficient.

iii. Next, 100 μL of stabilized methylene blue is added and homogenized, and 4 mL of chloroform is added. After vigorous stirring for 30 s and then 5 min at rest in the same cuvette, without filtering, the solution is measured for absorbance at 650 nm against air.

5.4.5.3.5 Phenols

5.4.5.3.5.1 General discussion. Phenol is an aromatic organic compound with the molecular formula C_6H_5OH. It is a white crystalline solid that is volatile. It is a component of industrial paint strippers used in the aviation industry and other chemically resistant coatings. Phenol derivatives are also used in the preparation of cosmetics.

Phenol and its derivates are serious water pollutants. Small amounts of phenols directly influence taste and smell of water. The environmental pollution proceeds from industrial sources by the manufacture of dyes, papers, plastics, drugs, and antioxidants or from the use of phenols as pesticides and insecticides. Laws in most countries limit the concentration of phenols in drinking water. The upper limit for total phenol in drinking water is fixed at 0.5 pg/L by the EU-Directive 80/778, in bathing water was fixed at 50 pg (pg)/L by 76/160/CEE and surface water intended for the abstraction of drinking water was fixed at 1–100 pg/L by 75/440/CEE [1].

A rapid, accurate and sensitive procedure is required for phenol determination in water. For this purpose, two methods are discussed below.

5.4.5.3.5.2 Principle. The method discussed here is based on the oxidative coupling reaction with 4-aminoantipyrine (4-AAP) and absorbance measurement at 510 nm. This reagent has some disadvantages as the para-substituted phenols cannot be determined and it also requires preliminary distillation of the sample to separate the interferents. The detection limit is 0.5 pgL with chloroform extraction method and 1 pg/L with direct photometric method.

One reported procedure is based on the blue-colored reaction between phenol, sodium nitroprusside and hydroxylamine hydrochloride in a buffer medium. Procedure with this method is discussed below.

5.4.5.3.5.3 Analytical procedure. All the solutions need to prepare from analytical reagents *(A.R.)* grade reagents and distilled water without phenol. Follow the following method for standard solution preparation and measurement.

i. Prepare the stock standard solution of phenol by dissolving 1.0 g of phenol in 1000 mL of distilled water (1000 mg/L). The solution should be stable for at least 30 days and kept in a refrigerator.

ii. Prepare the working standard solutions of 20 mg/L by appropriate dilution of the stock standard solution with distilled water and use within 1 day.

iii. Prepare the sodium nitroprusside solution 1.0×10^{-2} mol/L by dissolving 0.7450 g of sodium nitroprusside $\{Na_2[Fe(CN)_5NO].2H_2O\}$ in 250 mL of distilled water.

iv. Prepare the hydroxylamine hydrochloride solution 4.0×10^{-2} mol/L by dissolving 0.6950 g of hydroxylamine hydrochloride in 250 mL of distilled water.

v. Prepare the buffer solution (pH 12) by adding 500 mL of NaH_2PO_4 solution (0.1 mol/L) and 17.5 mL of NaOH solution (6 mol/L) to a 1000-mL volumetric flask and diluting it to the mark with distilled water.

vi. From the above stock solution, put the 0.25, 0.5, 1.0, 2.0, 2.5, 5.0, 7.5 mL of the working standard solution, respectively, 50-mL volumetric flask with matching corks and dilute to the mark with distilled water. The series of prepared solutions will contain phenol in the range of 0.05–5.0 mg/L. Next, add 1.0 mL of 1.0×10^{-2} mol/L sodium nitroprusside solution, 1.0 mL of 4.0×10^{-2} mol/L hydroxylamine hydrochloride solution, and 3.0 mL of buffer solution in each standard solution. After mixing well, allow to stand for 15 min and then measure absorbance at 700 nm in 2-cm cells against a distilled water blank.

5.4.5.3.6 Determination of phenol (alternative method)

5.4.5.3.6.1 General discussion. Phenol yields the peak value (wavelength) within 270–296 nm. You can double confirm by using UV–Vis Spectroscopy where the maximum absorption indicated the peak value that you are looking for. The most frequently used peak for quantification is 270 nm, where you have a good linear correlation between absorbance and concentration, provided you don't have any interfering compounds. You can use water as reference. Ensure that the lamp that is using the spectrophotometer is deuterium and the sample holder is of quartz cuvette.

Water samples showing contamination of phenols are best examined by extracting the phenol into the organic solvent; tri-n-butyl phosphate is considered very suitable for this. It requires alkaline condition which is maintained by addition of tetra-n-butylammonium hydroxide.

5.4.5.3.6.2 Reagents

i. *Stock phenol solution.* Weigh out 0.5 g phenol, dissolved in distilled water and make upto the mark in a 500 mL graduated flask: it is recommended that freshly boiled and cooled distilled water be used.

ii. *Standard phenol solution (0.025 mg L^{-1}).* Dilute 25 mL of stock solution to 1L using freshly boiled and cooled distilled water be used.

iii. *Tetra-n-butyl ammonium hydroxide (0.1 M solution in methanol).* Dissolve 20 g of tetra-n-butyl ammonium iodide in 100 mL of dry methanol and pass this solution through the column at a rate of about 5 mL/min and collect the effluent in a vessel fitted with a "Carbosorb" guard tube to protect if from atmospheric carbon dioxide. Then pass 200 mL of dry methanol through the column. Standardize the methanoic solution by potentiometric titration of an accurately weighed (0.3 g) of benzoic acid. Calculate the molarity of the solution. And add sufficient methanol to make it approx. 0.1 M.

iv. HCl (5 M), tri-n-butyl phosphate.

Procedure.

i. From the above stock solution of phenol, prepare 4 standard solutions of phenol by placing 200 mL of boiled and cooled distilled water in each of four stopped, 500 mL bottles, and adding to each 5 g of NaCl (this assists the extraction procedure by salting out the phenol).

ii. Add respectively, 5.0, 10.0 mL, 15 mL and 20 mL of the standard phenol solution to the four bottles, then adjust the pH of the solution to about 5 by the careful addition of 5 M HCl (check with pH meter).

iii. Add distilled water to each bottle to make a total volume of 250 mL and then add 20 mL tri-n-butyl phosphate. Then close tightly with stopper and shake it well with mechanical shaker for 30 min. Transfer to separatory funnel and when the phases have settled, run off and discard the aqueous layers.

iv. Prepare an alkaline solution of the phenol concentrate by placing 4 mL of a tri-n-butyl phosphate layer in a 5 mL graduated flask and then adding 1 mL of Tetra-n-butyl ammonium hydroxide: do this for each standard solution. The reference solution consists of 4 mL of the organic layer and 1 mL of methanol.

v. Measure the absorbance of each of the extracts from the four test solutions and plot a calibration curve.

vi. The unknown solution (in the range of 0.5–2 mg phenol/L) is also treated in the manner described above and measure the absorbance. With the aid of slope of the calibration, evaluate the concentration of the unknown.

Notes: sections (5.18–5.23), which are associated with application of UV–Vis spectroscopy for quantitative analysis of primary amines, carbonyl group, anionic detergent, phenol, were, discussed on the basis of Ref. [12,24,37,42–45] and are as follows:

- G.H. Jeffery, J. Bassett, J. Mendham, R.C. Denney, Vogel's Textbook of Quantitative Chemical Analysis sixth ed., Longman Scientific & Technical, England. Longman Group UK Ltd.,1989.
- Kang, Y. Wang, R. Li, Y. Du, J. Li, B. Zhang, L. Zhou, Y. Du, Microchemical Journal 64 (2000) 161–171.

- J. Mendham, R.C. Denney, J.D. Barnes, M Thomas, B. Sivasankar, Vogel's Textbook of Quantitative Chemical Analysis sixth ed., Pearson, (2009) pg. 668−670.
- Kang, Y. Wang, R. Li, Yaoguo Du, J. L., B. Zhang, L. Zhou, Y. Du, Microchemical Journal 64 (2000) 161,171.
- Jurado et al./Chemosphere 65 (2006) 278−285
- http://chemguide.co.uk/organicprops/amines/background.htm
- https://en.wikipedia.org/wiki/Amine

5.4.5.3.7 Enzymatic analysis

5.4.5.3.7.1 General discussion. Routine enzyme analysis is required in biochemical and clinical analysis. These methods have been adapted for food analysis and bioanalysis. Compounds that occur in nature, e.g. sugars, acids and their salts and alcohols, can be analyzed enzymatically since living cells contain enzymes capable of synthesizing and decomposing these substances. Thus if such an enzyme and a suitable measuring system can be made available, the compound can be determined enzymatically.

Many of the ultraviolet methods of analysis are based on the increase or decrease in absorbance of the coenzymes NADH (nicotinamide adenine, dinucleotide, reduced form) or NaDPH (nicotinamide adenine, dinucleotide phosphate, reduced form) which absorbs light at $\lambda_{max} = 340$ nm.

Spectroscopic and colorimetric methods are based on the formation of light absorbing dye in the visible spectrum. Oxidases react with a substrate (analyte) to form hydrogenperoxide as an intermediate. This can react with a leuco-dye in the presence of a peroxidase enzyme and a dye is formed which can be measured in the visible region of the spectrum. Two examples of the use of the test kits for the analysis of foodstuffs are discussed.

5.4.5.3.8 Glycol in fruit juice

5.4.5.3.8.1 Principle. Glycerol is converted by adenosine-5′-triphosphate (ATP) to L-glycerol-3-phospate in the reaction catalyzed by glycerokinage (GK):

$$\text{Glycerol} + \text{ATP} \rightarrow \text{L-glycerol-3-phosphate} + \text{ADP}$$

The adenosine-5′-diphosphate (ADP) formed in this reaction is reconverted by phosphoenolpyruvate (PEP) with the aid of pyruvate kinage (PK) into ATP with the formation of pyruvate:

$$\text{ADP} + \text{PEP} \rightarrow \text{ATP} + \text{pyruvate}$$

In the presence of the enzyme lactate dehydrogenase (L-LDH) pyruvate is reduced to L-lactate by reduced nitonamide adenine dinucleotide (NADH) with the oxidation of NADH to NAD.

$$\text{Pyruvate} + \text{NADH} + \text{H}^+ \rightarrow \text{L-Lactate} + \text{NAD}^+$$

The amount of NADH oxidized in this reaction is stoichiometric to the amount of glycerol. NADH is determined by means of its light absorption at 340 nm.

5.4.5.3.8.2 Reagents. Most enzymatic methods use test kits supplied by manufacturers e.g. Boehringer Mannheim, which are sufficient for a given number of determinations of the analyte (s). Here, the test combination obtained from Boehringer Mannheim also contains enough reagents for 30 determinations. It comes in five bottles, three of type 1 and one each of type 2 and 3: Each contains reagents of the following compositions:

Bottle 1: Contains approximately 2g coenzyme/buffer mixture, consisting of glycyllglycine buffer (pH 7.4), NADH-7 mg, ATP ~22 mg, PEP-CHA ~11 mg, magnesium sulfate, stabilizers.

Bottle 2: Contains approximately 0.4 mL suspension, consisting of pyruvate kinage and lactate dehydrogenase.

Bottle 3: Contains approximately 0.4 mL glycerokinase suspension.

Solutions for ten determinations is solved the contents of one of the coenzyme/buffer bottles in 11 mL or redistilled water. Allow the solution to stand for about 10 min before use. The contents of the bottle 2 and bottle 3 are used undiluted.

Procedure:

i. Dilute the fruit juice sample to yield a glycerol concentration of less than 0.4 g/L.
ii. If the juice is turbid, filter and use the clear solution for the essay.
iii. When analyzing strongly colored juices, decolorize the sample as follows: Take 10 mL of the juice and add about 0.1 g of polymide powder or polyvinylpolypyrolidine. Stir for 1 min and filter. Use the clear solution (which may be slightly colored for the determination).
iv. Pipette into a cuvette 1 mL of diluted bottle 1 solution, 2 mL of distilled water, 0.1 mL sample solution and 0.01 mL of bottle 2 suspension.
v. In a separate cuvette prepare a blank using the same reagents as for the sample, but do not add the sample solution.
vi. Mix each solution well and record its absorbance at 340 nm against either an air or distilled water reference when the reaction is complete (5–7 min).
vii. Start the reaction by adding 0.01 mL of bottle 3 suspension to both cuvettes. Mix well and wait for completion of the reaction (about 5–10 min).
viii. Measure the absorbance of the sample and blank immediately one after the another and at 340 nm. If the reaction has not stopped after 15 min, continue measurement at 2 min intervals until the absorbance decreases constantly over 2 min intervals.
ix. Extrapolate the absorbance to the time the bottle 3 suspension was added.

Calculation:

Let the absorbance before adding bottle 3 be A_1 he and the absorbance after adding the bottle 3 be A_2. Determine the absorbance differences (A_1 - A_2) for both the blank and sample:

$$\Delta A = (A_1 - A_2)_{sample} - (A_1 - A_2)_{blank}$$

The measured absorbance difference should be at least 0.1 to achieve the sufficiently accurate results. If ΔA is larger than 1.0, the concentration of glycol in the sample solution is too high and should be diluted.

The concentration c (g/L) is given by

$$c = \frac{V \times MW}{1000\varepsilon dv} \times \Delta A$$

where V = final volume (3.02 mL); V, sample volume (0.1 mL); MW, relative molecular weight of the glycol (92.1); d, path length (1 cm); ε, is the absorption coefficient (6.3 L mmol^{-1}cm^{-1}). i.e., $c = 0.414\ \Delta A$.

5.4.5.3.9 Cholesterol in mayonnaise

5.4.5.3.9.1 Principle. Cholesterol is oxidized by cholesterol oxidase:

$$\text{cholesterol} + O_2 \rightarrow \Delta^4\text{-cholesterone} + H_2O_2$$

In the presence of catalase, the H_2O_2 produced in this reaction oxidizes methanol to formaldehyde.

$$CH_3OH + H_2O_2 \rightarrow CH_2O + H_2O$$

The formaldehyde reacts with acetylacetone to form yellow lutidine dye in the presence of ammonium ions:

$$CH_2O + NH_4^+ + 2 \text{ acetylacetone} \rightarrow \text{lutidine dye} + 3H_2O$$

The concentration of the lutidine dye formed is stoichiometric to the amount of cholesterol and is measured by the increase of light absorbance in the visible range at 405 nm.

5.4.5.3.9.2 Reagents. The test combination obtained from Boehringer Mannheim also contains enough reagents for 25 determinations. It comes in five bottles, three of type 1 and one each of type 2 and 3:

Bottle 1: Contains approximately 95 mL solution consisting of ammonium phosphate buffer (pH 7.0), methanol 2.6 mol/L catalase and stabilizers.

Bottle 2: Contains approximately 60 mL solution consisting of acetylacetone 0.3 mol/L, 0.3 mol/L and stabilizer.

Bottle 3: Contains approximately 0.8 mL cholesterol oxidase suspension.

Cholesterol reagent: Mix three parts of the solution from bottle 1 with two of the solution from bottle 2 in a brown bottle adjusted to room temperature. Allow the mixture to stand at room temperature for 1 h before use.

Solution 3 Use contents of bottle undiluted.

Other reagents: Freshly prepared 1.0 M methanoic KOH.

Procedure:

  i. Weigh accurately 1 g of mayonnaize and 1 g of sea sand into a 50 mL RB flask.

 ii. Add 10 mL of methanoic KOH and heat under reflux for 25 min, stirring continuously. Transfer the supernatant solution into a 25 mL graduated flask with a pipette.

iii. Boil the residue twice with portions of 6 mL propan-2-ol each under reflux for 5 min. Collect the solution in the volumetric flask and allow to cool.

 iv. Dilute the contents of the flask to the mark with propan-2-ol and mix. If the solutions are turbid, filter through a fluted filter paper. Use the clear solution for the assay. The sample solution should contain between 0.07 and 0.4 gL^{-1} cholesterol.

  v. Pipette a sample blank by pipetting 5.0 mL of the cholesterol reagent and 0.4 mL of the sample solution; mix thoroughly.

 vi. Into a separate test tube pipette 2.5 mL of the sample blank and add 0.02 mL of solution 3. This is the sample. Mix thoroughly.

vii. Cover both tubes and incubate in a water bath at 37—40 °C. For 60 min. Allow to cool to room temperature.

viii. Read the absorbance of the sample blank and sample one after the other in the same cuvette against and air reference at 405 nm.

ix. Subtract the absorbance of the blank from the absorbance of the sample (ΔA). In order to achieve sufficient accuracy, the measured absorbance difference ΔA should be at least 0.1 to achieve the sufficiently accurate results.

Calculation:

Let the absorbance before adding bottle 3 be A_1 he and the absorbance after adding the bottle 3 be A_2. Determine the absorbance differences ($A_1 - A_2$) for both the blank and sample:

$$\Delta A = (A_1 - A_2)_{sample} - (A_1 - A_2)_{blank}$$

The measured absorbance difference should be at least 0.1 to achieve the sufficiently accurate results. If ΔA is larger than 1.0, the concentration of glycol in the sample solution is too high and should be diluted.

The concentration c (g/L) is given by

$$c = \frac{V \times MW}{1000 \varepsilon dv} \times \Delta A$$

where V, final volume (5.4 mL); V, sample volume (0.4 mL); MW, relative molecular weight of the glycol (386.64); d, path length (1 cm); f, dilution factor $= 2.52/2.5 = 1.008$; ε is the absorption coefficient (7.4 L mmol^{-1}cm^{-1}). i.e. $c = 0.711\Delta A$.where w is the weight of the mayonnaize sample in grams.

5.4.5.3.10 Aromatic hydrocarbons and binary mixtures

5.4.5.3.10.1 Principle. Measurements of absorption spectra of typical hydrocarbon by UV Visible spectrophotometry allow one to determine the mixtures of hydrocarbons.

Reagents: Methanol and benzene toluene Avoid inhalation or skin contact with benzene; it is carcinogenic. Use analytical grade reagents for all the solution preparation.

Procedure:

i. Place 0.05 mL of benzene in a 25 mL graduated flask and prepare a stock solution by diluting to the mark with methanol Using a 0.1 mL capillary micropipette (Work in a fume cupboard).

ii. Prepare a series of five dilutions of the stock solutions; use 2 mL graduated pipette to transfer 0.25, 0.5, 0.75, 1 and 1.5 mL of the solution into a series of 10 mL graduated flasks then make up to the mark with methanol.

iii. Using stoppered quartz cells, use solution 5 (i.e., the most concentrated of the test solutions) to plot an absorption curve using pure methanol as the blank.

iv. Take absorbance reading over the wavelength range 200–300 nm using a spectrophotometer. Make a note of the λ_{max} values for the peaks observed in the curve. There should a well developed peak at approximately 250 nm, and using each of the test solutions in turn, measure the absorption at the observed peak wavelength and test the validity of Beer's law (that is, the least square fit on the measured absorbance data should yield follow linear trend).

v. Now starting with 0.05 mL toluene, repeat the procedure to obtain five working solutions, 1' (' represents 'dash') to 5'. Use solution 5' to plot the absorption curve of toluene; again record the λ_{max} values for the peaks of the curves. There should be a well developed peak at about 270 nm, and using the five new test solutions, measure the absorbance of each one at the observed peak wavelength and test the application of Beer's law.

vi. Next measure solution 5′ at the wavelength used for benzene, and solution 5 at the wavelength used for toluene.

vii. Prepare a benzene-toluene mixture, take 0.05 mL of each liquid in a 25 mL graduated flask then make upto the mark with methanol. Take 1.5 mL of this solution, place in a 10 mL graduated flask and dilute to the mark with methanol; this solution contains benzene at the same concentration as solution 5, and toluene at the same concentration as solution 5'.

viii. Measure the absorbance of this solution at two wavelengths where toluene and benzene show maximum absorbance. Then use the procedure detailed in previous section (section above section for Cr and Mn) to evaluate the composition of the solution and compare it with the composition calculated from the amounts of benzene and toluene taken.

Notes: sections (5.24−5.27), which are associated with application of UV−Vis spectroscopy for quantitative determination of glycol, cholesterol and aromatic hydrocarbon, were discussed on the basis of Ref. [12,24,37,42−45].

Problems:

1. Gaseous ozone has a molar absorptivity of 2700 $M^{-1}cm^{-1}$ at the absorption peak near 260 nm in the spectrum at the beginning of this chapter. Find the concentration of ozone (mol/L) in air if a sample has an absorbance of 0.23 in a 10.0 cm cell. Air has negligible absorbance at 260 nm.

2. (a) What is the absorbance of a 2.33×10^{-4} M solution of a compound with a molecular absorptivity of $1.05 \times 10^3 M^{-1}cm^{-1}$ in a 1.0 cm cell? (b) What is the transmittance of the solution in (a)? (c) Find A and % T when the path length is doubled to 2.0 cm. (d) What would be the absorbance in (a) for a different compound with twice as great a molar absorptivity $(\varepsilon = 2.1 \times 10^3 M^{-1}cm^{-1})$? The concentration and path length are unchanged from (a).

3. How do we measure the benzene in Hexane in the following conditions?

(a) The solvent hexane has negligible UV absorbance above a wavelength of 200 nm. A solution prepared by dissolving 25.8 mg of benzene (C_6H_6) in hexane and diluting to 250.0 mL has an absorption peak at 256 nm, with an absorbance of 0.266 in a 1.0 cm cell. Find the molar absorptivity of benzene at this wavelength.

(b) A sample of hexane contaminated with benzene has an absorbance of 0.07 at 256 nm in a cell with a 5.0 cm path length. Find the concentration of benzene.

4. What is a figure of merit?

5. A 25.0 mL aliquot of an aqueous quinine solution was diluted to 50 mL and found to have an absorbance of 0.656 at 348 nm when measured in a 2.5 cm cell. A second 25.0 mL aliquot was mixed with 10.0 mL of a solution containing 25.7 ppm of quinine; after dilution to 50.0 mL, this solution has an absorbance of 0.976 (2.5 cm cell). Calculate the concentration in parts per million in the sample (ppm).

6. A 0.599 g pesticide sample was decomposed by wet ashing and then diluted to 200 mL in a volumetric flask. The analysis was completed by treating aliquots of this solution as indicated.

Sample(mL)	2.75 ppm Cu^{2+}	Lignad	H_2O	Absorbance(A)
5	0	20	25	0.723
5	1	20	24	0.917
Calculate the percentage of copper in the sample.				

7. What types of energy levels can be involved in the measurement of analytes by molecular spectroscopy? List one specific type of molecular spectroscopy that makes use of each of these changes in energy levels.

8. Explain how UV–Vis spectroscopy can be used for either the measurement or identification of a molecule.

9. What is the standard addition method? Under what circumstances is this method typically used for analysis?

10. A solution has an analyte concentration of 5.7×10^{-3} M that gives a transmittance of 43.6% at 480 nm and when measured in a 5 cm cuvette. Calculate the molar absorptivity of the analyte. What is the expected lower limit of detection for this analyte if the smallest absorbance that can be measured by the instrument is 0.001?

11. What is the upper limit of the linear range for the analyte in problem 10 if deviations for Beer's law are found to occur at an absorbance of approximately 1.0?

References

[1] EPA, How EPA Regulates Drinking Water Contaminants? 2018. https://www.epa.gov/dwregdev/how-epa-regulates-drinking-water-contaminants#decide.

[2] F.W. Billmeyer, M. Saltzman, Principles of Color Technology, John Wiley & Sons, Inc., New York, 1981.

[3] R.S. Hunter, R.W. Harold, The Measurement of Appearance, John Wiley & Sons, Inc., New York, 1981.

[4] M. Roger, Ground Water and the Rural Homeowner, Pamphlet, U.S. Geolgoical Survey, 1982.

[5] https://water.usgs.gov/edu/groundwater-contaminants.html, 2018.

[6] https://www.epa.gov/dwstandardsregulations, 2018.

[7] https://quizlet.com/220963261/calibration-of-instrumental-methods-flash-cards/, 2018.

[8] APHA Method 4500-NO_3: Standard Methods for the Examination of Water and Wastewater, eighteenth ed., 1992.

[9] M. McCasland, N.M. Trautmann, K.S. Porter, R.J. Wagenet, Nitrate: Health Effects in Drinking Water, Pesticide Safety Education Program (PSEP), Kornell University, 2019. http://psep.cce.cornell.edu/facts-slides-self/facts/nit-heef-grw85.aspx.

[10] F.A.J. Armstrong, Determination of nitrate in water ultraviolet spectrophotometry, Anal. Chem. 35 (9) (1963) 1292–1294.

[11] J. Patton, J.R. Kryskalla, Colorimetric Determination of Nitrate Plus Nitrite in Water by Enzymatic Reduction, Automated Discrete Analyzer Methods, Book Chap. 8 Section B, Methods of the National Water Quality Laboratory, Laboratory Analysis. Techniques and Methods 5–B8, USGS. Virginia, 2011.

[12] A. Skoog, F.J. Holler, S.R. Crouch, Instrumental Analysis, Cengage Learning, New Delhi, 2007, p. 420 (Reprint).

[13] D.S. Hage, J.D. Carr, Analytical Chemistry and Quantitative Analysis, Pearson Publication (Int. Ed.), 2011, p. 442.

[14] N.A. Zatar, M.A. Abu-Eid, A.F. Eid, Spectrophotometric determination of nitrite and nitrate using phosphomolybdenum blue complex, Talanta 50 (1999) 819–826.

[15] A. Shrivastava, V.B. Gupta, Methods for the Determination of Limit of Detection and Limit of Quantitation of the Analytical Methods, Chronicles of Young Scientists, 2011, p. 2.

[16] S. Lim, Determination of Phosphorus Concentration in Hydroponics Solution, 2011.

[17] EPA Method 365.1, in: J.W. O'Dell (Ed.), Revision 2.0: Determination of Phosphorus by Semi-automated Colorimetry, U.S. ENVIRONMENTAL PROTECTION AGENCY, 1993 (EPA.

[18] Environmental Protection Agency, 1978. Water.*epa.gov*.

[19] Method 4500-P, APHA "Standard Methods for the Examination of Water and Wastewater", 1992.

[20] https://archive.epa.gov/water/archive/web/html/vms56.html.

[21] A.T. Palin, J. Am. Water Work. Assoc. 49 (1957) 873.

[22] D.L. Harp, Current Technology of Chlorine Analysis for Water and Wastewater, Technical Information Series of Hach Company — Booklet No.17, Lit. No. 7019, L21.5, 2002. Printed in U.S.

[23] APHA 4500-Cl G (Chlorine residual/Colorimetric Method), Standard Methods for the Examination of Water and Wastewater, eighteenth ed., Inorganic, nonmetals, 1992, 4000.

[24] G.H. Jeffery, J. Bassett, J. Mendham, R.C. Denney, Vogel's Textbook of Quantitative Chemical Analysis, fifth ed., Bath Press Ltd., Great Britain, 1994, pp. 480—481. Reprint.

[25] Flinn Scientific Spectrophotometer Laboratory Manual, Flinn Scientific, Batavia, IL, 1994, pp. 55—60.

[26] APHA, 3500-Fe B. Phenanthroline Method, "Standard Methods for the Examination of Water and Wastewater".

[27] CFR Part 136 Appendix B — Definition and Procedure for the Determination of the Method Detection Limit.

[28] The World Health Organization (WHO), Ammonia, Environmental Health Criteria, Geneva, 1986, p. 54.

[29] International Organization for Standardization, Water Quality—Determination of Ammonium. Geneva, 1984, 1986 (ISO5664:1984; ISO6778:1984; ISO7150-1:1984; ISO7150-I7152:1986.

[30] APHA, Standard Methods for the Examination of Water and Wastewater, seventeenth ed., American Public Health Association/American Water Works Association/Water Pollution Control Federation, Washington, DC, 1989.

[31] P.J. Craig, Organometallic Compounds in the Environment, John Wiley & Sons, Inc., New York, 1986.

[32] WHO, Arsenic in Drinking Water, Geneva, vol. 210, 1999.

[33] J.C. Ng, Environmental contamination of arsenic and its toxicological impact on humans, Environ. Chem. 2 (3) (2005) 146—160.

[34] J.K. Thakur, K.R. Thakur, A. Ramanathan, M. Kumar, S.K. Singh, Arsenic contamination of Groundwater in Nepal- an overview, Water 3 (2011) 1—20.

[35] WHO, Arsenic Contamination in Groundwater Affecting Some Countries in the South-East Asia Region, WHO, Washington, DC, USA, 2001.

[36] FAO, Arsenic Threat and Irrigation Management in Nepal, 2004. Rome.

[37] J. Mendham, R.C. Denney, J.D. Barnes, M. Thomas, B. Sivasankar, Vogel's Textbook of Quantitative Chemical Analysis, sixth ed., Pearson, 2009, pp. 668—670.

[38] V.D. Lenoble, B. Serpaud, J.C. Bollinger, Arsenite oxidation and arsenate determination by the molybdene blue method, Talanta 61 (2003) 267.

[39] https://www.shimadzu.com/an/uv/support/uv/ap/apl.html.

[40] M. Nakahara, The Science of Color, Baifukan, 2002, p. 108.

[41] https://www2.chemistry.msu.edu/faculty/reusch/VirtTxtJml/Spectrpy/UV-Vis/spectrum.htm.

[42] C. Kang, Y. Wang, R. Li, Y. Du, J. Li, B. Zhang, L. Zhou, Y. Du, Microchem. J. 64 (2000) 161—171.

[43] E. Jurado, et al., Chemosphere 65 (2006) 278—285.

[44] http://chemguide.co.uk/organicprops/amines/background.html.

[45] https://en.wikipedia.org/wiki/Amine.

Introduction to nanomaterials and application of UV–Visible spectroscopy for their characterization

6.1 Nanomaterials

6.1.1 Introduction

The word "nano" comes from the Latin "nanus" meaning literally dwarf (very small) and has found an ever-increasing application to different fields of science since 2005. Below we will learn well accepted definitions of nanoparticles:

According to IUPAC, particles of any shape with dimensions in the range 1 and 100 nm (1×10^{-9} and 1×10^{-7} m) are called nanoparticles (also called "ultrafine particles"): In other word, they encompass systems whose size is above molecular dimensions and below macroscopic ones (generally >1 nm and <100 nm). The basis of the 100-nm limit is the fact that novel properties that differentiate particles from the bulk material typically develop at a critical length scale of under 100 nm.

According to (ISO/TS 27,687: 2008), a nanoparticles is a nano object with all three external dimensions in the nanoscale", although in practice the term is often used to refer to particles larger than 100 nm (up to 500 nm). The reason for this is that the behavior of nanoparticles and the applicability of measurement techniques vary with size and environment, to extent that 500 nm particles can either be considered very large or very small depending on the frame of reference.

Nanoparticles can be both natural and man-made entities, and are widely found in the environment as well as at the laboratory (some of the synthesis methods are discussed in Section (6.1)). Their origins and properties are highly varied, making study a rich branch of analytical science.

Applications: Much of the interest in nanoscale materials arises from both an understanding of the physical, chemical, and size-dependent phenomena on the nanometer length scale and the development and beneficial uses of these materials in a wide-range of applications. Some of them are listed as: Materials of these dimensions have been widely used in the design and development of new devices for various applications such as optical, photonics, optoelectronics, magnetic, photovoltaics, sensors, catalysis, biological labeling, biomedical, and pharmaceutical applications. Some of the synthesis methods, progression of optical properties with sizes and application in solar cells are discussed in the consecutive section.

Chemical Analysis and Material Characterization by Spectrophotometry. https://doi.org/10.1016/B978-0-12-814866-2.00006-3

6.1.2 General properties of nanoparticles (with a focus on the wide bandgap metal oxides)

Fig. 6.1 shows some of the important and, in some cases, interrelated physicochemical properties of nanoparticles that control (influence) their microscopic as well as their macroscopic behaviors (bulk properties). These properties include size shape, surface composition, aggregation, concentration, and their ability to be active, i.e., to have changing properties as a function of time or some other variables. These properties impact and dictate the most fundamental characteristics of nanomaterials including their ability to absorb solar radiation and carrying photo-generated charges, emit radiation (when used in LED), and get into living cells (for drug delivery application).

6.1.2.1 Size and shape

Size: One of the most interesting aspects of nanoscience is that properties of nanoparticles change with size. It has been demonstrated that many fundamental properties are size dependent on the nanoscale. For example, the most stable crystalline phase of a material is size dependent. From thermodynamic considerations, the total free energy ($G_{nanoparticle}$) is a sum of the free energy of the bulk (G_{bulk}) and the surface ($G_{surface}$) of the nanoparticle. That is,

$$G_{nanoparticle} = G_{surface} + G_{bulk} \tag{6.1}$$

For nanoparticles, $G_{surface}$ is no longer a minor component but in fact becomes a large component of the total free energy.

Surface free energies and surface stress are important components to the overall phase stability of nanoparticles. For example, Titanium dioxide (TiO_2) in anatase phase is more stable than in rutile for a

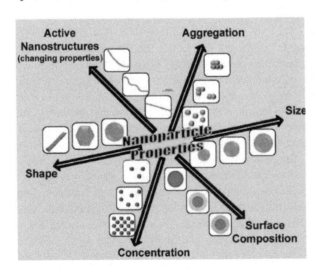

FIG. 6.1

Progression of microscopic (smaller) to macroscopic (larger-bulk) system. Microscopic and macroscopic behaviors of nanoparticles depend on number of important features (e.g., shape, size, concentration, surface composition, and aggregation) nanostructures (see in the text).

V. H. Grassian, J. Phys. Chem. C, 112 (2008), adapted with permission from, Copyright (2008), American Chemical Society.

particle size below 14 nm. However, it has been recently shown that the stability of rutile nanoparticles increases relative to anatase and brookite at low pH due to surface charges.

Electronic properties are size dependent on the nanoscale. The electronic band gap, E_g, for semiconductor nanoparticles can be estimated for a spherical particle as:

$$E_g = \frac{\pi^2 h^2}{2R^2} \frac{1}{\mu} - \frac{1.8e^2}{\varepsilon R}, \tag{6.2}$$

where R is the particle radius, μ is the reduced mass of the exciton or the electron-hole pair, ε is the dielectric constant of the semiconductor and h is Planck's constant. These changes in the bandgap will influence many properties of nanoparticles including surface reactivity.

For pure nanocrystalline metal oxide (cupper oxide, for example) powders, the ability to reflect near-infrared light (NIR) (750–2500 nm) was found to be much higher (+20%) relative to larger common macro-crystalline powders and minerals due to the smaller crystallite sizes and smaller mean aggregate sizes in accordance with Kubelka-Munk theory.

Also, the *localized surface Plasmon resonance* (LSPR) of Cu nanoparticles deposited on a substrate is significantly affected by the presence of copper oxides and the removal of the oxide species yields a dramatic difference in the observed LSPR. For noble metals, LSPR depends on both metal and size. For the two most common metals used, Ag and Au, LSPR can occur throughout the visible region of the spectrum depending on the size of the nanoparticles (Fig. 6.2).

Shape: Also shape of the nanoparticles found to play a crucial role in materials' properties (e.g., shape dependence on absorption spectrum of spherical noble metal nanoparticles). For nanorods, theoretical calculation of the surface plasmon resonance (SPR) showed there are two controlling factors: the bulk plasma wavelength, a property dependent on the metal itself, and the aspect ratio of the nanorods (a geometrical parameter). By changing the shape of Au and Ag nanoparticles, the SPR can be red-shifted into the near-infrared region of the electromagnetic spectrum. As shown by Haes et al., the LSPR for Ag nanoparticles shifts across the electromagnetic spectrum as the shape of the particle changes from a sphere to cylinder to cube to prism to pyramid (**see** Fig. 6.3).

Size- and Shape-Tunable LSPR spectra of Ag nanoparticles: Plasmonics is an emerging branch of nanophotonics that examines the properties of the collective electronic excitations in noble metal films or nanoparticles commonly known as surface plasmons. In a typical NSL process, the deposition of 50 nm (thickness) of Ag over a single-layer mask self-assembled from nanospheres with $D = 400$ nm produces nano-triangles with an in-plane width $a \approx 100\,nm$, height $b \approx 50$ nm, and inter-particle separation distance dip $\approx 230\,nm$.

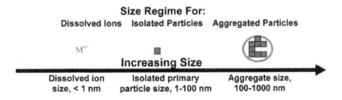

FIG. 6.2

Nanoparticles can be present as isolated particles or they can form aggregates up to microns in size *or* they can dissolve into ions in solution.

V. H. Grassian, Nanoscience and Nanotechnology: Environmental and Health Impacts, Reprinted with permission of (Copyright ©2008) John Wiley & Sons Inc.

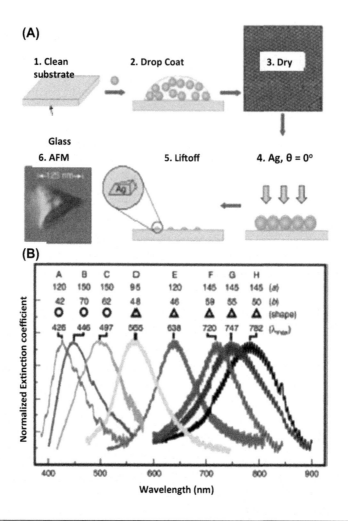

FIG. 6.3

(A: top) Schematic representation of the nanosphere lithography (NSL) fabrication process. The AFM image in step 3 is 5 μm × 5 μm. (B: bottom). Size- and shape-tunable localized surface plasmon resonance spectra, LSPR, of various Ag nanoparticles (labeled A–H) fabricated by NSL. The wavelength of maximum extinction, λ_{max}, is changed by varying the (A) in-plane width and (B) out-of-plane height of the nanoparticles.

Surface Plasmon resonance (in UV Vis spectra), have been utilized to determine the size and heterogeneity of nanoparticles. Specifically, for noble metals, nanoparticle size can be determined from the maximum in the UV absorption spectrum (which is produced due to the surface Plasmon) and the width of the extinction band is a measure of the sample heterogeneity with broader peaks (that is caused by plasmon resonance) appearing for samples that are more heterogeneous. Below we discuss a typical example of LSPR spectra of Ag nanoparticles.

NSL-derived nanoparticles exhibit intense UV–visible extinction (i.e., the sum of absorption and scattering) bands that are not present in the spectrum of the bulk metal. **Figure (bottom)** shows that the

LSPR spectra can easily be tuned all the way from the near-UV through the visible spectrum and even into the mid-IR by changing the size or shape (triangle or hemisphere) of the nanoparticles.

Additionally, several other surprising LSPR optical properties have been discovered for NSL-derived Ag nanoparticles: (1) λ_{max} shifts by 2−6 nm per 1 nm variation in nanoparticle width or height, (2) the molar decadic (tenfold) extinction coefficient is $3 \times 10^{11}\,M^{-1}\,cm^{-1}$, and (3) the LSPR oscillator strength per atom is equivalent to that of atomic silver in gas or liquid phases.

6.1.2.2 Surface properties: composition, charge and functionalization

Although there is growing evidence to suggest that the surface properties of nanoparticles, including surface reactivity, are distinctly different from larger particles, often it is difficult to get a quantitative understanding of surface composition, charge, and functionalization for nanoparticles. Surface functionalization will impact everything from secondary size (through aggregation) to water solubility and the ability of nanomaterials to get into cells (for drug delivery applications). Thus, there is a great deal of interest in gaining a more quantitative understanding of the surface properties of nanoparticles. It is understood that structural disorder and unusual surface relaxation are two fundamental differences between the surface structure of nanoparticles and larger, bulk materials.

6.1.2.3 Aggregation of nanoparticles

In aqueous environments and as aerosol, there is a tendency for nanoparticles to form aggregates that are much larger than the primary size of the nanoparticle. The tendency for nanoparticles to aggregate depends on a number of factors including surface functionalization, nanoparticle concentration, pH, and ionic strength.

6.1.2.4 Concentration

Because of the small mass of individual nanoparticles, an important consideration is how to standardize nanoparticle concentrations. The important metric whether mass density, number density or some other unit, such as surface area, will in part depend on the application under consideration.

Consider spherical particles suspended in a volume, V, a comparison of mass and number density for particles ranging from 1 to 1000 nm. Specifically, it is shown that, for constant particle number density concentrations (number of particles/cm^3), the concentration in mass density ($\mu g/m^3$) changes by orders of magnitude. In particular a comparison between 5, 50, and 500 nm particles, the mass concentration changes by a million-fold over this range.

Typically for particulate matter in the air (such as air pollution regulations of particulate matter), it is mass concentrations. For example, if mass concentrations are kept constant at 1 $\mu g/m^3$, it can be seen that there are orders of magnitude more 5 nm nanoparticles relative to larger particles (e.g., 50 and 500 nm).

6.2 Classification of nanomaterials

Nanomaterials may be categorized in various ways such as chemical composition, properties, shape and size, and dimensions. Here, we discuss classification based on the number of dimensions, which are not confined to the nanoscale range. According to modified classification scheme by Pokropivny and Skorokhod, nanomaterials are classified into 0D, 1D, 2D, and 3D nanostructured materials (NSMs).

6.3 Classification of nanomaterials (according to their dimensions)
6.3.1 Zero-dimensional (0-D) nanomaterials

Materials wherein all the dimensions are measured within the nanoscale (no dimensions, or 0-D. Example: spheres and nanoclusters. See Fig. 6.4). The electrons of this kind of nanomaterials are confined along all three directions. Therefore, the electrons are not allowed to move anywhere in the system.

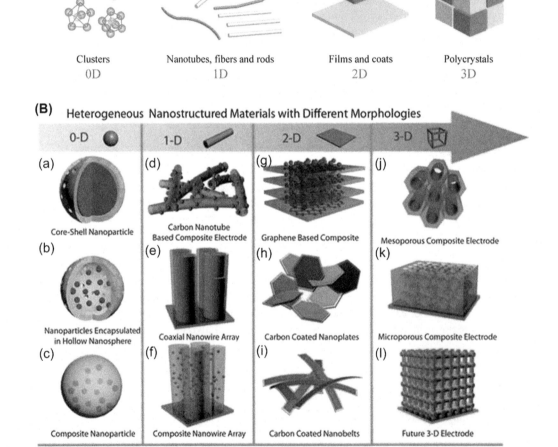

FIG. 6.4

(A) Classification of materials according to their dimensions. (B) Specific examples of materials with various sizes and dimensions.

The effect of confinement on the resulting energy states can be calculated by quantum mechanics, as the "particle in a box" problem. An electron is considered to exist inside of an infinitely deep potential well (region of negative energies), from which it cannot escape and is confined by the dimensions of the nanostructure. The energy states in such confinement are given by Eq. (6.3).

$$0-D: \quad E_n = \left[\frac{\pi^2 \hbar^2}{2mL^2}\right]\left(n_x^2 + n_y^2 + n_z^2\right) \tag{6.3}$$

where $\hbar \equiv h/2\pi$, h is Planck's constant, m is the mass of the electron, L is the width (confinement) of the infinitely deep potential well, and n_x, n_y, and n_z are the principal quantum numbers in the three dimensions x, y, and z. The smaller the dimensions of the nanostructure (smaller L), the wider is the separation between the energy levels, leading to a spectrum of discrete energies.

The 0-D nanomaterials can be of various kinds which can be stated as: (a) Amorphous or crystalline. (b) Single crystalline or polycrystalline, (c) Composed of single or multi-chemical elements, (d) Exhibit various shapes and forms. (e) Exist individually or incorporated in a matrix, (f) Metallic, ceramic or polymeric.

The most common representation of 0-D nanomaterials are quantum dots (nanoparticles with small diameter). Particles of this dimension such as uniform particle arrays of quantum dots, heterogeneous particle arrays, core shell quantum dots, onions, hollow spheres, and nanolenses have been synthesized by several research groups and for the application in light-emitting diodes, solar cells, single-electron transistors, and lasers.

6.3.2 One-dimensional nanomaterials

Materials aligned (expanded) only in one dimension that is outside the nanoscale and the electrons are free to travel in one direction and confined in the other two directions (x&y or x&z or y&z), as explained by Eq. (6.4).

$$1-D: \quad E_n = \left[\frac{\pi^2 \hbar^2}{2mL^2}\right]\left(n_x^2 + n_y^2\right) \tag{6.4}$$

This leads to needle like-shaped nanomaterials. 1-D materials include nanotubes, nanorods, nanowires and nanofibers. 1-D nanomaterials nanomaterials can be: (a) Amorphous or crystalline, (b) Single crystalline or polycrystalline, (c) Chemically pure or impure, (d) Stand-alone materials or embedded within another medium, (e) Metallic, ceramic, or polymeric.

For example, carbon nanotubes (CNTs) (Fig. 6.4D) are hexagonal networks of carbon atoms, 1 nm in diameter and 100 nm in length, as a layer of graphite rolled up into a cylinder. CNTs are of two types, single-walled CNTs and multiwalled CNTs. The small dimensions of CNTs, combined with their remarkable physical, mechanical, and electrical properties, make them unique materials. They display metallic or semiconductive properties, depending on how the carbon leaf is wound on itself. The current density that a nanotube can carry is extremely high and can reach 1 billion A/m^2, making it a superconductor. The mechanical strength of CNTs is 60 times greater than that of the best steels. CNTs have a great capacity for molecular absorption and offer a 3D configuration. Moreover they are chemically very stable.

Due to these unique features, one dimensional device such as nanowires and nanotubes have received extensive attention in both fundamental and applied studies. They play a crucial role not only in important future optoelectronic devices but also in data storage and biochemical and chemical sensors, and they also can be used to enrich our understanding of basic quantum mechanics.

6.3.3 Two-dimensional nanoparticles

In 2-D system, the electrons are confined in only two directions. That is, the electrons are allowed to move in any two directions in the system and explained by Eq. (6.5). E.g., nanosized thin films, nanoplates and branched structures. 2-D nanomaterials exhibit plate-like shapes. Two-dimensional nanomaterials include nanofilms and nanolayers. 2-D nanomaterials can be: (a) Amorphous or crystalline, (b) Made up of various chemical compositions, (c) a single layer or multilayer structures, (d) Deposited on a substrate, (e) Integrated in a surrounding matrix material, (f) Metallic, ceramic, or polymeric.

$$2-D: \quad E_n = \left[\frac{\pi^2\hbar^2}{2mL^2}\right](n_x^2) \tag{6.5}$$

These two-dimensional nanomaterials with certain geometries exhibit unique shape-dependent characteristics and subsequently are used as building blocks for the key components of nano-devices. In addition, 2D NSMs are particularly interesting not only for the basic understanding of the mechanism of nanostructure growth, but also for the investigation and development of novel appli-cations in sensors, photocatalysts, nanocontainers, nanoreactors, and templates for 2D structures of other materials.

6.3.4 3-D system (three-dimensional nanoparticles)

Bulk nanomaterials are materials that are not confined to the nanoscale in any dimension. In 3-D system (bulk), the electrons are free to move in all three directions and there are no confinement and limitations. Example: powders, multilayer, fibrous and poly crystalline materials wherein nano-structural elements of 0-D, 1-D and 2-D are closely related with each other and form interfaces. These materials are thus characterized by having three arbitrary dimensions above 100 nm. Materials possess a nanocrystalline structure and in this structure, bulk nanomaterials can be composed of a multiple arrangement of nanosize crystals, most typically in different orientations. With respect to the presence of features at the nanoscale, 3-D nanomaterials can contain dispersions of nanoparticles, bundles of nanowires, and nanotubes as well as multi-nanolayers (Also see examples in Fig. 6.4B).

6.3.4.1 Summary

Quantum effects: The overall behavior of bulk crystalline materials changes when the dimensions are reduced to the nanoscale.

For 0-D nanomaterials, where all the dimensions are at the nanoscale, electrons are fully confined in 3-D space. No electron delocalization occurs.

For 1 -D nanomaterials, electron confinement occurs in 2-D, whereas delocalization takes place along the long axis of the nanowire/rod/tube.

In the case of 2-D nanomaterials, the conduction electrons will be confined across the thickness but delocalized in the plane of the sheet.

Note that, in 1-D and 2-D nanomaterials, electron confinement and delocalization coexist.

For 3-D nanomaterials the electrons are fully delocalized.

6.4 Approaches for synthesis of nanomaterials

Synthesis of nanomaterials with strict control over size, shape, and crystalline structure has become very important for the applications of nanotechnology in numerous fields including thin film based solar cell technology, electronics, photo-catalysis (e.g., photocatalytic degradation of organic or organometallic pollutant in contaminated/waste water), medicine (for drug delivery), and electronics. There exist a number of methods to synthesize the nanomaterials which are categorized in two techniques: "Top down" and "Bottom up".

6.4.1 (I) top-down approach

This Top-down approach of nanoparticles synthesis refers to slicing or successive cutting of a bulk material to get nano sized particle, while retaining the original integrity (see Fig. 6.5). In other words, top-down is when the structure is cut out from a bigger piece "manually" or by a kind of self-structuring process. In some sense, all current microelectronics is fabricated using this approach. One of the practices of such approaches, for example, is the synthesis of nanostructures by etching out crystal planes (removing crystal planes) which are already present on the substrate.

Disadvantages and advantages: The biggest problem with top-down approach is the imperfection of surface structure and significant crystallographic damage to the processed patterns. These imperfections which in turn leads to extra challenges in the device design and fabrication. However, top-down approach typically provides better control, but is limited to "countable" number of structures (though this number may be billions and billions) and therefore, leads to the bulk production of nano material. It is considered that regardless of the defects produced by top-down approach, they will continue to play an important role in the synthesis of nano structures.

6.4.2 (II) bottom-up approach

Contrary to top-down approach, the "Bottom-up approach" refers to the buildup of a material from the bottom: Atoms, molecules and even nanoparticles themselves can be used as the building blocks

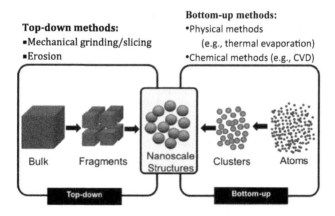

FIG. 6.5

Schematics of the "Top-Down" and "Bottom-up" approaches.

for the creation of complex nanostructures. In other words, in bottom-up method the structure is created from smaller building blocks (wet chemical methods such as hydrothermal, sol-gel, co-precipitation, etc of nanoparticles is a typical example of it). The useful size of the building blocks depends on the properties to be engineered. For example, synthesis of nanostructures onto the substrate by stacking atoms (or molecules) onto each other, which gives rise to crystal planes, crystal planes further stack onto each other, resulting in the synthesis of nanostructures.

Though the bottom-up approach often referred in nanotechnology, it is not a new concept. All the living beings in nature observe growth by this approach only and also it has been in industrial use for over a century. Examples include the production of salt and nitrate in chemical industry.

Advantage: The advantage of bottom-up is the cost (less expensive), scalability and in general better uniformity of the product.

Both approaches play very important role in modern industry and most likely in nanotechnology as well.

6.5 Thin solid films
6.5.1 Introduction

A thin film is a layer of material ranging from fractions of a nanometer (monolayer) to several micrometers in thickness. Thin-film deposition is the act of applying a thin layer (film) onto a surface (called substrates) or onto previously deposited layers. The controlled synthesis of materials as thin films (a process referred to as deposition) is a fundamental step in many applications. A familiar example is the household mirror, which typically has a thin metal coating on the back of a sheet of glass to form a reflective interface. The process of silvering was once commonly used to produce mirrors. Advances in thin film deposition techniques during the 20th century have enabled a wide range of technological breakthroughs in areas such as electronic semiconductor devices, LEDs, optical coatings (such as antireflective coatings), hard coatings on cutting tools, and for both energy generation (e.g., thin-film solar cells) and storage (thin-film batteries).

6.5.2 Methods for thin film fabrication

Deposition techniques fall into two broad categories, depending on whether the process is primarily chemical or physical. The physical methods of thin film preparation comprise vapor deposition method and sputtering method. Whereas, the chemical methods of thin film preparation comprise vapor deposition method and sputtering method (Fig. 6.6).

6.5.2.1 Physical methods
6.5.2.1.1 Resistance heating evaporation

One of the common methods of Physical Vapor Deposition (PVD) is Thermal Evaporation. In this technique, the pure source material is evaporated in a vacuum. The vacuum allows vapor particles to travel directly to the target object (substrate), where they condense back to a solid state. The deposited layer, also called films, is usually in the thickness range of usually nanometer to microns and can be a single material, or can be multiple materials in a layered structure. The schematic view of the PVD process is shown in Fig. 6.7.

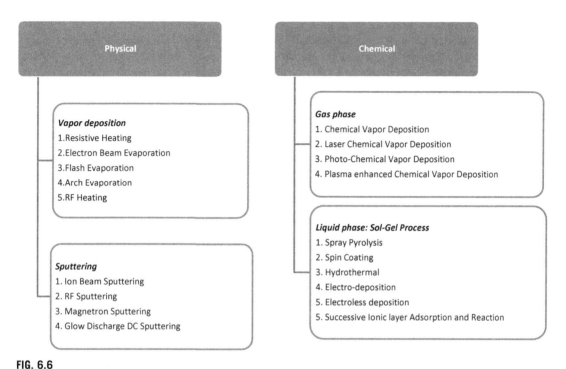

FIG. 6.6

Methods for thin film preparation.

Resistance heating evaporation

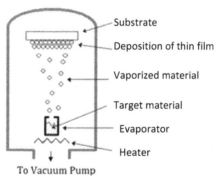

FIG. 6.7

Schematic view of thermal evaporation process by "Resistance Heating Evaporation" method.

One of the primary means of heating the source material is called simple electrical resistive heating. It is often referred to as resistive heating (also called Filament Evaporation) or filament. There are numerous different physical configurations of these resistive evaporation filaments, including many known as "boats" - essentially thin sheet metal pieces of suitable high temperature metals (such as tungsten) with formed indentations or troughs into which the material is placed.

The materials to be applied with PVD techniques can be pure atomic elements including both metals and nonmetals, or can be molecules such as oxides and nitrides. The material on to which coating is applied referred to as the substrate, and can be any of a wide variety of things such as: semiconductor wafers, solar cells, optical components, or many other possibilities.

Since, in most instances of "Thermal Evaporation" processes the material is heated to its melting point and is liquid, it is usually located in the bottom of the chamber, often in some sort of upright crucible. The vapor then rises above this bottom source, and the substrates are held inverted in appropriate fixtures at the top of the chamber. The surfaces intended to be coated are thus facing down toward the heated source material to receive their coating.

Steps may have to be taken to assure film adhesion, as well as control various film properties as desired. Fortunately, Thermal Evaporation system design and methods can allow adjustability of a number of parameters in order to give process engineers the ability to achieve desired results for such variables as thickness, uniformity, adhesion strength, stress, grain structure, optical or electrical properties, etc.

Similar to Resistive Heating, E-Beam Evaporation PVD systems also utilizes quartz crystal deposition control (whereby real time deposition rate monitoring and control takes most of the work out of achieving the right thickness).

Accessories such as residual gas analyzers (RGA's), and other custom features and custom automation are also available to achieve perfection in thermal evaporation techniques and methods. Cryogenic pumps are the most popular type of high vacuum pump for evaporation, but other options (such as turbo pumps and diffusion pumps) are also available, to customize the system as per need. Regardless of which options are selected, 'Thin Film Evaporation systems' can offer the advantages of relatively high deposition rates, real time rate and thickness control. It also offers good evaporation stream directional control for such purposes as 'Lift Off' processing to achieve direct patterned coatings.

6.5.2.1.2 Inert-gas condensation

(IGC) is a bottom-up approach to synthesize nanostructured materials, which involves two basic steps. The first step is the evaporation of the material and the second step involves a rapid controlled condensation to produce the required particle size (see Fig. 6.8).

6.5.2.1.3 Electrical arc discharge

As depicted on the **upper part of** Fig. 6.8, in this technique, an electric arc is used to vaporize the material directly. About 10% of the clusters formed are ions, and this avoids the need for a separate cluster ionization stage. Generally, cluster sizes of up to around 50 atoms are deposited on the target. Nanoshel uses the arc discharge heating to produce ultrafine powders: For example, Fe, Si, SiC, and Al_2O_3. Gas pressure and arc current are the critical parameters that need to be controlled during evaporation to obtain the desired particle size.

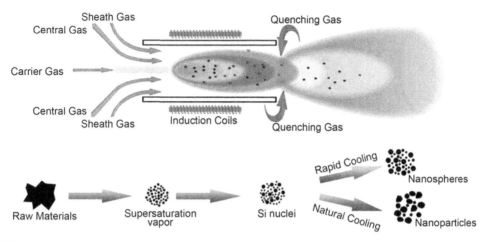

FIG. 6.8

Methods for thin film preparation: Electrical arc discharge.

6.5.2.1.4 RF plasma method

Nanoparticles are also synthesized by *RF plasma method*. This plasma is generated by RF heating coils. The starting metal is contained in a pestle and this pestle is contained in an evacuated chamber. The metal is heated above its evaporation point by using high voltage RF coils wrapped around the evacuated chamber. Helium gas is allowed to enter the system and this gas forms high temperature plasma in the region of the coils. The metal vapor nucleates on the helium gas atoms and diffuses up to a cold collector rod, where nanoparticles are collected. Finally these nanoparticles are passivated by the introduction of appropriate gas (Oxygen).

6.5.2.1.5 Sputtering

Over 150 years ago, the first sputtering was observed in a discharge tube by Bunsen and Grove. Sputtering is the second type of physical vapor deposition process that uses active radiation kinds, which has several different systems that are used for thin-film deposition including ion-beam sputtering, radio frequency (rf) sputtering (called radio frequency diode), E-Gun Sputtering, Laser Gun, magnetron Sputtering, and DC diode (also known as dc sputtering and diode sputtering). The simplest model among these sputtering systems is the DC diode (see Fig. 6.9).

As shown in Fig. 6.9, in the DC diode sputtering system, the source material (a solid surface called target) and substrate are placed in the vacuum chamber. The target is connected with the cathode plate ($-$ ve) and the substrate located on the anode plate, The chamber is filled with sputtering gas such as argon gas (Ar), which is an inert gas, at low pressure (4×10^{-2} torr). In this type of system of deposition, ions of argon (Ar^+) that produced in the radiance discharge are accelerated at high speed by an imposed electric field toward the cathode (target), and sputter the target. Sputtering of the target produces positive, negative, as well as neutral charged clusters. The clusters produced are hot and cool down during flight (on the way to substrate) inside the chamber and comes in contact on the surface of the target, resulting in thin films. Below we discuss ion Beam Sputtering (IBM) method.

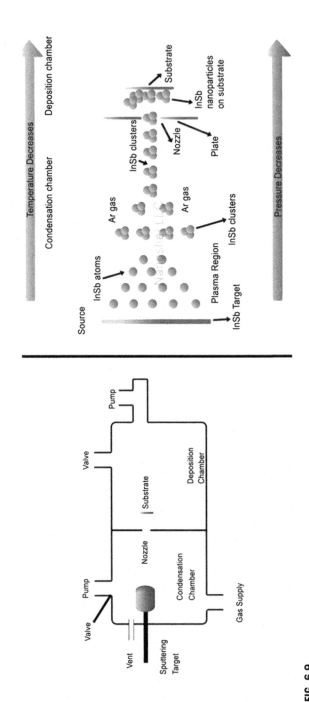

FIG. 6.9

DC diode sputtering system which contain a sputtering source material (material to be deposited), a carrier gas, deposition chamber containing a (substrate in which a thin film is to be deposited) and pumping system for low vacuum.

6.5.2.1.5.1 Ion Beam Sputtering method.
Ion Beam Sputtering, also called Ion Beam Deposition (IBD), is a thin film deposition process that uses an ion source to sputter a target material (metal or dielectric). The typical configuration of an IBD system consists ion source, a target, and a substrate. Fig. 6.10 shows a simple pictorial view of IBD process, and also included is an advanced ion beam sources that uses an electron gun. The electron beam strikes on the surface of the target and produces free ions or clusters (they can be both the ionic and neutral). As shown in Fig. 6.10 (right), these ionic or neutral particles evaporate and deposit on the substrate which is lying just above the target.

1. The ion beam of Ar gas is produced by applying high voltage between two electrodes at the cathode and anode: The cathode electrode the source material (target is attached) whereas at anode the substrate is housed.
2. Then the ion beam (Ar^+) is focused on a target material, and
3. The sputtered species (cations, anions or neutral particles) cool down during flying around the vacuum chamber and deposit onto a substrate to create either a metallic or dielectric film.
4. It may be desirable to heat the substrates (with an electric heater) during deposition to improve overall performance.

IBD Advantages: Because the ion beam is monoenergetic (ions possess the equal energy) and highly collimated, it enables extremely precise thickness control and deposition of very dense, high quality films as compared to other PVD (physical vapor deposition) technologies.

6.5.2.1.5.2 Laser ablation (LA).
It is a process in which a laser beam is focused on a target surface to remove material from the irradiated zone. Besides the production of nanomaterials, deposition of thin metallic and dielectric films, laser ablation has also been used for many technical applications, including: fabrication of superconducting materials, routine welding and bonding of metal parts, and micromachining of MEMS structures.

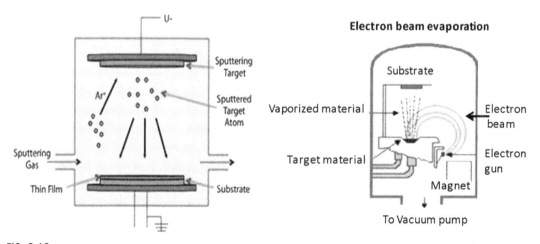

FIG. 6.10

Schematic view of thin film deposition system, using ion Gun (Ar^+ ion beam: Left) and electron gun (right) as a sputtering sources which bombard at the surface of the target producing ions or clusters. These ionic or neutral particles deposit as a thin film on the heated substrates.

6.5.2.2 Chemical methods

Chemical Vapor Deposition (CVD) is a process in which gaseous compounds which do not react at room temperature are passed over a substrate heated to a temperature at which either the reactants decompose or combine with other constituents to form a layer on the substrate. This technology has been widely used for material-processing technology: The majority of its applications involve depositing solid thin-film to surfaces. The majority of the elements in the periodic table have been deposited by CVD techniques, some in the form of the pure element, but more often combined to form compounds. Below we summarize the major steps for thin film preparation with this technique.

The basic steps involved in CVD process are summarized below (also illustrated in Fig. 6.11A).

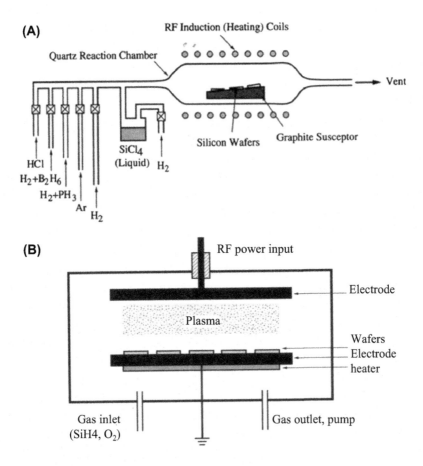

FIG. 6.11

(A). Atmospheric cold-wall system used for deposition of silicon. Carrier gas (e.g., Ar) and fuel gas (e.g., H_2) and precursor (in the form of aerosol: Here $SiCl_4$ is mixed with H_2 gas) are allowed to flow with the aid of controlled valves from the respective storage tanks to the deposition chamber. (B). Plasma assisted chemical vapor deposition system for thin film preparation.

1. *Transport (in the chamber):* Transport of reactants to the deposition region (inside the deposition chamber).
2. *Transport (at the substrate):* Transport of reactants from the main gas stream through the boundary layer to the wafer surface housed in the deposition chamber.
3. *Adsorption* of reactants on the wafer surface.
4. *Surface reactions,* including: chemical decomposition or reaction, surface migration to attachment sites (kinks and ledges); site incorporation; and other surface reactions (for example, emission and re-deposition). Depending upon the method employed, some of the chemical reaction undergoing during the deposition process are illustrated below.

Thermal decomposition: Molecule AB (g) decomposes to A (in the solid form, which deposits on the substrate) and B (g):

$$AB(g) -> A(s) + B(g)$$

Example: Si deposition from Silane at 650 °C: $SiH_4(g) \rightarrow Si(s) + 2H_2(g)$.
Reduction (using H_2): In the presence of hydrogen gas, the reduction process (at the heated substrate) may be generalized as follows:

$$AX(g) + H_2(g) \rightarrow A(s) + HX(g).$$

Example: $SiCl_4$ precursor accompanied with H_2 gas loaded in the deposition (reaction) chamber (see Fig. 6.11):

$$SiCl_4(g) + 2H_2(g) \rightarrow Si(s) + 4HCl \ (1200 \ °C).$$

Tungsten (W) deposition at 300 °C: $WF_6(g) + 3H_2(g) \rightarrow W(s) + 6HF(g)$

Oxidation (using O_2): In the presence of oxygen, the oxidation process (at the heated substrate) may be generalized as follows:

$$AX(g) + O_2(g) \rightarrow AO(s) + [O]X(g)$$

Example: SiO_2 deposition from silane and oxygen at 450 °C (lower temp than thermal) is proceeded with the following reaction:

$$SiH_4(g) + O_2(g) -> SiO_2(s) + 2H_2(g).$$

Desorption of byproducts. Mainly following two steps are occurred: (i) Transport of byproducts through boundary layer. (ii) Transport of byproducts away from the deposition region.
Steps 2−5 are most important for film growth rate.
CVD advantages and disadvantages (as compared to physical vapor deposition).
Advantages:

➢ High growth rate and good reproducibility are possible to achieve.
➢ Materials which are hard to evaporate can be deposited.

➢ Films can be grown also with the epitaxial growth (which is termed as "vapor phase epitaxy, VPE)". For instance, MOCVD (metal-organic CVD).
➢ Generally better film quality is achieved.

Disadvantages:

➢ It requires high process temperatures.
➢ The processes are complex, toxic. Corrosive gasses are also produced.
➢ Also the film may not be pure (hydrogen incorporation ...).

Some of the type of CVD techniques are discussed below.

6.5.2.2.1 APCVD (atmospheric pressure CVD)

A thin film synthesis method where the substrate is exposed to one or more volatile precursors, at atmospheric pressure, which react or decompose on the surface to produce a deposit. For example below we discuss preparation of cadmium telluride (CdTe) thin film for solar cell applications.

Although there are many demonstrated methods for producing high-efficiency CdTe solar cells, large scale commercial production of thin-film CdTe PV modules has not yet been realized. APCVD combines proven reaction chemistry with state-of-the-art engineering principles to enable design of thin film deposition reactors for the manufacturing environment.

Advantages and disadvantages of APCVD.

Advantages:

• Low equipment cost compared to vacuum processing because equipment will need neither the structural strength nor the pumping systems of a vacuum chamber.
• Large area uniformity achieved through control of temperature and gas flow - both of which are subject to rigorous engineering design.
• Simplified process control and source replenishment because the source gas generation is physically separated from the deposition chamber.

Presently, APCVD is used commercially to deposit transparent conducting oxide (TCO) films commonly used in CdTe solar cells. In fact, the processing sequence: deposition of thin films of various metal oxides for TCO, deposition of CdS, CdTe thin films and metalorganic CVD of electrodes could be performed in a single continuous process.

Disadvantages:

1. APCVD is extremely susceptible to oxidation due to the greater gas density and residence times. As such vacuum-caliber reactor construction is required, eliminating many of the purported advantages of APCVD.
2. In addition, tremendous quantities of liquid nitrogen would be required to operate APCVD, representing a utility cost that is substantially greater than operating a vacuum pump.
3. APCVD has inherently poor utilization; for example during the CdTe film preparation, approximately 90% of the CdTe passes by the substrate. Although in principle this material could be recovered and reused, from a practical standpoint this is a major drawback.

6.5.2.2.2 LPCVD (low pressure CVD)

- LPCVD features low deposition rate limited by surface reaction, so uniform film thickness (*many* wafers stacked *vertically* facing each other; in APCVD, wafers have to be laid horizontally side by side.
- Gas pressure is maintained around $1-1000$ mTorr (lower P $=>$ higher diffusivity of gas to substrate).
- This technique yields better film uniformity & step coverage and fewer defects.
- Required operating temperature (i.e., process temperature) is $\geq 500\,°C$.

6.5.2.2.3 Plasma-enhanced chemical vapor deposition (PECVD)

As shown in Fig. 6.11B, a beam of ion current (plasma jet) is produced by applying RF between two electrodes, which initiates chemical reaction among reactants. The energy from plasma is transferred into the reactant gases, forming radicals that are very reactive. (RF: radio-frequency, typically 13.56 MHz for PECVD).

Advantages:

1. As thermal energy is less critical when RF energy exists, we can prepare films at relatively low temperature ($<300\,°C$).
2. At such low temperature, surface diffusion is slow, so one must supply kinetic energy for surface diffusion, plasma (ion bombardment) provides that energy and enhances step coverage.
3. Moreover, plasma (i.e. ion bombardment on the surface of the film) increases film density, composition, and step coverage.

Therefore, this method is used for depositing film on metals (Al ...) and other materials that cannot sustain high temperatures. At such low temperatures APCVD and LPCVD give increased porosity and poor step coverage.

Disadvantages: Often plasma damages films, and also the film is contaminated with hydrogen.

There are numerous advantages of using chemical methods. In some cases, nanomaterials are obtained as colloidal particles in solutions, which can be filtered and dried to obtain powder. We can obtain thin films or nanoporous materials by electro-deposition and etching. Advantages of chemical synthesis are manifold. In many cases very well known chemical reaction route can be optimized to obtain nanoparticles. Particles of different shapes and sizes are possible depending upon the chemicals used and reaction conditions.

6.5.2.2.3.1 CVD sources and substrates. Types of sources: (i) Gasses (easiest), (ii) Volatile liquids *(iii)* Sublimable solids *(iii)* Combination.

Choice of source materials: The source materials should be: (i) Stable at room temperature, (ii) Sufficiently volatile, (iii) High enough partial pressure to get good growth rates, (iv) Reaction temperature < melting point of substrate, (iv) Produce desired element on substrate with easily removable by-products, (v) Possesses low toxicity.

Substrates: The chosen substrate must possess good adsorption properties and should have the capacity to initiate surface reactions for the material to be coated. For example, WF_6 deposits on Si but not on SiO_2.

6.5.2.2.4 Laser beam chemical vapor deposition (LCVD)

Like all the other CVD processes, LCVD is another approach to grow materials using the unique properties of laser outlined earlier. As will be clear in the following, LCVD promises various advantages over other CVD techniques especially in the field of microelectronics and general film growth. In some aspects there are qualitative similarities with some variants of CVD and in others the mechanism is quite different. Both attributes of the laser beam, viz. heat and light, have been exploited in the deposition of materials by this variant of CVD. In this method, the laser is focused through a transparent window and the transparent reactants onto an absorbing substrate, creating a localized hot spot at which the reaction takes place. The absorptivities of the reactants and the substrate determine the laser wavelength which is used.

Thus LCVD can be broadly classified into two categories: pyrolytic LCVD (thermal effect) and photolytic LCVD (light-wavelength effect).

6.5.2.2.4.1 Pyrolytic LCVD. Fig. 6.12 shows that in pyrolytic LCVD the laser light impinges upon the desired substrate, heating it locally over the beam area. Owing to the incident thermal profile of the

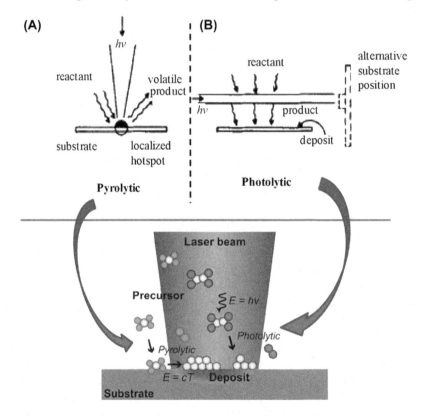

FIG. 6.12

Schematic illustration of basic laser deposition mechanisms. Process (a: pyrolytic) leads to heating of localized area of the substrate followed by the decomposition of the reactant, whereas in the process (b: photolytic) the laser beam directly decomposes the reactant (bond breaking), then, one of the product deposits on the substrate. The lower schematic drawing is included to further clarify these processes.

From: S. K. Roy, Bull. Mater. Sci., Vol. 11, 1988.

beam, the surrounding reactant gas undergoes thermal decomposition and deposition takes place in a manner analogous to that in conventional CVD or LPCVD. Correct choice of laser can help avoid absorption of the laser photon by the reactant gases and aid heat the substrate surface. It has been possible to deposit various materials, metals, semiconductors and insulators by this thermo-chemical heterogeneous reaction.

6.5.2.2.4.2 *Photolytic LCVD.* As Fig. 6.12 shows, in the photolytic LCVD technique the reactant gas molecules are photo-dissociated owing to absorption of photons of appropriate wavelength of the laser light. The beam, either by single or multi-photon absorption, excites particular vibrations of the reactant molecules and raise the internal energy to induce dissociation. The reactive species produced by such a process of bond breaking interact with substrate giving rise to the deposited film. Evidently, bond breaking is intrinsically efficient since energy is not randomly distributed throughout the internal degree of freedom of the molecules. This technique has also been used in deposition of different materials in thin film form.

   Advantages: The use of a laser as the heat source for chemical vapor deposition offers several distinct advantages: (a) It allows better control of the coating (coating within a specified area more accurately); (b) limited distortion of the substrate due to heat; (c) the possibility of cleaner films due to the small area heated; and (d) availability of rapid, i.e., non-equilibrium, heating and cooling rates. It should be possible to generate deposits of almost any material that can be deposited by conventional CVD and probably some which cannot. Possible applications include: one-step ohmic contacts; localized protective coatings; localized coatings for waveguide optics; and localized wear and corrosion resistant coatings. [S.D. Allen, Ref. J. Appl. Phys. 52, 6501 (1981)]

6.5.2.2.5 Atomic layer deposition (ALD)

ALD is a thin-film deposition technique based on the sequential use of a gas phase chemical process; it is a subclass of chemical vapor deposition. As shown in Fig. 6.13, during atomic layer deposition a film is grown on a substrate by exposing its surface to alternate gaseous species (typically referred to as precursors. The majority of ALD reactions use two chemicals as precursors). In contrast to chemical vapor deposition, the precursors are never present simultaneously in the reactor, but they are inserted as a series of sequential, non-overlapping pulses. In each of these pulses the precursor molecules react with the surface in a self-limiting way, so that the reaction terminates once all the reactive sites on the

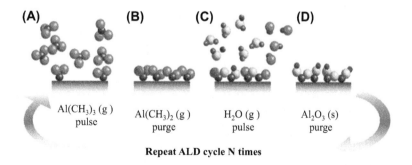

(A) $Al(CH_3)_3$ (g) pulse   **(B)** $Al(CH_3)_2$ (g) purge   **(C)** H_2O (g) pulse   **(D)** Al_2O_3 (s) purge

Repeat ALD cycle N times

FIG. 6.13

Schematic of ALD process for depositing Al_2O_3 thin film.

surface are consumed (that is, these precursors react with the surface of a material one at a time in a sequential, self-limiting, manner). Consequently, the maximum amount of material deposited on the surface after a single exposure to all of the precursors (a so-called ALD cycle) is determined by the nature of the precursor-surface interaction. By varying the number of cycles it is possible to grow materials uniformly and with high precision on arbitrarily complex and large substrates.

ALD is considered one deposition method with great potential for producing very thin, conformal films with control of the thickness and composition of the films possible at the atomic level. A major driving force for the recent interest is the prospect seen for ALD in scaling down microelectronic devices according to Moore's law.

A basic schematic of the atomic layer deposition process is shown in Fig. 6.13: In "Frame A"precursor 1 (in blue) is added to the reaction chamber containing the material surface to be coated. After precursor 1 has adsorbed on the surface, any excess is removed from the reaction chamber. Precursor 2 (red) is added ("Frame B") and reacts with precursor 1 to create another layer on the surface ("Frame C"). Precursor 2 is then cleared from the reaction chamber and this process is repeated until a desired thickness is achieved and the resulting product resembles "Frame D".

Advantage: The process of ALD is often performed at lower temperatures, which is beneficial when working with substrates that are fragile, and some thermally unstable precursors can still be employed with ALD as long as their decomposition rate is slow.

A wide range of materials can be deposited using ALD, including oxides, metals, sulfides, and fluorides, and there is a wide range of properties that these coatings can exhibit, depending on the application.

Applications of ALD method: The application range of atomic layer deposition is vast, and that is why it has become a popular tool to develop nano coatings and thin films.

One of the most popular applications is the use of ALD thin films in the semiconductor manufacturing industry as electronics become miniaturized. The thin films and coatings produced using ALD help to make these products even smaller yet and maintain a high standard of performance we demand in our consumer electronics.

6.5.2.2.6 Sol-gel process

6.5.2.2.6.1 (I) Spray pyrolysis. **Spray pyrolysis** is a **process** in which a thin film is **deposited** by **spraying** a precursor solution on a heated surface, where the chemical constituents react to form a chemical compound. The chemical reactants are selected such that the products other than the desired compound are volatile at the temperature of **deposition**. As shown in the (Fig. 6.14), the precursor solution is atomized in a droplet generating apparatus. In spray pyrolysis, the reaction takes place from the vapor phase at moderate high temperature, and can be performed in air for depositing transparent thin films of oxides such as TiO_2, ZnO, F-doped SnO_2, graphene oxide, etc compact layers on the surface of the transparent conductive. Spray pyrolysis method has been adopted due to its simplicity and relatively cost-effective processing method (especially with regard to equipment cost). Below an example is given how F-doped SnO_2 films can be deposited.

The setup consists of a chamber of dimension in which a heater with a substrate holder (metal plate) is kept and the temperature of the substrate was controlled by a temperature controller. The pressure required to generate erosol is created by the nebulizer with maximum pumping capacity of pressure 22,556 Kpa which can spray precursor solution 0.33 mL = min. The precursor solution was kept in a small plastic vessel of capacity ca. 10 mL, which was in direct connection with the nebulizer's

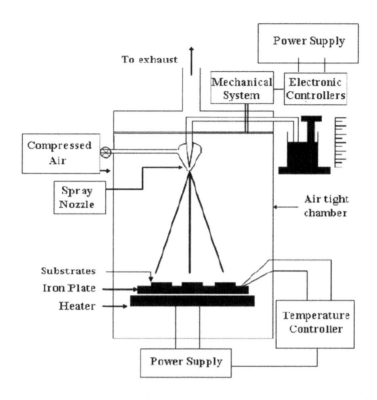

FIG. 6.14

Schematic diagram of spray pyrolysis technique.

outlet through which pumped air comes in contact with the solution and carries all the way to the glass substrate residing on the heater. The noise level of nebulizer was 60 db around one meter area.

The host precursor solution consisted of dehydrate Stannous chloride ($SnCl_2.2H_2O$) dissolved in.deionized water and ethanol (volume ratio 1:9) and the precursor concentration was 1.8 M (90.5 gm $SnCl_2.2H_2O$ dissolved in 5 mL of concentrated Hydrochloric acid (HCl) was heated at 90 °C for 10 min. Then 20 mL of distilled water was added to it and again stirred for another 15 min). The Ammonium fluoride (NH_4F) with molar concentration 0.85 M was added in above precursor and the mixture was stirred for 60 °C at 15 min. The dopant precursor NH_4F was added maintaining the mass ratio of F:Sn in the solution equal to 0.85:1.81 (From our earlier study, it was found that the films with FTO precursor molar concentration ratio of F:Sn = 0.85:1.81 yields higher transparency and low electrical resistivity. Then the mixture was continuously stirred for 1 h at a temperature of 60 °C and allowed to remain for 24 h for aging. Before FTO films deposition, Corning 7059 glass substrates of dimension ($2.5 \times 2.5 \times 0.1$) cm^3 were ultrasonically cleaned using detergent and distilled water and then washed with alcohol followed by drying in hot air oven. The substrates were heated to the $450 \pm 5)$ ^{o}C temperature before commencing spray. Films of various thicknesses were deposited by spraying the solution of fixed composition in different times. After deposition, the coated substrates were allowed to naturally cool down to room temperature before being taken out from the spray chamber for calcination under hot air at 500 °C followed by further heat treatment under N_2 gas at the same temperature (500 °C). To note, the actual incorporation of Fluorine into the film was expected to be very low due to the high volatility of fluorine compounds produced during the film deposition, as reported in earlier studies.

With this technique, the effect of different deposition times (i.e. films' thickness), different temperature and calcination environment (in presence of air or inert gas in the muffle furnace) on the FTO thin films' structural, optical and electrical properties were measured with XRD, SEM, UV−Vis spectroscopy, and 4-Probe measurements, respectively. The FTO films fabricated by spraying the precursor solution for 30 min with F to Sn to ratio of 0.85:1.81 and calcined in the presence of air followed by low temperature plasma treatment yielded the lowest electrical resistance of 17 Ωcm^{-1} and maximum transmittance of ca. 80 %. Also, the effect of calcination environment.

6.5.2.2.6.2 Spin coating. Spin coating is a procedure used to deposit uniform thin films to flat substrates with the help of spin coated (see Fig. 6.15). Usually a small amount of coating material is applied on the center of the substrate, which is either spinning at low speed or not spinning at all. The substrate is then rotated at high speed in order to spread the coating material by centrifugal force. A machine used for spin coating is called a spin coater, or simply spinner.

Rotation is continued while the fluid spins off the edges of the substrate, until the desired thickness of the film is achieved. The applied solvent is usually volatile, and simultaneously evaporates. The higher the angular speed of spinning, the thinner the film. The thickness of the film also depends on the viscosity and concentration of the solution, and the solvent.

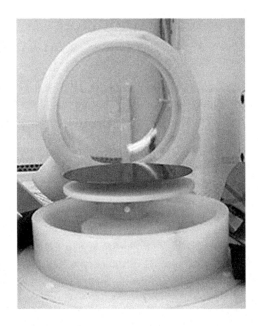

FIG. 6.15

Spin coater (Laurell Technologies WS-400) which has been used to apply photo-resist to the surface of a silicon wafer.

Spin coating is widely used in microfabrication of functional oxide layers on glass or single crystal substrates using sol-gel precursors, where it can be used to create uniform thin films with nanoscale thicknesses. It is used intensively in photolithography, to deposit layers of photoresist about 1 µm thick (Photoresist is typically spun at 20 to 80 revolutions per second for 30−60 s). It is also used to deposit thin transparent films of several hundred nanometer to micrometer thick films for solar cell and light emitting diode, and bio-sensing applications. Below an example is given how Fe doped ZnO films can be deposited.

Preparation of solution For preparing ZnO transparent films, zinc acetate dihydrate $Zn(CH_3COOH)_2.H_2O$, 2-methoxy ethanol and monoethanolamine (MEA) were used as a starting material, solvent, and stabilizer, respectively. $Zn(CH_3COOH)_2.H_2O$ was dissolved in a mixture of MEA and 2-methoxy ethanol solution at room temperature. The molar ratio of MEA to zinc acetate was maintained at 1.0 where the concentration of zinc acetate was 0.35 M. Then, the solution was stirred at 50 C for 1 h to yield a clear homogeneous solution, which served as coating solution after cooling to room temperature. MEA controls the size of sol particles and provides the required viscosity for the diffusion of the precursor on the glass substrate during the coating process. The coating was usually performed after aging the precursor solution for 24 h. For doping Fe ions, $FeSO_4.7H_2O$ was mixed in appropriate molar fraction by atomic mass in percentage (from 1% to 4%).

Substrate preparation The ordinary glass and FTO coated glass with the size of $\sim 2.5 \times 2.5 \times 0.25$ cm were used as substrates. The substrates were dipped into a beaker filled with ethanol and were ultrasonically treated for 15 min to remove oil and dirt from them followed by rinsing with distilled water and oven drying at 100 °C.

Preparation of ZnO film The aged precursor solution was dropped onto the surface of ordinary glass substrate or a conducting side of FTO glass substrate and placed on the holder of the spin coater set at a rotating speed of 3000 rpm for 60 s. After deposition, the substrate was dried at 350 C for 15 min in a muffle furnace to completely evaporate solvent and organic residuals. The coating and drying process were repeated for ten cycles for each sample. Then, finally, the sample was annealed at a temperature of 450 C for 1 h. Figure 1 shows the flowchart of preparation of transparent ZnO film by *spin coating* process.

Then the structural, morphological, and optical properties of Fe-doped ZnO transparent thin films spin coated on ordinary and FTO glass substrates using X-ray diffraction (XRD), scanning electron microscopy (SEM), and spectrometric reflectometry (SR) techniques were investigated. The XRD spectra of the Fe-doped ZnO films showed the polycrystalline films with a preferential orientation along the c-axis and without any impurity phases SEM images revealed that the randomly oriented matrix shaped nanoparticles of ZnO doped with 1% Fe (with respect to Zn) transformed into cylindrical long grains at higher dopant concentration (4% Fe). The surface reflectance of the Fe-doped ZnO film, particularly in the wavelength range of 400−650 nm, was as high as 85% and decreased when the concentration of Fe increased. Moreover, the band gap (3.26 eV), derived with the aid of measured transmittance curves of un-doped ZnO thin film was decreased to ca. 3.21 eV when Fe ions were doped into the film and was independent of the dopant concentration.

6.5.2.2.6.3 Hydrothermal. Hydrothermal synthesis a unique method for crystallizing substances from high-temperature aqueous solutions at high vapor pressures; also termed "hydrothermal method." The term "hydrothermal" is of geologic origin.

Hydrothermal synthesis can be defined as a method of synthesis of single crystals that depends on the solubility of minerals in hot water under high pressure. The crystal growth is performed in an apparatus consisting of a steel pressure vessel called an autoclave (see Figs. 6.16 and 6.17), in which a

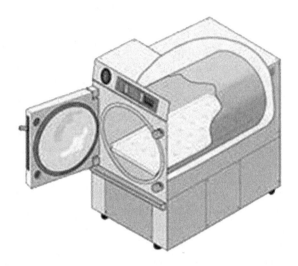

FIG. 6.16

Cutaway illustration of a cylindrical-chamber autoclave.

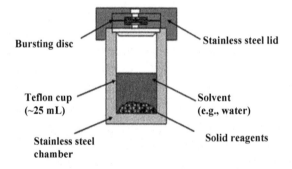

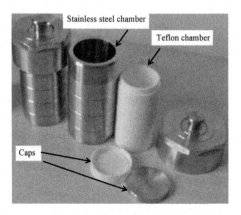

FIG. 6.17

Image of Teflon lined stainless steel autoclaves with inner components (upper). The Teflon chamber is housed inside the stainless steel chamber in which reaction is carried out. Lower image is of the commercially available Teflon lined stainless steel chamber.

nutrient is supplied along with water. A temperature gradient is maintained between the opposite ends of the growth chamber. At the hotter end the nutrient solute dissolves, while at the cooler end it is deposited on a seed crystal, growing the desired crystal.

6.5.2.2.6.4 Temperature-difference method.
This is the most extensively used method in hydrothermal synthesis and crystal growing. Supersaturation is achieved by reducing the temperature in the crystal growth zone. The nutrient is placed in the lower part of the autoclave filled with a specific amount of solvent. The autoclave is heated in order to create two temperature zones. The nutrient dissolves in the hot zone and the saturated aqueous solution in the lower part is transported to the upper part by convective motion of the solution. The cooler and denser solution in the upper part of the autoclave descends while the counterflow of solution ascends. The solution becomes supersaturated in the upper part as the result of the reduction in temperature and crystallization sets in.

Advantages and Disadvantages: Advantages of the hydrothermal method over other types of crystal growth include the ability to create crystalline phases which are not stable at the melting point. Also, materials which have a high vapor pressure near their melting points can be grown by the hydrothermal method. The method is also particularly suitable for the growth of large good-quality crystals while maintaining control over their composition. Disadvantages of the method include the need of expensive autoclaves, and the impossibility of observing the crystal as it grows. Various techniques are used to grow nanocrystal using the concept of concept of hydrothermal. One of the most commonly used is Temperature −difference method and discussed below.

In a typical synthesis, 0−60 mL of deionized water was mixed with 0−60 mL of concentrated hydrochloric acid (36.5%s38% by weight) to reach a total volume of 60 mL in a Teflon-lined stainless steel autoclave (125 mL volume, Parr Instrument Co.). The mixture was stirred at ambient conditions for 5 min before the addition of 1 mL of titanium butoxide (97% Aldrich). After stirring for another 5 min, substrates (for example, FTO coated glass, silicon wafers or glass slides) ultrasonically cleaned for 60 min in a mixed solution of deionized water, acetone, and 2-propanol with volume ratios of 1:1:1, were placed at an angle against the wall of the Teflon-liner with the conducting side facing down. The hydrothermal synthesis is conducted at 80−220 °C for 1−24 h in an electric oven. After synthesis, the autoclave is cooled to room temperature under flowing water, which takes approximately 15 min. The substrate is was taken out, rinsed extensively with deionized water and allowed to dry in ambient air.

6.5.2.2.6.5 Electro-deposition.
Electro-deposition is a long-established way to deposit metal layers on a conducting substrate. Ions in solution are deposited onto the negatively charged cathode, carrying charge at a rate that is measured as a current in the external circuit. The process is relatively cheap and fast and allows complex shapes. The layer thickness simply depends on the current density and the time for which the current flows. The deposit can be detached if the substrate is chosen to be soluble by dissolving it away.

Electroplating is a process that uses electric current to reduce dissolved metal cations so that they form a thin coating on an electrode. Besides, unlike other CVD methods, electro-deposition doesn't rely on a vacuum atmosphere which greatly reduces the cost and increases scalability. In addition, relatively high rates of deposition can be achieved with electro-deposition.

Principle Electro-deposition reverses the galvanic cell by supplying energy to drive non-spontaneous redox reactions to plate an electrode with a thin film. The anode is made of the metal to be plated and is oxidized by supplying direct current. As demonstrated in Fig. 6.18, this oxidation at the anode creates ions which dissolve and flow through the electrolytic solution, which contains metal salts and other ions to permit the flow of electricity. The dissolved ions are reduced and plated onto the cathode.

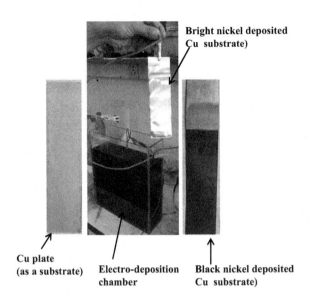

FIG. 6.18

Electrochemical deposition of Nickel on cupper (Cu) substrate: Left: Cu substrate, electrochemical bath for bright Ni deposition (Middle) with Ni layer deposited Cu substrate, and Black Ni deposited Cu substrate (right). Black Ni thin film has been used for solar energy harvesting (as a solar thermal collector).

The process of electroplating requires both materials used in the process to be conductive, as such metals and metallic compounds are primarily used. In order for the plating to be successful, the surface of the material that will be plated needs to be completely clean. Surface cleanliness is ensured by dipping the material in a strong acid or briefly connecting the electroplating circuit in reverse - if the electrode is clean, atoms from the plating metal will bond to it effectively. Even if the surface is clean, ineffective plating can result when components have complex geometries, which lead to an uneven distribution of plating thickness. Plating thickness can be controlled by varying the duration of electric current between the metals and the strength of the current applied between metals. Increasing either or both of these will increase the thickness of the plating. In other words, the film thickness can be controlled by monitoring the amount of charge delivered, whereas the deposition rate can be followed by the variation of the current with time. By controlling the thickness of the plating, plating issues resulting from complex geometries can be avoided.

The final films can range in thickness from a few nanometers to tens of microns and can be deposited onto large specimen areas of complex 3s6hape, making the process highly suitable for industrial use.

6.5.2.2.6.6 Electroless deposition. Electroless plating, also known as chemical or auto-catalytic plating, is a non-galvanic plating method that involves several simultaneous reactions in an aqueous solution, which occur without the use of external electrical power. For example, for Ni deposit in a metal substrate, the process involves placing the part in an aqueous solution and depositing nickel, creating a catalytic reduction of nickel ions to plate the part without any electrical energy dispersal. Unlike electroplating, this is a purely chemical process, with no extra machines or electrical power necessary. Below method for Ni deposition on aluminum substrate is explained.

Firstly a nickel layer (catalytic sites) is formed on the aluminum substrate via galvanic displacement method for the initiation of electroless nickel plating. Galvanic displacement reaction takes place due to the difference in the standard redox potentials between the solid material to be displaced and the ions of the source material. Since nickel has a higher redox potential compared to aluminum, nickel ions in solution will displace aluminum from the surface. After the formation of this initial layer, electroless nickel plating is performed to deposit nickel on the surface. For electroless nickel plating, 30 g/L $NiCl_2.6H_2O$, either 0.7 or 2.1 g/L $NaBH_4$, 60 g/L $NH_2(CH_2)_2NH_2$ and 40 g/L $NaOH$ is used. The plating solution is adjusted to a pH of 13 and plating done at either 70 °C or 80 °C. The nickel layer thus formed is cleaned with distilled water and stored in a dust free environment after drying with air or nitrogen.

6.6 Material properties, analytical (characterization) techniques and measurement methods for nanomaterials based thin films

6.6.1 Material properties

As discussed in section (1), chemical properties of interest for those studying nanoparticles include total chemical composition, mixing state (internal/external), surface composition, electrochemistry and oxidation state. Physical properties of interest include number and mass concentration, size, surface area, total mass, morphology and optical properties. Because of their very high surface area to mass ratio (see Table 6.1), and high surface curvature, nanoparticles may be particularly chemically active.

Nanoparticles are of current interest because of an emerging understanding of their possible application in human health improvement (drug delivery) and environmental sustainability (wastewater treatment, solar energy harvesting). Nanoparticles are used in many different applications and created by many different processes. Their characterization pose interesting analytical challenges.

Table 6.2 compares the properties of nominally spherical nanoparticles. Particle size has a dramatic effect on the physical properties of a collection of particles. Mass-based measurements are heavily weighted toward the largest particles, whereas smaller particles have a much larger surface area per unit mass.

Table 6.1 Types of vapor deposition methods, operating conditions, usages, and their advantage and disadvantages.

Type	Advantage	Disadvantage	Usage	Pressure/temp
APCVD atmospheric pressure CVD	Simple, fast	Poor step coverage	Low temp oxides	10−100 kPa 350−1200 °C
LPCVD low pressure CVD	Excellent cleanness, conformity and uniformity	High temp, low deposition rate	Polysilicon, nitride, oxide	100 Pa 550−600 °C
PECVD plasma enhancened CVD	Low temp	Rise for particle and chemical contamination	Low temp oxides, passivation nitrides	200−600 Pa 300−400 °C

Table 6.2 Correlation between relative mass per particle and relative surface area per unit mass.

Particle diam. (nm)	Relative Mass per particle	Relative surface area per unit Mass
10,000	8,000,000	1
2,500	125,000	16
1000	8000	100
50	1	40,000

Source: The characterization of nanoparticles, amc technical briefs, Ed. Michael Thompson, Analytical Committee AMCTB 48 Dec. 2010.

Moreover, comparison of results between phases is very difficult, and matrix effects can be significant due to the high surface area to mass ratio of nanoparticles. The techniques presented below give a general overview of common measurements made on nanoparticles (in solid state form, in particular thin film) for a range of applications.

The nanoparticles can take several forms including loose powders, wet or dry 'powder cakes' and thin solid film. As such, any analysis must take into account how the particles will eventually be used because this will affect their final agglomeration state and other properties.

Nanoparticles in the solid phase exist either as a powder or encapsulated or deposited in a solid medium (e.g., thin film). Listed below are some of the many techniques that can be applied to various nanoparticle (ENP) characterization or concentration measurement scenarios. Each has pros and cons. The choice of sampling and analytical options will depend on many factors including the particles and properties of interest, target nanoparticle size, sample matrix, cost, and required turnaround time.

Nowadays, many more types of nanomaterials are synthesized than only a decade ago, and in higher amounts than before, requiring the development of more precise and credible protocols for their characterization (see for example in Fig. 6.19). However, such a characterization is sometimes incomplete. This is because of the inherent difficulties of nanoscale materials to be properly analyzed, compared to the bulk materials (e.g. too small size and low quantity in some cases following laboratory-scale production). In addition, the unaffordable equipment (or analytical) cost do not permit every research team to have easy access to a broad range of characterization facilities. In fact, quite often a wider characterization of NPs is necessary, requiring a comprehensive approach, by combining techniques in a complementary way. In this context, it is desirable to know the limitations and strengths of the different techniques, in order to know if in some cases the use of only one or two of them is enough to provide reliable information when studying a specific parameter (e.g. particle size).

Herein we describe extensively the use of different methods for the characterization of NPs. These techniques are sometimes exclusive for the study of a particular property, while in other cases they are combined. We discuss all these techniques in a comparative way, considering factors such as their availability, cost, selectivity, precision, non-destructive nature, simplicity and affinity to certain compositions or materials.

There are microscopy-based techniques (e.g., TEM, HRTEM, and AFM) which provide information on the size, morphology and crystal structure of the nanomaterials. Many other techniques provide further information on the structure, elemental composition, optical properties and other common and more specific physical properties of the nanoparticle samples. Examples of these

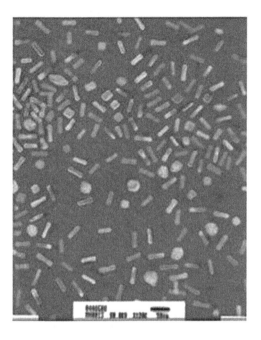

FIG. 6.19

Nanoparticles differ in their physical properties such as size, shape, and dispersion, which must be measured to fully describe them.

techniques include X-ray, spectroscopy and scattering techniques. Size, size distribution and organic ligands present on the surface of the particles may affect other properties and possible applications of the NPs. In addition, the crystal structure of the NPs and their chemical composition are thoroughly investigated as a first step after nanoparticle synthesis.

Until now, there were no standardized protocols for this aim. Credible and robust measurement methods for NPs will greatly affect the uptake of these materials in commercial applications and allow the industry to comply with regulations. Nevertheless, there are important challenges in the analysis of nanomaterials because of the interdisciplinary nature of the field, the absence of suitable reference materials for the calibration of analytical tools, the difficulties linked to the sample preparation for analysis and the interpretation of the data. In addition, there are unmet challenges in the characterization of NPs such as the measurement of their concentration in situ and on-line, especially in a scaled-up production, as well as their analysis in complex matrices.

For this reason, it is crucial to characterize the nanomaterials prepared in several ways to the maximum extent. We do not only focus on the characterization of the nanoparticle core, but also on the surface ligands that influence the physical properties.

6.6.2 Analytical techniques (overview)

Nanoparticles are commonly made up of inorganic substances such as gold, silver, or iron oxide, but can also comprise of liposomes and organic structures such as carbon nanotubes. Nanoparticles can be characterized in two ways. The first method of which are methods aiming to determine the physical properties of the nanoparticle itself, such as size, shape, monodispersity, crystal structure or optical characteristics. Other characterization methods aim to determine chemical characteristics of the particles, including the presence and method of bonding of ligands or other conjugated molecules, and the zeta potential.

Nanoparticles differ in their physical properties such as size, shape, and dispersion, which must be measured to fully describe them. Nanoparticles are unlike conventional chemicals in that their chemical composition and concentration are not sufficient metrics for a complete description, because they vary in other physical properties such as size, shape, surface properties, crystallinity, and dispersion state. The characterization of nanoparticles is a branch of nanometrology that deals with the characterization of physical and chemical properties of nanoparticles. Nanoparticles have at least one primary external dimension of less than 100 nm, and are often engineered for their unique properties.

A list of techniques commonly employed to characterize physical aspects (their size, morphology, and surface charge, using such advanced microscopic techniques) of nanoparticles is given below:

- UV–Visible Spectroscopy
- Scanning Electron Microscopy (SEM)
- X-Ray Diffraction (XRD)
- Transmission Electron Microscopy (TEM)
- Atomic Force Microscopy
- Dynamic Light Scattering (DLS)
- Elipsometry
- Nanoparticle Tracking Analysis (NTA)

Several of these techniques, including XRD, DLS, DCS, NTA and TEM focus on determining the nature of crystallinity, dimensions of a particle, and the degree of homogeneity between the particles in a colloid. The average particle diameter, their size distribution, and their charge affect the physical stability and the in vivo distribution of the nanoparticles. Electron microscopy techniques are very useful in ascertaining the overall shape of polymeric nanoparticles, which may determine their toxicity. The surface charge of the nanoparticles affects the physical stability and redispersibility of the polymer dispersion as well as their in vivo performance. Several techniques have been used to characterize the size, crystal structure, elemental composition and a variety of other physical properties of nanoparticles. In several cases, there are physical properties that can be evaluated by more than one technique. Different strengths and limitations of each technique complicate the choice of the most suitable method, while often a combinatorial characterization approach is needed. In addition, given that the significance of nanoparticles in basic research and applications is constantly increasing, it is necessary that researchers from separate fields overcome the challenges in the reproducible and reliable characterization of nanomaterials, after their synthesis and further process (e.g. annealing) stages.

The properties of nanoparticles often bridge the microscopic and macroscopic regimes, meaning that conventional theories do not necessarily allow us to predict their behavior.

6.6.3 Materials properties and measurement methods (tools)

6.6.3.1 Particle size and dispersions

Particle size is the external dimensions of a particle, and dispersity is a measure of the heterogeneity of sizes of molecules or particles in a mixture. A collection of objects is called **uniform** if the objects have the same size, shape, or mass. A sample of objects that have an inconsistent size, shape and mass distribution is called **non-uniform.** If the particle is elongated or irregularly shaped, the size will differ between dimensions, although many measurement techniques yield an equivalent spherical diameter. There are many methods of measuring particle size (or, more correctly, size distribution). It can be calculated from physical properties such as settling velocity, diffusion rate or coefficient, and electrical mobility. Size can also be calculated from microscope images using measured parameters such as Feret diameter, Martin diameter and projected area diameters; electron microscopy is widely used for this purpose for nanoparticles, in conjunction with other measurements. In particular, for solid nanoparticles Transmission electron microscopy (TEM), High resolution transmission microscopy (HRTEM) and atomic force microscopy (AFM) and also X-ray diffraction (XRD) are often used.

As depicted in Fig. 6.20, nanoparticles with different particle sizes can have different physical properties. For example, gold nanoparticles of different sizes appear as different colors. Size measurements may differ between methods because they measure different aspects of particle dimensions, they average distributions over an ensemble differently, or the preparation for or operation of the method may change the effective particle size. One of the methods used includes UV−Visible spectroscopy. As discussed in the previous chapters, this technique uses light in the visible and

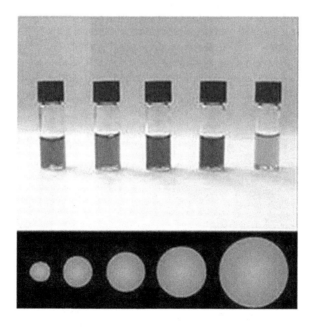

FIG. 6.20

Nanoparticles with different particle sizes can have different physical properties. For example, gold nanoparticles of different sizes appear as different colors.

adjacent ranges. The absorption or reflectance in the visible range directly affects the perceived color of the chemicals involved. In this region of the electromagnetic spectrum, atoms and molecules undergo electronic transitions.

6.6.3.2 Morphology (shape)

Morphology refers to the physical shape of a particle (as indicated in Fig. 6.19), as well as its surface topography, for example, the presence of cracks, ridges, or pores. Morphology influences dispersion, functionality, and toxicity, and has similar considerations as size measurements. Evaluation of morphology requires direct visualization of the particles through techniques like scanning electron microscopy (SEM), TEM and AFM. Fig. 6.21 demonstrates a TEM image of star-shaped gold nanoparticles. Because microscopy involves measurements of single particles, a large sample size is necessary to ensure a representative sample and orientation and sample preparation effects must be accounted for.

6.6.3.3 Surface area

Surface area is an important metric for nanoparticles because it influences reactivity and surface interactions with ligands. The most common technique is the *nitrogen adsorption technique* based on the BET isotherm, and is routinely carried out in many laboratories.

Also, the solid nanoparticles can be collected onto a substrate and their external dimensions can be measured using electron microscopy, and then converted to surface area using geometric relations.

6.6.3.4 Composition

Bulk chemical composition refers to the atomic elements of which a nanoparticle is composed, and can be measured in ensemble or single-particle elemental analysis methods.

Suitable surface techniques include *X-ray photoelectron spectroscopy* (and other X-ray spectroscopy methods) and *secondary ion mass spectrometry*. Composition has also been measured, utilizing elemental detectors such as energy-dispersive X-ray analysis or electron energy loss spectroscopy while using scanning electron microscopy or transmission electron microscopy.

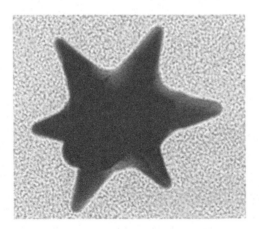

FIG. 6.21

Nanoparticles can take on a non-spherical shape, such as this star-shaped gold nanoparticles.

Bulk techniques generally use digestion followed by conventional wet chemical analyses such as mass spectrometry, atomic emission spectroscopy and ion chromatography.

6.6.3.5 Crystallography

As shown in Fig. 6.22, the arrangement of elemental atoms in a nanoparticles may be organized into a crystal structure or may be amorphous. Crystallinity is the ratio of crystalline to amorphous structure. Crystallite size, the size of the crystal unit cell, can be calculated through the Scherrer equation.

The Scherrer equation, in X-ray diffraction and crystallography, is a formula that relates the size of sub-micrometer particles, or crystallites, in a solid to the broadening of a peak in a diffraction pattern. It is named after Paul Scherrer. It is used in the determination of the size of particles of crystals in the form of powder.

$$B(2\theta) = \frac{K\lambda}{L\cos\theta} \quad B(2\theta) = \frac{0.94\lambda}{L\cos\theta} \tag{6.6}$$

Generally, a crystal structure is determined using powder X-ray diffraction, or selected area electron diffraction using a transmission electron microscope, though others such as Raman spectroscopy exist. X-ray diffraction requires on the order of a gram of material, whereas electron diffraction can be done on single particles.

6.6.3.6 Optical properties (of nanomaterials based thin films)

The past few decades have witnessed the use of various nanoparticles in different applications such as electronics, solar cells, sensors, drug delivery, etc. The properties and applications of the final products depend on the physicochemical properties of the nanoparticles and hence their properties have to be examined before usage in different applications.

crystalline

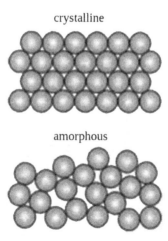

amorphous

FIG. 6.22

The atoms in a nanoparticle may be arranged in a crystal structure, may be amorphous, or may be intermediate between the two.

For example, thin films of metal and metal oxide have been widely used in solar cells, light emitting diode as metal contact and semiconductors. These devices consist of films with thicknesses of about 1 μm and it is important to know the refractive index, thickness and absorption coefficient as a function of wavelength to predict the photoelectric behavior of a device. Knowledge of these optical constants is also necessary to determine the optical gap.

Numerous techniques like scanning electron microscopy (SEM), X-ray-diffreaction, X-ray diffraction XRD crystallography, and Raman (light-scattering methods) spectroscopy have been employed to characterize nanoparticles. However, most of these techniques are time consuming, have very high cost, and most importantly need a large concentration of nanoparticles to be analyzed.

Role of absorption spectroscopy in material characterization: The unique optical properties of metal nanoparticles are a consequence of the collective oscillations of conduction electrons, which when excited by electromagnetic radiation are called surface plasmon polariton resonances. Those changes have an influence on the refractive index next to the nanoparticle's surface; thus, it is possible to characterize nanomaterials using UV–Vis spectroscopy. Besides, nanoparticles have optical properties that are very sensitive to nanoparticles size, shape, aggregation (agglomeration) problems, the structure of the particles, concentration changes, etc. Hence a method that is versatile, reliable, and economical, like UV–Vis spectroscopy, can be used for the in situ characterization of nanoparticles even at very low concentrations.

Absorption spectroscopy uses electromagnetic radiation between 190 and 800 nm and is divided into the ultraviolet (190–400 nm) and visible (400–800 nm) regions. Because the absorption of ultraviolet or visible radiation by a molecule leads to a transition among electronic energy levels of the molecule, it is also often called electronic spectroscopy. The information provided by this spectroscopy when combined with other sources of spectral data provides clues to valuable structural information of various molecules.

In UV–Vis spectroscopy, the intensity of light that is passing through the sample is measured. UV–Vis spectroscopy is a fast and easy-to-operate technique for nanoparticles characterization, especially for colloidal suspensions. There are several advantages of UV–Vis techniques, such as simplicity, sensitivity, and selectivity to nanoparticles and short time of measurement, and what is more, calibration is not required. Therefore, these techniques are increasingly used for nanoparticle characterization in many fields of science and industry.

Measurement of optical properties of nanomaterials based thin films by means of **UV–Vis spectroscopy**.

UV–Vis spectroscopy, specifically, allows us to understand the following optical properties of thin films:

(i) Optical transmittance *(T)* of semiconductor thin films,
(ii) Refractive Index *(n)*,
(iii) Film thickness *(d)*,
(iv) Absorptance *(k)*, absorption coefficient *(ε)*,
(V) Band gap *(E$_g$)* of thin films
(Vi) Nanoparticles' size and shape dependent optical properties (e.g., Tunable Localized Surface Plasmon Resonance Spectra)

Exercise 1. How does the refractive index of material depends on the property of two materials in contact?

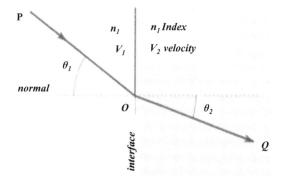

FIG. 6.23

Refraction of light at the interface between two media of different refractive indices, with $n_2 > n_1$. Since the phase velocity is lower in the second medium ($v_2 < v_1$), the angle of refraction θ_2 is less than the angle of incidence θ_1; that is, the ray in the higher-index medium is closer to the normal.

When light moves from one medium to another, it changes direction, i.e. it is refracted. If it moves from a medium with refractive index n_1 to one with refractive index n_2, with an incidence angle to the surface normal of θ_1, the refraction angle θ_2 can be calculated from Snell's law:

$$n_1 Sin\theta_1 = n_2 Sin\theta_2; \quad n_2 = \frac{n_1 Sin\theta_1}{Sin\theta_2} \tag{6.7}$$

The refractive index n of an optical medium is defined as the ratio of the speed of light in vacuum, $c = 299{,}792{,}458$ m/s, and the phase velocity v of light in the medium.

$$n = c/v \tag{6.8}$$

The phase velocity is the speed at which the crests or the phase of the wave moves, which may be different from the group velocity, the speed at which the pulse of light or the envelope of the wave moves.

When light enters a material with higher refractive index, the angle of refraction will be smaller than the angle of incidence and the light will be refracted toward the normal of the surface. The higher the refractive index, the closer to the normal direction the light will travel (See below in Fig. 6.23). When passing into a medium with lower refractive index, the light will instead be refracted away from the normal, toward the surface.

Exercise: 2. How does interference related with the film's thickness and radiation wavelength ?

An optical coating consists of a number of thin layers that create interference effects used to enhance transmission or reflection properties for an optical system. How well an optical coating performs is mainly dependent upon the following factors:

(i) the number of layers (of different materials),
(ii) the thickness of each layer and
(iii) the differences in refractive index at the layer interfaces.

The transmission properties of light are predicted by wave theory. One outcome of the wave properties of light is that waves exhibit interference effects. Light waves that are in phase with each other undergo constructive interference, and their amplitudes are additive. Light waves exactly out of phase with each other (by 180°) undergo destructive interference, and their amplitudes cancel. It is through the principle of optical interference that thin film coatings control the reflection and transmission of light.

When designing a thin film, though the wavelength of light and angle of incidence are usually specified, the index of refraction and thickness of layers can be varied to optimize performance. As refraction and thickness are adjusted these will have an effect on the path length of the light rays within the coating which, in turn, will alter the phase values of the propagated light. As light travels through an optical component, reflections will occur at the two interfaces of index change on either side of the coating. In order to minimize reflection, ideally there should be a 180° phase shift between these two reflected portions when they recombine at the first interface. This phase difference correlates to a $\lambda/2$ shift of the sinusoid wave, which is best achieved by adjusting the optical thickness of the layer to $\lambda/4$. Fig. 6.24 shows an illustration of this concept.

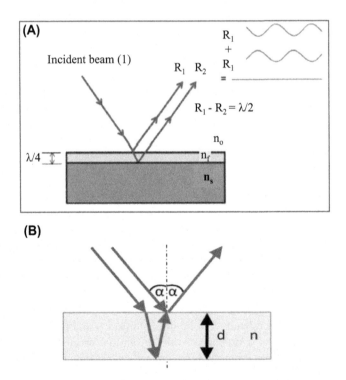

FIG. 6.24

(A). For an air (n_0)/film (n_f) interface, illustrated is the 180° phase shift between the two reflected beams (R_1, R_2), resulting in destructive interference of the reflected beams. (B). For an air (n_0)/film (n_f) interface, illustrated is the two reflected beams (R_1, R_2) in same phase, resulting in constructive interference of the reflected beams.

Index of refraction influences both optical path length (and phase), but also the reflection properties at each interface. The reflection is defined through Fresnel's Equation, which provides the amount of reflection that will occur from the refractive index change at an interface at normal incidence:

$$R = \left(\frac{n_p - n_m}{n_p + n_m}\right)^2 \tag{6.9}$$

The intensity of reflected light is not only a function of the ratio of the refractive index of the two materials, but also the angle of incidence and polarization of the incident light. If the incident angle of the light is altered, the internal angles and optical path lengths within each layer will be affected, which also will influence the amount of phase change in the reflected beams. It is convenient to describe incident radiation as the superposition of two plane-polarized beams, one with its electric field parallel to the plane of incidence (p-polarized) and the other with its electric field perpendicular to the plane of incidence (s-polarized). When a non-normal incidence is used, s-polarized and p-polarized light will reflect differently at each interface which will cause different optical performances at the two polarizations.

Determining the refractive index, n, and the absorptance (absorption coefficient), k, of a coating are two important parameters in thin film research. In real materials, the polarization does not respond instantaneously to an applied field. This causes dielectric loss, which can be expressed by the complex index of refraction that can be defined (Eq. 6.10):

$$n' = n + ik \tag{6.10}$$

where: $n*$ = the complex refractive index.

i = the square root of -1
n = the refractive index

Here, the imaginary part κ indicates the amount of absorption loss (or the absorption index) when the electromagnetic wave propagates through the material. The term k is often called the extinction coefficient in physics. Both n and k are dependent on the wavelength. In most circumstances $k > 0$ (light is absorbed).

The attenuation coefficient, α, and k are related as: $\alpha = 4\pi\kappa/\lambda_0$.

In this chapter we will show how the absorptance, refractive index, and film thickness of thin films can be calculated from the spectral data.

Measurement with Transmission/Reflection Spectroscopy: To determine n and k a number of optical measurements are needed which require accessories to be added to the UV/Vis spectrometer. To calculate the absorption coefficient (re: absorptance, extinction) of a thin film the transmission and absolute reflectance spectra of the material needs to be acquired (it follows that the material cannot be opaque). Absorptance is light that is not transmitted or reflected by a material, but is absorbed. The equation,

$$T + R + A = 1, \tag{6.11}$$

describes the theory, where T = transmittance, R = reflectance, and A = absorptance.

6.6.4 Methods for measuring optical properties of thin films

Below we discuss the measurement methods for following optical properties of the thin transparent films:

- Transmittance (T) of semiconductor thin films,
- Refractive Index (n),
- thickness (d),
- absorptance (A),
- absorption coefficient (ε),
- Band gap (Eg) of thin films: with Swanepoel method)

6.6.5 Methods for measuring optical properties of thin films

6.6.5.1 Transmittance (T) of semiconductor thin films

Measuring the transmittance and absorptance (A) of thin solid Materials.

A picture of the installed Spectroscopic Reflectometer is shown in Fig. 6.25, with a top down view of the control system of UV−Vis spectrophotometer.

For transmission and reflection spectroscopy we mostly use a commercial spectrophotometer that are capable of recording spectra in the visible range as well as in the near infrared and UV. To compensate for the complicated intensity distribution of the light source, we should measure and compare the signal from the sample to a reference beam. Additionally, a baseline need to be recorded prior to the actual measurements to calibrate the instrument. However the transmission and reflection measurements are limited to wavelengths at which the sample has an average absorption coefficient.

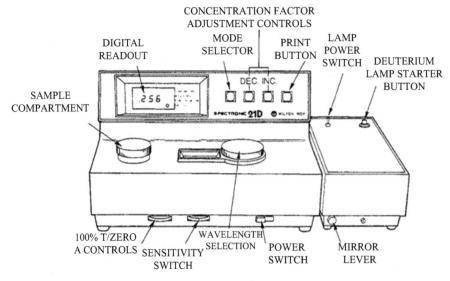

FIG. 6.25

Control panel of UV−Vis spectrophotometer.

It should be noted that for high absorption there is virtually no transmission. As both reflection and transmission spectra are required to calculate the absorption coefficient α this method is no more applicable then.

If α is too low however thin film interference effects will appear because light waves that are reflected on the two sides of the sample film will overlay. Therefore, spectrophotometry has to be used in combination with other techniques in order to obtain a complete absorption spectrum for a sample.

Film thickness, d, is determined from the wavelength between peaks and valleys in interference patterns.

The thickness of films can be determined using a surface profiling stylus or by various interferometric methods. But here, we will discuss optical method (an alternative method) first proposed by Swanepoel in 1983. This optical method allow to determine the thickness from the interference fringes of the transmission spectrum to an accuracy of better than 1% provided the films are of good quality.

To determine refractive index $n(\lambda)$ and absorption coefficient $\alpha(\lambda)$ by simple straightforward calculations using the transmission spectrum alone. The accuracy is also of the order of 1% which is even better than the accuracy of the elaborate iteration methods. This method considers only the spectrum in the optical region (electromagnetic radiation in the wavelength range between 100 nm and 1 mm).

A large number of components are coated by a thin layers of material, e.g. thin films for solar cells and light emitting diode, glass panes, headlights, wafers, foils. Determining the optical constants such as refractive index, absorption coefficient and thickness of such layers is an important task during manufacturing and for final quality control. As long as the layer dimensions are in the range of few nanometers to about 100 μm, optical spectroscopy can be applied as a non-contact, non-destructive and fast method to determine the film thickness.

There are two main principles used to determine the layer thickness of thin films: white light interference/minimum reflectance and modeling/pattern comparison. Both techniques require back reflected light from the film to be spectrally analyzed. At the boundaries of thin film, one against the substrate, the other against air, light is reflected due to a change in the index of refraction, n. At the photodetector, the reflected light interferes and causes a modulation of the overall reflected signal across the wavelength axis, λ. This intensity modulation over λ is analyzed can be analyzed to determine film thickness.

Analytical method: White-light interference method: At large optical layer thickness, the back-reflected spectrum shows a fringe pattern, and the distance of the peak positions can be used to determine the thickness. As the film thickness is increased, the spacing between the fringes decreases. Therefore, the maximum thickness of the film is given by the resolution of the spectrometer. The minimum thickness limit of this method is given by only a fraction of a fringe detected by a spectrometer. To analyze very thin films, a wide-range spectrometer is necessary. An additional consideration is the film has to be transparent over this range.

Metal oxide nanomaterials based thin films (such as tin oxide, SnO_2, titanium oxide, TiO_2) have been the subject of intense study during the last three decades, and great efforts have been to develop the mathematical formulation describing the transmittance and reflectance of different optical systems. These metal oxide and Chalcogenide compound semiconductor thin films are well-known IR-transmitting materials. They exhibit a wide range of photo-induced effects accompanied by significant changes in their optical constants that enable them to be used as opto-electronics, light emitting diodes, solar cells, bio-sensors and many other optical elements.

The precise knowledge of the optical properties (refractive index, absorption coefficient, dispersion parameter, film thickness, etc) of such thin films are obviously necessary for exploiting these materials in very interesting potential technologies.

Great efforts have been made to develop the method for determining the refractive index n (the real part of the complex refractive index), absorption coefficient, dispersion parameter etc. of the films. Many methods have developed including.

 (i) ellipsometry,
 (ii) fitting method (fitting analytical model in transmittance curve), and
(iii) reflection spectrum method (Cauchy relation).

The most common method was developed by Swanepoel. Swanepoel extracted the refractive index, film thickness and absorption coefficient of the films only from their transmittance curve. However, Swanepoel method would lead to obvious abnormal values of the film thickness in the region of strong absorption which were rejected by Swanepoel. Therefore, very recently, Jin and co-workers improved the model proposed by Swanepoel for accurately estimate those optical parameters from the measured transmittance spectrum. Moreover, Shaban and co-workers effectively employed the Swanepoel method for the thin films of the up to the thickness of 2.34 µm. Readers are referred to these literatures for more details. But here we will discuss the basic principles prescribed by Swanepoel. Besides, an alternative method, which employs Cauchy relation to fit in the measured reflection curve will also be discussed.

Estimation of refractive index: Here, as an example, we consider metal oxide thin film deposited on a thick transparent substrate (e.g., glass substrate). Schematic view of the thin film deposited on the glass substrate is shown in Fig. 6.26. Assuming normal incident and taking into account interference due to multiple refractions at air/film and film/substrate interfaces, Swanepoel has shown that the total transmission T is given by,

$$T\left(\lambda\right)=\frac{Ax}{B-CxCos\varphi+Dx^2} \tag{6.12}$$

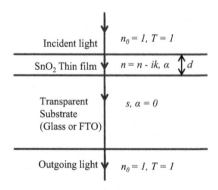

FIG. 6.26

Swanepoel model: Estimation of refractive index, film thickness, absorption coefficient and band gap.

Here.

$$A = 16n^2 s \tag{6.13a}$$

$$B = (n+1)^3 (n+s^2) \tag{6.13b}$$

$$C = 2(n^2 - 1)(n^2 - s^2) \tag{6.13c}$$

$$D = (n-1)^3 (n - s^2) \tag{6.13d}$$

$$\varphi = (4\pi nd)/\lambda \tag{6.13e}$$

$$x = exp(-\alpha d) \tag{6.13f}$$

Where n ($n' = n + ik$, where k is the extinction coefficient and equal to $\alpha\lambda/4\pi$) is the index of refraction of the film, s is the refractive index of the substrate ($s = 1.52$ for glass). Actually, the values of the refractive index of the substrate, s are obtained from the transmission spectrum of the substrate T_s using the well-known equation;

$$s = \frac{1}{T_s} + \sqrt{\left(\frac{1}{T_s} - 1\right)} \tag{6.14}$$

The thickness of the film is represented by the symbol d, φ is the phase difference between the direct and the multiple reflected beams, x is the absorbance which depends on absorption coefficient α and d of the film. The thickness of the substrate is several orders of magnitudes higher than d. The system is surrounded by air with refractive index $n_o = 1$. There are only two independent variables n and x in the above equation. In order to get these two variables two independent equations are required. The extremes (maxima and minima) of the interference fringes can be expressed as

$$T_M = \frac{Ax}{B - Cx + Dx^2} \tag{6.15}$$

$$T_m = \frac{Ax}{B + Cx + Dx^2} \tag{6.16}$$

Where T_M and T_m are the peaks and valleys of the envelope (upper and lower tangent) envelopes of the transmission spectrum, respectively.

Following the concept of Swanepoel method, which is based on creating the upper and lower envelopes of the measured transmittance spectrum, the refractive index of the film n_1 in the spectral region of transparent, weak and medium absorption regions at any wavelength can be calculated by the expression:

$$n_1 = \sqrt{N_1 + \sqrt{N_1^2 - S^2}} \qquad (6.17)$$

$$\text{where } N_1 = \frac{2S}{T_m} + \frac{S^2 + 1}{2} \qquad (6.18a)$$

for transparent region. And for weak and medium absorption regions N_1 becomes

$$N_1 = 2S \frac{T_M - T_m}{T_M \, T_m} + \frac{S^2 + 1}{2} \qquad (6.18b)$$

Additionally, the well-known basic equation for the interference fringe is:

$$2nd = m\lambda \qquad (6.19)$$

where m is the interference order (Note: Eq. (6.19) contains information on the product of n and d (film thickness) and there is no way of obtaining information on either n or d separately using this equation only).

If n_1 and n_2 are the refractive indices at two adjacent maxima (or minima) at λ_1 and λ_2, it follows that the film thickness d is given by expression:

$$d = \frac{\lambda_1 \lambda_1}{\lambda_1 n_2 - \lambda_2 n_1} \qquad (6.20)$$

Where n_1 and n_2 are the refractive index at two adjacent maxima (or minima) whose interference order is an integer (at maxima) or half-integer (at minima) at λ_1 and λ_2. This value of m can now be used, along with n_1 to calculate order number m_o.

To note, the thickness obtained from Eq. (6.20) is very sensitive to the uncertainty in the value of the refractive index which is derived from equation (6.17).

Now, the accuracy of d can be significantly increased by taking the corresponding exact integer or half integer values of m associated to each extreme (Fig. 6.27) and deriving a new thickness, d_2, from Eq. (6.20), again using the values of n_1, the values of d_2 found in this way have a smaller dispersion ($\sigma_1 > \sigma_2$).

With the exact value of m and the very accurate value of d, Eq. (6.19) can then be solve for n at each λ, and thus, the final values of the refractive index n_2 are obtained. Fig. 6.28 depicts the extracted refractive index as a function of wavelength for cobalt doped (0.25% by wt.) tin oxide film.

Absorption coefficient: For the strong absorption region, where interference fringes disappear, the absorption coefficient α, is given by:

$$\alpha(\lambda) = \frac{1}{d} ln \left[\frac{16n^2 s}{(n+1)^3 (n+s^2) T_0} \right] \qquad (6.21)$$

in which $T_o = A_x/B$, and d is the film thickness. The n and s are the refractive indices of the film and substrate, respectively.

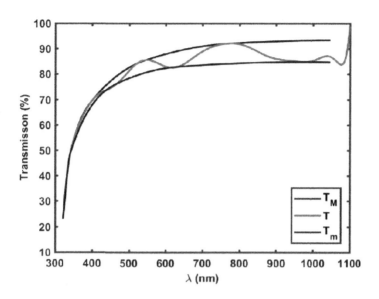

FIG. 6.27

Transmittance spectrum of cobalt doped (0.25% by wt) tin oxide film. Envelopes were created to calculate refractive index.

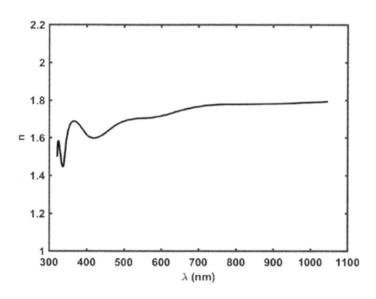

FIG. 6.28

Refractive index as a function of wavelength for cobalt doped (0.25% by wt.) tin oxide film. Above 400 nm, the trend shows refractive index increases with decreasing the energy of the radiation.

In the spectral region of medium absorption where interference fringes appear distinctly in the transmission spectra, α can be obtained by

$$\alpha(\lambda) = \frac{1}{d} ln \left[\frac{(n-1)^3(n-s^2)}{E_m - \sqrt{\left[E_m^2 - (n^2-1)^3(n^2-s^4) \right]}} \right] \tag{6.22}$$

$$E_m = \frac{8n^2 s}{T_m} - (n^2-1)(n^2-s^2)$$

It is noticeable that the wavelike patterns due to interaction of probe beam reflected from different interfaces will be appeared in the optical absorption coefficient which is evaluated from the transmittance data using the known relation

$$\alpha = \frac{1}{d} ln \left[\frac{1}{T(\lambda)} \right] \tag{6.23}$$

The absorption curve for cobalt doped (0.25% by wt.) tin oxide film, shown in Fig. 6.29, reflects the trend presented in the transmittance curve, higher absorption coefficient in the UV region and lower in the visible and near IR region with monotonic decrease toward lower energy (higher wavelength) side.

Optical energy gap (band gap): The optical Energy gap E_g is another important quantity that characterizes semiconductors and dielectric materials since it has a very important role in the design and modeling of such materials. An approximate functional dependency of α on the photon energy, in the strong absorption region, is given by the expression:

$$\alpha(h\nu) = \frac{k(E - E_g)^2}{E} \tag{6.24}$$

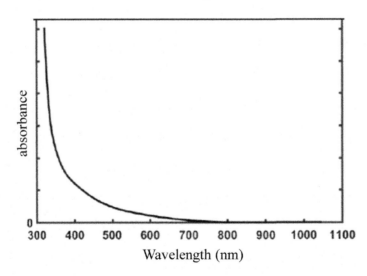

FIG. 6.29

Absorption coefficient (relative) as a function of wavelength for cobalt doped (0.25% by wt.) tin oxide film.

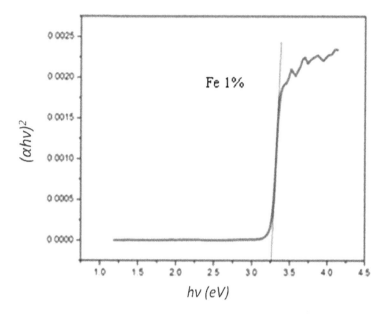

FIG. 6.30

An example of Tauc plot: The $(\alpha h\nu)^2$ versus photon energy ($h\nu$) curve for iron doped zinc oxide (Fe:ZnO, Fe concentration w.r.to Zn was only 1%). One can clearly see the E_g value for Fe doped znO film (Fe 1%) is 3.25 eV.

where k is a constant which is almost independent of the chemical composition of the film, but depends on the transition probability. $E\ (=h\nu)$ is the photon energy and E_g is the optical band gap. The optical band gap is defined as the E axis (abscissa) intercept of the linear part of the plot of $\sqrt{\alpha E}$ versus E. Through the linear fitting of the experimental data the following equation can be obtained:

$$(\alpha E)^{1/2} = k(E - E_g) \tag{6.25}$$

For example, the derived band gap from the linear fit in the curve shown in Fig. 6.30 yields band of about 3.255 eV.

Nanomaterials' size- and shape dependent properties: Here, we will discuss some of the size-dependent properties and surface chemistry of metal and metal oxide nanoparticles in liquid and solid phase environments. During their minute size below 100 nm, nanomaterials have properties quite different from materials of the same chemical composition but of macroscopic size. These size dependent properties relate to color, melting point, electronic and magnetic properties, chemical bond formation, surface hydrophilicity/hydrophobicity, catalysis, and many more. Controlling the size permits tuning these properties in a large range and has led to important new developments in materials science, comparable to a third dimension of the periodic table. The section aims at an understanding of the fundamentals of the size-dependence of materials properties, to learn how to synthesize and characterize these materials and also to learn how to apply them in the areas of energy materials and catalysis.

For example, nanomaterial-based solar cell incorporates sun light-absorbing materials such as nanomaterials of PbS or perovskite. Fig. 6.31 shows basic arrangement of the components in the solar cell, which uses natural dye molecules attached to the TiO$_2$ nanoparticles. Fig. 6.32 displays absorption spectra on TiO$_2$ nanoparticles (NPs) film of the PbS quantum dots (QDs), Lead iodide perovskite and Lead iodide perovskite with PbS QDs. The figure clearly demonstrates that the absorption nature and values are superior when PbS and methyl ammonium leadhalide perovskite are combined together.

Moreover, this absorption property critically depends on the size of perovskite particles. Fig. 6.33 displays the typical crystal structure of a generic ABX$_3$ perovskite (left) and color of colloidal solutions of CsPbX$_3$ (X = Cl, Br, I) under UV light. The Fig. 6.33 also displays MAPbX$_3$ solutions (prepared by dissolving nanomaterials of different sizes) and thin films from nanomaterials of different sizes.

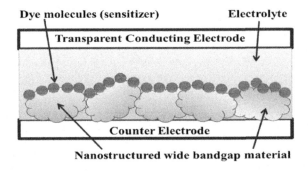

FIG. 6.31

Scheme of the dye sensitized solar cell (or perovskite solar cell). As shown in the figure, light harvesting dye or metal halide perovskite is adsorbed on the film of nanostructured wide bandgap materials such as TiO$_2$ NPs, ZnO, PbS quantum dots (QDs).

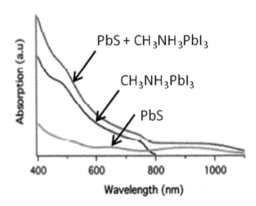

FIG. 6.32

Absorption spectra on TiO$_2$ NPs film of the PbS quantum dots (QDs), Lead iodide perovskite and Lead iodide perovskite with PbS QDs.

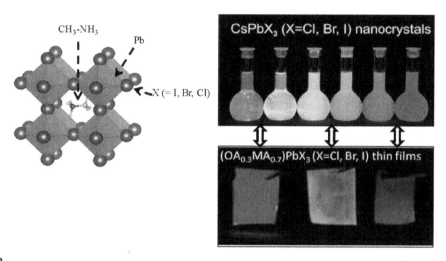

FIG. 6.33

Band gap tenability for ABX$_3$ structures ranging from 1.15 to 3.06 eV. Color of colloidal solutions of CsPbX$_3$ (X = Cl, Br, I) under UV light as reported by Protesescu et al.

Moreover, Fig. 6.34 **shows the representative** PL spectra of CsPbX$_3$ extended toward MAPbCl$_3$ perovskites. The PL peaks for the colloidal CsPbCl$_3$ and CsPbI$_3$ are at 405 nm (3.06 eV) and 700 nm (1.77 eV), respectively. For the Sn/Pb metal, the peaks for MAPbI$_3$ and MASnPbI$_3$ are at 780 nm (1.59 eV) and 960 nm (1.29 eV), respectively. Interestingly, due to the bandgap anomaly of these

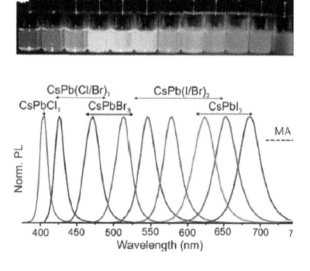

FIG. 6.34

Cesium lead halide perovskite size dependence of color and their associated photoluminescence, PL, spectra. The shape of these PL spectra is simulated using Gaussian functions in order to resemble the colloidal PL data.

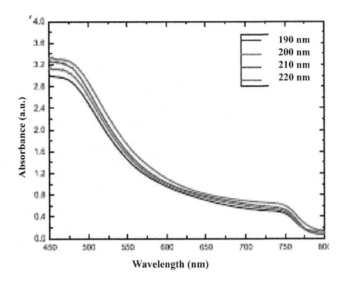

FIG. 6.35

Absorbance spectra of the $CH_3NH_3PbI_3$ perovskite films with different thicknesses on glass substrates.

compounds, the most red-shifted peak has been reported for $MASn_{0.8}Pb_{0.2}I_3$ at 1080 nm (1.15 eV). The shape of these PL spectra is simulated using Gaussian functions in order to resemble the colloidal PL data illustrating the versatile shifting of the PL position from $CsPbCl_3$ (3.06 eV) to $MASn_{0.8}Pb_{0.2}I_3$ (1.15 eV). The large bandgap range has made perovskites attractive for various applications outside of photo-voltaic ranging from lasing, light emitting devices, sensing, photo-detectors as well as X-ray and particle-detection.

Besides size of nanomaterials, also absorption depends on film thickness. Fig. 6.35 plots the UV−visible absorption measurements of the $CH_3NH_3PbI_3$ perovskite films with different thicknesses on glass substrates. The $CH_3NH_3PbI_3$ perovskite film with 190-nm thickness shows lower absorption than that of the film with 220-nm thickness (as discussed before, the perovskite's shape reported to be cuboid-like crystals with the average $CH_3NH_3PbI_3$ crystal size from about 200 nm to about 600 nm). As a consequence, when the thickness increases, more sunlight can be absorbed to generate excitons in the perovskite film.

6.7 Questions

1. In *resistive heating physical vapor* thin film deposition technique (in particular, sputtering), among the following steps which one is the mandatory.
   a. The inert gas collides with sputtering target and releases the sputtered target atom which deposit on the substrate
   b. The precursor solution is directly deposited on the substrate
   c. The precursor solution is spin casted on the substrate
   d. A vapor of sample is created by heating which deposits on the substrate

2. The major steps in chemical vapor deposition that causes to deposit good thin films are

3. Estimation of thickness from measured transmittance curve by using Swanepoel method requires the following input parameters:

a. Maxima and minima of a transmittance curve and associated wavelength

b. Transmittance values at the onset and off set of the curve

c. Refractive indices of the analyte and substrate

d. The parameters given in (a) and (c)

4. In plasma enhanced chemical vapor deposition, the role of plasma is:

a. To produce temperature

b. To produce vapor of the sample

c. To helps to break up gas molecules

d. None of all

5. Major steps involved in spray pyrolysis deposition are

6. What is the difference between top down and bottom up methods for creating nano-structures?

7. What are the practical applications of thin solid films from nanomaterials?

8. Write stepwise methods for Physical vapor deposition, with an example of resistive heating with figure.

9. Distinguish between the following: i) Spin coating and Spray pyrolysis. ii) Single crystalline, polycrystalline and amorphous materials.

10. How can you derive a band gap of a thin film from the measured transmittance curve? Explain with necessary equation and well labeled diagram.

Further reading

[1] V.H. Grassian, When size *really* matters: size-dependent properties and surface chemistry of metal and metal oxide nanoparticles in gas and liquid phase environments, J. Phys. Chem. C 112 (2008) 18303−18313.

[2] P.O. Oviroh, New development of atomic layer deposition: processes, methods, and applications, Sci. Technol. Adv. Mater. 20 (ja) (2019), https://doi.org/10.1080/14686996.2019.1599694 (inactive 5 April 2019). open access.

[3] R.L. Puurunen, Surface chemistry of atomic layer deposition: a case study for the trimethylaluminum/water process, J. Appl. Phys. 97 (12) (June 15 , 2005), 121301−121301−52.

[4] http://www.forgenano.com/uncategorized/atomic-layer-deposition/.

[5] https://www.ttu.ee/public/m/Mehaanikateaduskond/Instituudid/Materjalitehnika_instituut/MTX9100/Lecture5_NanomatFundamentals.pdf (assessed on: 8/17/2019).

[6] https://www.globalspec.com/reference/50219/203279/chapter-3-zero-dimensional-nanostructures-nanoparticles.

[7] https://www.rsc.org/publishing/journals/prospect/ontology.asp?id=CMO:0001316&MSID=c3nr00723e.

[8] M. Vert, Y. Doi, K.-H. Hellwich, M. Hess, P. Hodge, P. Kubisa, M. Rinaudo, F. Schue, Terminology for biorelated polymers and applications (IUPAC Recommendations 2012) (PDF), Pure Appl. Chem. 84 (2) (2012) 377−410.

[9] http://www.semicore.com/news/71-thin-film-deposition-thermal-evaporation.

[10] C.A. Silvera, R.G. Larson, N.A. Kotov, Nonadditivity of nanoparticle interactions, Science 350 (6257) (2015) 1242477.

[11] L.-C. Chen, J.-C. Chen, C.-C. Chen, C.-G. Wu, Fabrication and Properties of High-Efficiency, Perovskite/PCBM Organic Solar Cells, Chen *et al*, Nanoscale Res. Lett. 10 (2015) 312.

[12] D. Jung, Syntheses and characterizations of transition metal-doped ZnO, Solid State Sci. 12 (2010) 466–470.

[13] C. Wang, Z. Chen, Y. He, L. Li, D. Zhang, Structure, morphology and properties of Fe-doped ZnO films prepared by facing target magnetron sputtering system, Appl. Surf. Sci. 255 (2009) 6881–6887.

[14] I. Soumahoro, R. Moubah, G. Schmerber, S. Colis, M. AitAouaj, M. Abd-lefdil, N. Hassanain Berrada, A. Dinia, Structural, optical, and magnetic properties of Fe-doped ZnO films prepared by spray-pyrolysis method, Thin Solid Films 518 (2010) 4593–4596.

[15] Y. Natsume, H. Sakata, Zinc oxide films prepared by sol-gel spin-coating, Thin Solid Films 372 (2000) 30–36.

[16] R. Swanepoel, Determination of the thickness and optical constants of Amorphous silicon, J. Phys. 16 (1983) 1214–1224.

[17] L. El Mir, Z. Ben Ayadi, M. Saadoun, K. Djessas, H.J. von Bardeleben, S. Alaya, Preparation and characterization of n-type conductive (Al, Co) co-doped ZnO thin films deposited by sputtering from aerogel nanopowders, Appl. Surf. Sci. 254 (2007) 570–573.

[18] W.C. Hinds, Aerosol Technology: Properties, Behaviour & Measurements, second ed., Wiley-Blackwell, 1999.

[19] R. Hester, R.M. Harrison, Nanotechnology: Consequences for Human Health and the Environment, RSC Publishing, 2007.

[20] H.-G. Rubahn, Basics of Nanotechnology, third ed., Wiley VCH, 2008.

Infrared (IR) spectroscopy

7.1 Introduction

Definition/concept/applications: In absorption spectroscopy a beam of electromagnetic radiation passes through a sample. Much of the radiation passes through the sample without a loss in intensity. At selected wavelengths, however, the radiation's intensity is attenuated. This process of attenuation is called absorption. Similar to UV−Vis spectroscopy, IR spectroscopy is also based on absorption spectroscopy, on which a molecule or a functional group(s) in a molecule absorbs radiation in the IR range to excite (change) its vibrational motion.

Indeed, as indicated in Fig. 7.1, IR spectroscopy, is a very powerful tool that helps us in identification of chemical compounds (both the organic and inorganic), functional groups prevailing in a particular compound, bond length and also helps to learn more about its structure prevailing in all kinds of phases (gas, liquid and solid). Besides, gathering about the structural information of a compound, it is also useful to assess the purity of a compound. Therefore, it is considered a very important tool of the organic chemist and material scientists. Also, it is recommended that the teaching labs to have these instruments and undergraduate/graduate students are encouraged to acquire spectra of compounds made in the laboratory during their lab course.

In this analytical technique as well, spectra are acquired known as IR spectra: It basically comprises IR radiation source, monochromator, sample holder and an IR detector (discussed in Section 7.9). IR spectra are quick and easy to run.

7.2 IR radiation and relatio with molecular properties

IR radiation is that part of electromagnetic (EM) spectrum between the visible and microwave regions (i.e., it expands in the range from 10,000 to 100 cm^{-1}). The IR region of EM radiation is divided in three sub-regions called (1) Near-IR, (2) Mid-IR and (3) Far-IR. Of greatest practical use for material scientists and organic chemist is the limited portion between 4000 and 400 cm^{-1} (Near-IR).

Electromagnetic radiation consists of oscillating electric and magnetic fields that propagate through space along a linear path and with a constant velocity. The oscillations in the electric and magnetic fields are perpendicular to each other, and to the direction of the wave's propagation. Fig. 7.2 shows an example of plane-polarized electromagnetic radiation, consisting of a single oscillating electric field and a single oscillating magnetic field.

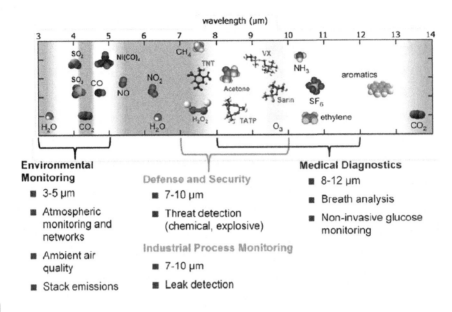

FIG. 7.1

IR wavelength ranges and their corresponding field of applications.

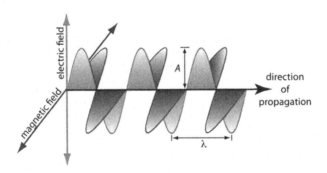

FIG. 7.2

Plane-polarized electromagnetic radiation showing the oscillating electric field (in red [gray in print version]) and the oscillating magnetic field (in blue [dark gray in print version]). The radiation's amplitude, A, and its wavelength, λ, are shown. Normally, electromagnetic radiation is unpolarized, with oscillating electric and magnetic fields present in all possible planes perpendicular to the direction of propagation (motion).

7.3 EM and molecular excitations

At a given temperature, a molecule (or functional groups bonded in a molecule) vibrates with its natural frequency. But, when it is exposed to IR radiation of matched frequency with a natural vibrational frequency of that molecule, it causes to change in the type of vibrations and rotations that are occurring in the molecule. To note, after absorption of a photon, the molecule begins to vibrate with a greater amplitude (but with the same frequency), and thus the molecule gains energy. In other words, when a photon of IR is absorbed by a molecule, it transfers (translate) the energy from photon to molecular vibration which may cause to excite one of the vibrational modes in the molecule and may bring to change in the dipole moment of the molecule.

7.3.1 EM radiation for rotational excitation

Infrared radiation of frequencies less than about 100 cm^{-1} (microwave region: The microwave region extends from 1000 to 300,000 MHz or 30 cm^{-1} m) is absorbed and converted by a molecule (or a functional group/groups present) into energy of molecular rotation. This absorption is also quantized; thus a molecular rotation spectrum consists of discrete lines.

7.3.2 EM radiation for vibrational excitation

Unlike rotational spectrum, vibrational spectra (vibrational states of an electronic state are shown in Fig. 7.2) appear as bands rather than lines because a single vibrational energy change is accompanied by a number of rotational energy changes. When the oscillating electric field of the EM radiation interacts with the molecule, EM couples with the alternating electric field of the molecule (which is produced due to the change of charge distribution accompanying a vibration of a molecule). The absorption of radiation frequency (or wavelength) depends on the relative masses of the atoms, the force constants of the bonds and the geometry of the atoms.

7.3.3 Relation between energy change (due to transition) and frequency of radiation frequency

As discussed before, depending on the energy of the electromagnetic radiation, it can bring about three kinds of transitions: rotational transitions, vibrational transitions, or electronic transitions (Overview of possible transitions is given in Fig. 7.3). Therefore, a molecule or a set of molecules can be investigated by the absorption of microwave radiation which provides transitions between rotational energy levels. In addition, if the molecules absorb infrared radiation it brings about provides the transitions between vibrational levels follows by transitions between rotational energy levels. Finally, when a molecule absorbs visible and ultraviolet radiation it brings about transitions between electronic energy levels follows by simultaneous transitions between vibrational and rotational levels (which is also called ro-vibronic transition: electronic transition followed by vibrational and rotational transitions, after absorption of photon).

For a molecule with known energy levels, along with wavelength, we can easily calculate the frequency of that molecule with the following relation:

$$\Delta E = E_u - E_l = h\nu \tag{7.1}$$

The above equation describes the energy change between upper state (E_u) and lower state (E_L) of energy.

- Frequency range between $10^9 - 10^{11}$ (which is in the microwave range) correlates to the rotation of polyatomic molecules.

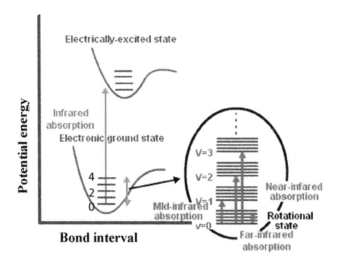

FIG. 7.3

Potential energy surface of a vibrating molecule bonded with a covalent bond. The vibrational levels of ground state and excited state are also shown. When we excite a molecule with IR radiation, it get excited to higher vibrational level (that is molecules vibrational level of that molecule is changed from ground state to higher vibrational state: For example, from $v = 0$ to $v = 1$. As shown in the right, each vibrational state have several rotational states).

- Frequency range between $10^{11}-10^{13}$ (which is in the far infrared range) correlates to the rotation of small molecules.
- Frequency range between $10^{13}-10^{14}$ (which is in the infrared range) correlates to the vibrations of flexible bonds.
- Frequency range between $10^{14}-10^{16}$ (which is in the visible and ultraviolet range) correlates to the electronic transitions.

Relation between frequency and wavenumber: The wavelength, λ, is defined as the distance between successive maxima of a electromagnetic radiation wave of a given wavelength. For ultraviolet and visible electromagnetic radiation the wavelength is usually expressed in nanometers ($1 \text{ nm} = 10^{-9} \text{ m}$), and for infrared radiation it is given in microns ($1 \text{ μm} = 10^{-6} \text{ m}$). The relationship between wavelength and frequency is:

$$\lambda = c/v \tag{7.2}$$

Another unit useful unit is the wavenumber, $\bar{v}$, which is the reciprocal of wavelength

$$\bar{v} = 1/\lambda. \tag{7.3}$$

Wavenumbers are frequently used to characterize infrared radiation, with the units given in cm^{-1}. Mostly, in IR spectra, optical transmittance (%) of a given sample is plotted as a function of wavenumbers.

Example 7.1. In 1817, Josef Fraunhofer studied the spectrum of solar radiation, observing a continuous spectrum with numerous dark lines. Fraunhofer labeled the most prominent of the dark lines with letters. In 1859, Gustav Kirchhoff showed that the D line in the sun's spectrum was due to the absorption of solar radiation by sodium atoms. The wavelength of the sodium D line is 589 nm. What are the frequency and the wavenumber for this line?

Solution: The frequency and wavenumber of the sodium D line are

$$N = c/\lambda = 3.00 \times 10^8 \text{ ms}^{-1} / 589 \times 10^{-9} \text{ m} = 5.09 \times 10^{14} \text{ s}^{-1}$$

$$v = 1/\lambda = (1/589 \times 10^{-9} \text{ m}) \times (1 \text{ m}/100 \text{ cm}) = 1.70 \times 10^4 \text{ cm}^{-1}$$

7.4 IR spectrum

Commonly, IR spectra are presented as a plot of transmittance (%T) versus wavenumber ($\bar{v}$), where % T provides band intensities and $\bar{v}$ represents the band positions whose unit is reciprocal of the centimetre (cm^{-1}). The symbol $\bar{v}$ is called "nu bar": Just for reminder the transmittance is the ratio of the radiant power transmitted by the sample to the radiant power incident of the sample. The result is that an IR spectrum has a very different appearance from the typical UV−Vis spectrum. The former typically show a large number of peaks, each of which is quite narrow compared to the entire range of wavelengths in the spectrum. In contrast to this UV−Vis spectra typically have only one or a few broad peaks for each analyte.

In the initial examination of the spectrum, the usual practice is to look for the presence of the various functional groups. In this way it may be possible to assign the compound to some particular structural class (or classes). Knowledge of the molecular formula will help to reject some of the alternatives. Identification of a compound is carried out by comparison with the published spectra (or with the spectrum of an authentic specimen).

7.5 Theory/concept of IR spectroscopy
7.5.1 Number of vibrational frequencies and IR spectra

A molecule has as many degrees of freedom as the total degrees of freedom of its individual atoms. Each atom has three degrees of freedom corresponding to the Cartesian co-ordinates (*x,y,z*) necessary to describe its position relative to other atoms in the molecule. A molecule of "*n*" atoms therefore has "*3n*" degrees of freedom. For nonlinear molecules, three degrees of freedom describes rotations and three degrees describe translation; the remaining "*3n-6*" degrees of freedom are vibrational degrees of freedom (also called *fundamental vibrations*). Linear molecules have "*3n − 5*" vibrational degrees of freedom, for only two degrees of freedom are required to describe rotation. To note, *fundamental vibrations* involve no change in the center of gravity of the molecule. However, these may not all be different, and may not all appear in the infrared absorption region. The actual number of fundamental frequencies depends largely on the symmetry of the molecule, and the less symmetrical molecule is, the larger is the number of different vibrational frequencies. Moreover, there may be present be vibrational frequencies other than the *fundamental* ones. These usually correspond to a little less than multiple of the *fundamental frequencies*, and are known as the *overtones or harmonics* (discussed in Section 7.6). Thus as "*n*" increases, the infrared spectrum becomes more and more complicated.

7.5.2 Criteria for a molecule to be IR active (dipole change)

It is the main criteria for a species to be IR active. Among the types of vibrations, only to those vibrations those results in change in the dipole moment of the molecule are observed in the IR. As an example, the symmetrical stretching of CO_2 is inactive in the IR because this vibration produces no change in the dipole moment of the molecule. In order to be IR active, a vibration must cause a change in the dipole moment of the molecule. Of the following linear molecules, carbon monoxide and iodine chloride absorb IR radiation, while hydrogen, nitrogen, and chlorine do not. *In general, the larger the dipole change, the stronger the intensity of the band in an IR spectrum*. $C≡O$, $I−Cl$, absorb and H_2, N_2 and Cl_2 in IR do not absorb

Mechanism of IR Absorption: The alternating electric field, produced by the changing the charge distribution accompanying a vibration within the molecule, the molecular vibration couples with the oscillating electric field of the electromagnetic radiation.

Carbon dioxide is an example of why one does not always see as many bands as implied by our simple calculation. Only two IR bands (2350 and 666 cm^{-1}) are seen for CO_2, instead of four corresponding to the four fundamental vibrations. In the case of CO_2, two bands are degenerate (two bending modes appear in the same frequency), and one vibration (the symmetric stretching one) does not cause a change in dipole moment. Other reasons why fewer than the theoretical numbers of IR bands are seen include: absorption is too weak to be observed; absorptions are too close to each other to be resolved on the instrument and absorption is not in the 4000−400 cm^{-1} range.

7.5.3 Types of vibrations

There are two kinds of molecular vibrations.

I. Stretching vibrations and
II. Bending vibrations

I. Stretching: The simplest kind of bond vibration involves stretching of a bond between two atoms. A stretching vibration is the rhythmical movement along the bond axis such that the inter-atomic distance is increasing or decreasing. It is the simplest kind of bond vibration (which involves stretching of a bond between two atoms). For example, water is a triatomic bent molecule with a bond angle of approximately 105° and the bond lengths of about 0.096 nm. It undergoes both the symmetric and asymmetric stretching vibrations (see in the consecutive section in detail) absorbing almost equal frequency. It also undergoes bending vibrations.

II. Bending vibration: A bending vibration may consist of a change in bond angle between bonds with a common atom. For example, in water molecule, oxygen is the common atom to both the hydrogen atoms. Scissoring, twisting, rocking are the kinds of bending vibrations (examples associated with these kinds of vibrations are given in the consecutive sub sections).

The stretching vibrations have higher frequencies (shorter wavelengths) than the deformation (bending) vibrations, and the intensities of the former (related to stretching) are much greater than those of the latter. *Although the masses of the bonded atoms predominantly influence the frequency of the absorption, other effects,* e.g., *environment* (i.e. the nature of neighboring atoms), *steric effects* etc also play a role (the steric effects seen in molecules that come from the fact that atoms occupy space). When atoms are put close to each other, this costs energy. The electrons near the atoms want to stay away from each other. This can change the way molecules want to react. Thus, in general, a particular group will not have a fixed maximum absorption wavelength, but will have a region of absorption: The maximum in this region depends on the rest of the molecule and also on the physical state of the compound gas, liquid and solid (For example in the form of thin film).

7.5.4 Models for understanding the molecular vibration

Usually two models are used to understand the molecular vibrations in IR spectroscopy:

(I) Static model and
(II) Spring model

I. Static model: This model implies that a water molecule is a static object (but in reality a water molecule is always undergoing some changes in its bond angles and bond lengths. Specifically, this model only depicts the average bond angle and bond length, which can have slightly higher or lower values at any given point in time. Let us consider, the various stretching and bending modes for a triatomic molecule such as H_2O molecule. The total modes of vibrations in this tri-atomic molecule ($n = 3$) are $3*3 - 6 = 3$: As shown in Fig. 7.4, this molecule absorbs three frequencies in IR range that are associated with two stretching modes (Symmetric and asymmetric stretching modes) and one bending mode. The symmetric and asymmetric stretching modes appear at 3657 cm^{-1} and 3756 cm^{-1}, respectively, whereas the bending mode appear at 1595 cm^{-1} in IR spectrum. Note symmetric and asymmetric stretching modes absorb radiation of almost equal frequencies, while for exciting bending mode requires only about a half of the values of stretching mode.

II. Spring model: This approach views these bonds in H_2O molecule as tiny springs, instead of fixed rods. Even at zero Kelvin temperature, the molecules will vibrate as these bonds undergo

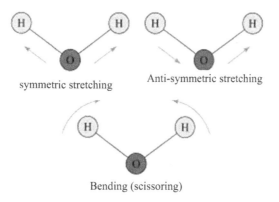

symmetric stretching   Anti-symmetric stretching

Bending (scissoring)

FIG. 7.4

Modes of vibrations in water molecules.

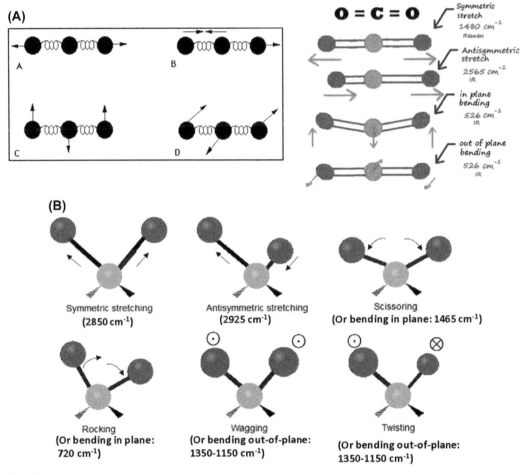

FIG. 7.5

(A) The symmetric stretching vibration mode in (Left: A) is IR inactive, as there is no change dipole moment of the molecule. The bending vibrations in (Left: C and D) are equivalent and are therefore, have the same frequency and are said to be doubly degenerated. Right panel shows alternative way of representation of the modes of vibrations A, B,C and D, and their associated frequencies. (B) Types of stretching and bending vibrations in CH_2 group appearing as a portion of a molecule (for e.g., in hydrocarbon molecule). The 3n-6 rule does not apply, since the CH_2 group represents only a portion of a molecule.

contraction and expansion. As depicted in Fig. 7.4, note the very close spacing in vibration frequency of the asymmetric and symmetric stretching, above, compared with the bending (scissoring) mode.

Let us also discuss the case of CO_2 molecule. The CO_2 molecule is linear and contains three atoms; therefore, it has four fundamental vibrations [$3 \times 3 - 5 = 4$] which are shown in Fig. 7.5.

These **vibrational modes**, shown in Fig. 7.5A, are responsible for the "greenhouse" effect in which heat radiated from the earth is absorbed (trapped) by CO_2 molecules in the atmosphere.

Vibrations in AX₂ functional group bonded in a molecule. The various stretching and bending modes for an AX_2 group appearing as a portion of a molecule, for example, the CH_2 group in a hydrocarbon molecule, are shown in Fig. 7.5B.

The types of vibration and associated frequency are given below:

1. Stretching (at 2850 cm^{-1})
2. Asymmetric stretching (at 2925 cm^{-1})
3. In-plane bending or scissoring (at 1465 cm^{-1})
4. Out-of-plane bending or wagging (1350 - 1150 cm^{-1})
5. Out-of-plane bending or twisting (1350 - 1150 cm^{-1})
6. In-plane bending or rocking (720 cm^{-1}).

The theoretical number of fundamental vibrations (absorption frequencies) will rarely be observed because overtones (transition due to absorption of radiation with multiples of given frequency: discussed in section 7.5.2) and also combination tones (sum of two other vibrations) increase the number of bands, whereas other phenomena reduce the number of bands. The following will reduce the theoretical number of bands.

1. The fundamental frequencies that fall outside of the 4000−400 cm^{-1} region^{-1}.
2. The fundamental bands that are too weak to be observed.
3. The fundamental vibrations that are so close that they coalesce (appear together or very close to each other).
4. The occurrence of a degenerate band from several absorptions of the same frequency in highly symmetrical molecules.
5. The lack of certain fundamental vibrations to appear in the IR because of the lack of change in molecular dipole.

The stretching frequencies of two atoms connecting by a bond can be approximately evaluated using **Hook's law**. This law considers two atoms and their connecting bond (covalent bond) are as a simple harmonic oscillator composed of two masses joined by a spring (see Fig. 7.6).

Specifically, according to this law the **frequency** of the **vibration** of spring is related to the mass and the force constant (f) of the spring. So, we can calculate the **wavenumber** or the **vibrational frequency** of these two atoms using the following relation:

$$\underline{v} = \frac{1}{2\pi c}\left[\frac{f}{M_1 M_2 /(M_1 + M_2)}\right]^{\frac{1}{2}} \quad \text{or} \quad \underline{v} = \frac{1}{2\pi c}\sqrt{\frac{f}{\mu}} \tag{7.4}$$

Atom 1   **Atom 2**

FIG. 7.6

Molecular model according to Hooke's law: According this model, the bond between two atoms is considered as a spring.

where $\underline{v}$ is the vibrational frequency (cm^{-1}), c = velocity of the light (cm/s), f = force constant of bond (dyne/cm). M_1 and M_2 are the masses of the bonded atoms x and y, respectively. And, the term $\left(\frac{M_1 M_2}{M_1 + M_2} = \mu\right)$ is known as reduced mass.

Application of Hooke's Law for frequency calculation: The value of f is approximately 5×10^5 dyne/cm for single bond and nearly two (10 $\times$ 10^5 dyne/cm) and three times (15 $\times$ 10^5 dyne/cm) this value for double and triple bonds, respectively. Calculation of frequency for C–H stretching vibrations employing Eq. (7.4), considering masses of 19.8 $\times$ 10^{-24} g and 1.64 $\times$ 10^{-24} g for C and H, respectively, gives the frequency of the C–H stretching vibrations at 3040 cm^{-1}. Actually, C–H stretching frequencies for methyl and methylene group generally observed in the region between 2960 and 2850 cm^{-1}. This indicates calculation using Hooke's law is not highly accurate because effects arising from the environment of the C–H within the molecule have been ignored. The frequency of IR absorption is commonly used to calculate the force constants of bonds. As indicated above, the region of an IR spectrum where bond stretching vibrations are seen depends primarily on whether the bonds are single, double, or triple or bonds to hydrogen. The following table shows frequency ranges where absorption by single, double, and triple bonds are observed in an IR spectrum.

Bond	Absorption region (cm^{-1})
C–C, C–O, C–N	800–1300
C=C, C=O, C=N, N=O	1500–1900
C≡C, C≡N	2000–2300
C–H, N–H, O–H	2700–3800

You should try calculating a few of these values to convince yourself that the Hooke's law approximation is a useful one.

To approximate the vibrational frequencies of bond stretching by Hook's law, the relative contribution of bond strength and atomic masses must be considered. For example, because C–H is lighter than F–H one may expect, the stretching frequency of F–H should appear in lower frequency, than that of C–H. However, the increase in the force constant from left to right across the first and second row of periodic table has a greater effect than the mass increase. Thus, the F–H group absorbs at a higher frequency (4134 cm^{-1}) than the C–H group (3040 cm^{-1}).

In summary, following are the factors that determine (affect) the frequency of absorption by a vibrating molecule:

1. Types of bonding
2. Types of vibrational mode
3. Masses of the bonded atoms in a molecule (or functional group)
4. Environment of vibrating species
5. Position in periodic table of involving atoms

7.6 Selection rules for infrared transitions

A **selection rule** describes how the probability of transitioning from one level to another cannot be **zero**. Thus, the selection rule for transition for a vibrating molecule (harmonic oscillator) is $\Delta v = \pm 1$. That is: v' = v – 1 OR v' = v + 1. Here, v' represents the new vibrational level of the molecule after emission or absorption of IR radiation.

The above mentioned criteria, for a particular vibration to be infrared active: the requirement of change in the dipole moment of the molecule during the vibration, means transition dipole moment must not be zero. Molecules can have a zero net dipole moment, yet still undergo transitions when stimulated by infrared light. Some examples will help here. Homo-nuclear diatomic molecules such as molecular hydrogen (H_2), nitrogen (N_2) and oxygen (O_2) have no net dipole moment. As the molecule stretches and compresses itself in its one vibrational mode, the symmetry is unchanged. Thus, infrared transitions are truly 'forbidden', since symmetry requires that the molecular dipole moment can never vary from zero.

On the other hand, heteronuclear diatomics such as HCl, CO and NO have a permanent dipole moment, which changes in magnitude as the molecule vibrates. Hence, HCl, CO and NO have electric dipole-allowed transitions (for which the transition matrix elements ($\mu_{v,v'}$ are non-zero) and, therefore, show infrared absorptions corresponding to vibrational excitation.

Linear symmetric molecules such as carbon dioxide, represented as O=C=O, have zero net dipole moments. The symmetric stretching mode doesn't change the dipole moment away from zero, so is infrared-inactive ($\mu_{0,1} = 0$). On the other hand, the anti-symmetric stretch and both bending modes create transient dipole moments, and so are infrared active (That is, $\Delta v = \pm 1$, transition can take place between adjacent vibrational levels, 0 to 1, 1 to 2, etc.).

7.6.1 Concept from quantum theory

The energy of infrared radiation produces a change in a molecule's or a polyatomic ion's vibrational energy. As shown in Fig. 7.3, vibrational energy levels are quantized; that is, a molecule may have only certain, discrete vibrational energies. From the quantum mechanical concept, the energy for an allowed vibrational mode,

$$E_v = \left(v + \frac{1}{2}\right)h v_0,$$

where v is the vibrational quantum number, which has values of *0, 1, 2, ...,* and v_0 is the bond's fundamental vibrational frequency. The value of v_0, which is determined by the bond's strength and by the mass at each end of the bond, is a characteristic property of a bond. For example, a carbon-carbon single bond (C—C) absorbs infrared radiation at a lower energy than a carbon-carbon double bond (C=C) because a single bond is weaker than a double bond.

At room temperature most molecules are in their ground vibrational state ($v = 0$). A transition from the ground vibrational state to the first vibrational excited state ($v = 1$) requires absorption of a photon with an energy of $h v_0$. Transitions in which Δv is ± 1 give rise to the fundamental absorption lines. Weaker absorption lines, called **overtones**, result from transitions in which Δv is *±2 or ±3* (see in the consecutive paragraphs).

IR spectrum shows bands rather than line spectrum due to coupling of various rotational transitions within a given vibrational transition IR spectrum is generally complex and contains many bands in addition to the ones corresponding to fundamental vibrational transitions. Even a relatively simple molecule, such as ethanol (C_2H_6O) (the IR spectrum shown in Fig. 7.7), has $3 \times 9-6$, or 21 possible normal modes of vibration, although not all of these vibrational modes give rise to an absorption.

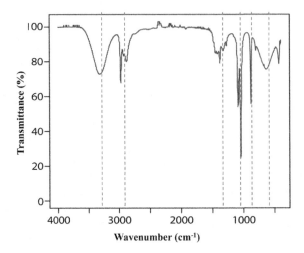

FIG. 7.7

Infrared spectrum of ethanol.

7.6.2 Overtones

Bands corresponding to integral multiple of fundamental vibration. They are due to transition from ground state to higher vibrational states. They are very weak bands. Specifically, overtones occur when a vibrational mode is excited from $v = 0$ to $v = 2$ which is called the first overtone, or $v = 0$ to $v = 3$, the second overtone. The fundamental transitions, $v = \pm 1$, are the most commonly occurring, and the probability of overtones rapid decreases as $v = \pm nv$ increases.

Based on the harmonic oscillator approximation, the energy of the overtone transition will be about n times the fundamental associated with that particular transition. For example, an absorption band at 1050 cm^{-1} may well have an accompanying band at 2100 ($2v$) and 3150 ($3v$) cm^{-1}.

The anharmonic oscillator calculations show that the overtones are usually less than a multiple of the fundamental frequency. Overtones are generally not detected in larger molecules. This is demonstrated with the vibrations of the diatomic HCl in the gas phase: (see Table 7.1).

Reasons for Overtones: In some special cases, higher-order electric multipole moments (such as the quadrupole tensor) can permit a transition to occur. Also, distortion of the potential energy curve away from the parabolic Harmonic Oscillator can make other transitions allowed, but with lower intensity. We can see this in the case for carbon monoxide that the v = 0 to v = 2, or 3 transitions, "forbidden" in the harmonic oscillator model, are much less intense (or much less probable) but are easily observed with the sensitivity of a modern Fourier-transform research spectrometer.

Table 7.1 HCl vibrational spectrum. We can see from Table, that the anharmonic frequencies correspond much better with the observed frequencies, especially as the vibrational levels increase.

Transition	$\bar{v}_{obs}$ [cm^{-1}]	$\bar{v}_{obs}$ [cm^{-1}] (Harmonic)	$\bar{v}_{obs}$ [cm^{-1}] (Anharmonic)
$0 \rightarrow 1$ (fundamental)	2885.9	2885.9	2885.3
$0 \rightarrow 2$ (first overtone)	5668.0	5771.8	5665.0
$0 \rightarrow 3$ (second overtone)	8347.0	8657.7	8339.0

7.6.3 Combination bands

Two vibrational frequencies in a molecule couple to give a new frequency within the molecule. This band is a sum of the two interacting bands.

7.6.4 Difference bands

Similar to combination bands. The observed frequency is the difference between the two interacting frequencies.

7.7 Isotope effects in IR spectrum

The different isotopes in a particular species may exhibit different fine details in infrared spectroscopy. For example, the O—O stretching frequency (in cm^{-1}) of oxyhemocyanin is experimentally determined to be 832 and 788 cm^{-1} for $v(^{16}O-^{16}O)$ and $v(^{18}O-^{18}O)$, respectively.

By considering the O—O bond as a spring, the wavenumber of absorbance, v can be calculated using Eq. (7.1).

$$\frac{\vartheta(O^{16})}{\vartheta(O^{18})} = \frac{\frac{1}{2\pi c}\sqrt{\frac{f^{16}}{\mu^{16}}}}{\frac{1}{2\pi c}\sqrt{\frac{f^{18}}{\mu^{18}}}} \quad or \quad \frac{832}{788} = \sqrt{\frac{9}{8}}.$$

By employing Eq. (7.4.), the reduced masses for $^{16}O-^{16}O$ and $^{18}O-^{18}O$ can be approximated as 8 and 9, respectively. Thus the effect of isotopes, on the vibration has been found to be stronger than previously thought.

7.8 Coupled interactions
7.8.1 Coupling interactions due to stretching vibrations
7.8.1.1 This results from a mechanical coupling interaction between the oscillators.- it appears when:

i) two bond oscillators share a common atom and ii) their individual oscillation frequencies are not widely different. For example in carbon dioxide molecule which consists of two bonds with a common carbon atom has two fundamental stretching vibrations an asymmetrical and a symmetrical stretching mode (N.B. It shows also bending vibrations at 667 cm^{-1} as has been mentioned)

(a) *The symmetrical stretching mode:* It consists of an in-phase stretching or contracting of the bonds.

$$\leftarrow O{=}C{=}O \rightarrow \text{ or } \rightarrow O{=}C{=}O \leftarrow$$

Absorption occurs at a frequency lower than that observed for the carbonyl group in an aliphatic acetone (which is observed at 1725–1705 cm^{-1}). The symmetrical stretching mode produces no change in the dipole moment of the molecule and it is therefore IR inactive and can't be observed in IR spectrum but it is easily observed in the Raman spectrum near 1340 cm^{-1} (N B. Band intensity in Raman spectra depends on bond polarizability rather than molecular dipole changes).

(b) *In the asymmetrical stretching mode:* The two C=O bonds stretch out of phase; one C=O bond stretches as the other contracts: ←O=C=O← .

The asymmetrical stretching mode is IR active, since it produces a change in the dipole moment. The absorption of asymmetrical stretching occurs at 2350 cm^{-1}, i.e., at a higher frequency than observed for a carbonyl group in aliphatic ketones. This difference in carbonyl absorption frequencies displayed by the CO_2 molecule results from strong mechanical coupling or interaction. In contrast, two ketonic carbonyl groups separated by one or more carbon atoms show normal carbonyl absorption near 1715 cm^{-1} because appreciable coupling is prevented by the intervening carbon atom (s).

Also, in primary amine and primary amide, the N–H stretching bands are observed in the $3497–3077 \text{ cm}^{-1}$ region, which is due to coupling. The observation of two C–H stretching bands in the $3000–2760 \text{ cm}^{-1}$ region for both the methylene and methyl groups are also accounted for coupling.

7.8.2 Coupling interactions due to bending vibrations

Vibrations resulting from bond angle changes frequently couple in a manner similar to stretching vibrations. For example, the ring out-of-plane bending frequencies of aromatic molecules depend on the number of adjacent hydrogen atoms on the ring: Coupling between the hydrogen atoms is affected by the bending of the C–C bond in the ring to which the hydrogen atoms are attached. Spectrum in Fig. 7.8 shows the IR spectrum of toluene. Note in the figure that the aromatic C–H stretches are to the left of 3000 cm^{-1}, and the methyl C–H stretches are to be right of 3000.

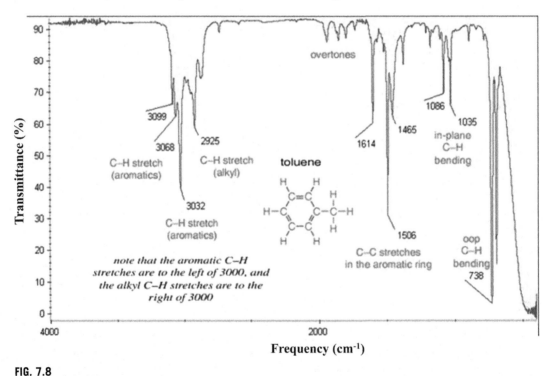

FIG. 7.8

IR spectrum of toluene.

7.8.3 Coupling due to interactions between stretching and bending vibrations

Coupling between stretching and bending vibrations can occur if the stretching bond forms one side of the changing angle. This type of interaction can be illustrated by the absorption of secondary acyclic amides which exist predominantly in the trans conformation. The structure of the secondary acylic amide is given in Fig. 7.9.

It shows strong absorption in the 1563-1515 cm^{-1}. This absorption involves coupling of the N—H bending and C—N stretching vibrations.

7.8.4 Coupling due to interactions between fundamental vibrations and overtones (fermi resonance)

Coupling may occur between fundamental vibrations and overtones or combination tone vibrations. Such interaction is known as Fermi resonance. Example of such interaction (Fermi resonance) is observation of in absorption peak of CO_2 at around 1340 cm^{-1}. As discussed before, the symmetrical stretching band of C=O appears in the Raman spectrum near 1340 cm^{-1}. Actually, there are two bands: one at 1286 and 1388 cm^{-1}. *The splitting results from coupling between the fundamental C=O stretching vibration near 1340 cm^{-1} and the first overtone of the bending vibration* (The fundamental bending vibration of CO_2 occurs near 666 cm^{-1}, the first overtone near 1234 cm^{-1}): The three fundamental vibrations are $v_1 = 1337$ cm^{-1} (frequency for symmetric stretching), $v_2 = 667$ cm^{-1}, $v_3 = 2349$ cm^{-1}. The first overtone of v_2 is $2v_2$ is $2(667) = 1334$ $cm^{-1}(v_2)$. Interaction occurs between $v_1 = 1337$ cm^{-1} and $2*v_2 = 1334$ cm^{-1}.

Another example of Fermi resonance in an organic structure is the doublet appearance of the C=O stretching of cyclo-pentanone under sufficient resolution of IR spectrometer. Fig. 7.10 shows the appearance of the spectrum of cyclo-pentanone under the usual conditions and the peak around the region where Fermi resonance of an α-methylene group shows two absorptions in the carbonyl stretch region.

Fig. 7.11 shows the zoomed spectrum of cyclo-pentane around 1740−1780 cm^{-1} region in four different solutions. The figure clearly shows two peaks due to Fermi resonance: Due to this kind of coupling, both shift in peak position and the peak intensities found to occur.

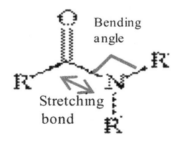

Organic amide

FIG. 7.9

Acylic amide in which coupling between stretching and bending vibrations occurs.

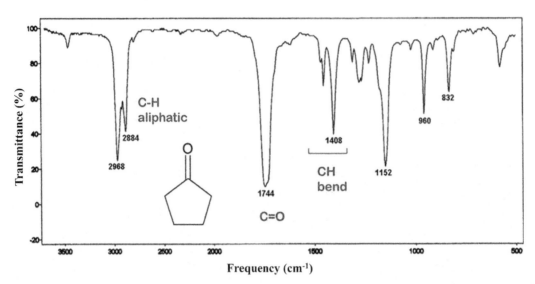

FIG. 7.10

Infrared spectrum of cyclopentanone (cyclic ketone with 5-member ring: see chemical structure inside the spectrum) thin film.

R. M. Silverstein, Francis X. Webster, Spectrometric Identification of Organic Compounds, 6th ed., Chap.3 pg. 75., Adapted with the permission from John Wiley & Sons, 1991.

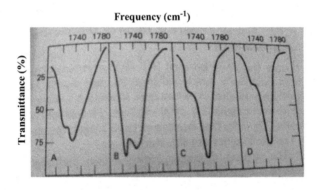

FIG. 7.11

Infrared spectrum of cyclopentanone in various media. (A) Carbon tetrachloride solution (0.15 M). (B) Carbon disulphide solution (0.023 M). (C) Chloroform solution (0.025 M). (D) Liquid state (thin films). Fermi resonance with overtone or combination band of an α-methylene group shows two absorptions in the carbonyl stretch region.

R. M. Silverstein, Francis X. Webster, Spectrometric Identification of Organic Compounds, 6th ed., Chap.3 pg. 75., Adapted with the permission from John Wiley & Sons, 1991.

It is useful to understand Fermi resonance because it helps assign and identify peaks within vibrational spectra (i.e., IR and Raman) that may not otherwise be accounted for, however, it should not be used lightly when assigning spectra.

Fermi resonance is a common phenomenon both in IR and Raman spectra. It requires that the vibrational levels be of the same symmetry species and that the interacting groups be located in the molecule so that mechanical coupling be noticeable.

7.8.5 The requirements for the effective coupling interaction may be summarized as follows

1. The vibrations must be of the same symmetry species if interaction is to occur.
2. Strong coupling between stretching vibrations requires a common atom between the groups.
3. Interaction is greatest when the coupled groups absorb, individually, near the same frequency.
4. Coupling between the stretching and bending can occur when the stretching bonds forms one side of the changing angle.
5. A common bond is required for coupling of bending vibrations.
6. Coupling is negligible when groups are separated one or more carbon atoms and the vibrations are mutually perpendicular.

7.9 Effect of hydrogen bonding in molecular vibration

Hydrogen bonding can occur in any system containing a proton donor group (X—H) and a proton acceptor (Y) if the s orbital of the proton can effectively overlap the p or π orbital of the acceptor group. Atoms X and Y are electronegative, with Y possessing lone pair electrons. The common proton donor groups in organic molecules are carboxyl, hydroxyl, amine, or amide groups. Common proton acceptor atoms are oxygen, nitrogen, and the halogens. Unsaturated groups, such as the C=C linkage, can also act the proton acceptors. Below is the summary of the influence of H-bonding in molecular vibrations.

Intermolecular hydrogen bonds gives rise to broad bands, while intramolecular hydrogen bonds give sharp and well defined bands. The inter- and intra-molecular hydrogen bonding can be distinguished by dilution: Intramolecular hydrogen bonding remains unaffected on dilution and as a result the absorption band also remains unaffected whereas in intermolecular, bonds are broken on dilution and as a result there is a decrease in the bonded O—H absorption.

The strength of hydrogen bonding is also affected by: i) Ring strain, ii) Molecular geometry, iii) Relative acidity and basicity of the proton donor and acceptor groups.

7.10 Instrument designs for infrared absorption

IR spectroscopy is similar to UV—Vis spectroscopy in that it requires a source of light, a means for separating this light into different, a sample holder and a detector. However, the specific instrument components in this kind of spectroscopy are made of different materials and often operate in different principles than that of UV—Vis spectroscopy.

For example, FT—IR, the monochromator is replaced with an interferometer (discussed in section 7.9). Rather than shining a monochromatic beam of light (a beam composed of only a single wavelength) at the sample, this technique shines a beam containing many frequencies of light at once and measures how much of that beam is absorbed by the sample. That is an FTIR instrument allows all wavelengths to fall on the sample simultaneously: In other word, an FTIR spectrometer simultaneously collects high-spectral-resolution data over a wide spectral range.

Below we discuss the components used in IR spectrophotometers followed by their types.

Instruments for measuring IR absorption require a continuous source of radiation and IR transducer. Below we briefly give overview about the major components and their unique properties which make them useful in IR spectroscopy: (i) light source, (ii) monochromator and (iii) transducers found in modern IR spectrophotometers.

7.10.1 Sample illumination system (light source)

The light source in IR spectroscopy is usually inert rod that is heated to a much lower temperature than is used for light source in UV–Vis spectroscopy. These sources produce continuous radiation approximating that of a black body radiation.

As glass and fused silica are opaque at wavelengths greater than 2.5 μm, the glowing source must not be in a glass bulb or in a casing that is made of these substances. The heated material is either silicon carbide (SiC, called Globar) or is a mixture of rare-earth oxides (producing a device known as a Nernst globar). For a globar heated at 1300–1500 K, useable light is provided at wavelengths of 0.4–20 μm.

(I) The Nernst Glower: The general construction of the Nernst glower is shown in Fig. 7.12A. This design includes the semiconducting material made of rare-earth oxide formed into a cylinder having diam. of 1–3 mm and length of 20–50 mm. Platinum leads are sealed to the ends of the cylinder to permit electrical connections. As current passes through the device, its temperature is increased to between 1200 and 2200 K. A heating source and a reflector is included to help pass the radiation coming out from the light source in the desired direction. Because the Nernst glower has a large negative temperature coefficient of electrical resistance (resistance ↓ as temperature ↑) thus circuit is current limited. Therefore, it must be heated externally to a dull red hot before the current is large enough to maintain the desired temperature.

This device produces adequate amounts of light at wavelengths ranging from 1 to 40 μm. Fig. 7.12B shows a typical spectral distribution, and shows the light produced by such a device closely matches what would be expected for blackbody radiation.

(II) Silicon carbide Globar

Introduction: Globar is a thermal light source that emits radiation in near IR region of the electromagnetic spectrum. Globar is a thermal radiator, meaning a solid which is heated to very high temperatures to produce light radiation. The name 'globar' is simply a combination of glow and bar, the heating element in the form of a bar is heated until it glows to emit radiation. Globar is a continuous radiation source that produces a broad, featureless range of wavelengths.

Construction of SIC globar: As shown in Fig. 7.13, the silicon carbide rod used as the heating element of the globar rod is of length 20–50 mm and width 5–10 mm. It is in the form of a U shaped rod, where the heating element is heated by an electric current. Globars that were used in the past required a separate cooling system in order to protect the electrical components. Modern globars, however, have the cooling system inbuilt in them by using ceramic technology.

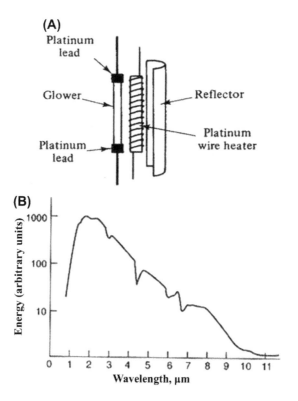

FIG. 7.12

(A) General design of a Nernst glower. This material must be preheated to achieve the conductance, which is done by using a separate platinum wire heater. A reflector helps to collect and direct the radiation that is given off by this source in the desired direction for use. (B) Spectral distribution of energy from a Nernst glower operated at approximately 2200 K.

The silicon carbide rod is heated electrically to temperatures of 1000–1650 °C. The emitted radiation from globar lies in the near IR range (The wavelength range of the emitted radiation is from 1 to 50 μm). This device is then combined with a downstream interference filter to produce radiation having wavelengths of 4–15 μm.

7.10.2 Monochromators

As glass and quartz absorbs IR radiation, these are not suitable materials for prism (or monochromators) to separate light of different wavelengths in IR spectroscopy. But, grating still can be used for this purpose. Besides, another device that can be used for this purpose is interferometer (This will be discussed later in this section). A grating in IR spectrometry works in the same manner as a grating that is used in UV–Vis spectroscopy (The working principle grating monochromator is similar in both of these kinds of spectroscopy). However, the spacing of the lines in these gratings is different.

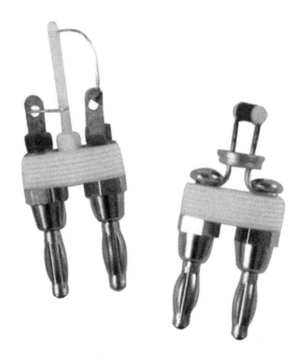

FIG. 7.13

Globar IR light source.

For UV−Vis spectroscopy, the spacing is typically 300−2400 grooves/mm. In IR spectroscopy, this spacing is 300/mm for 2−2.5 μm and 100 grooves/mm. for work with light at wavelengths of 5−16 μm.

7.10.3 Sample holding system (sample cells)

Infrared spectroscopy is routinely used to analyze gas, liquid, and solid samples and, therefore, suitable material should be chosen to allow to pass the IR radiation. Similar to light source and monochromator, sample cells made of glass and quartz are not suitable (as they are opaque) for IR spectroscopy. Instead, ionic salts such as NaCl, KBr, and CsBr, which are transparent to infrared radiation, are used. However, these materials are also not ideal, since they can be molded into desired shapes like glass and also they dissolve in water. To overcome the later problem, less soluble salts such as CaF_2 and AgCl can be used for the sample holder. Shape and method of loading samples in sample cell depends on the sample type (e.g., gas, liquid and solid) which are discussed below.

(I) **Gas samples:** Gases are analyzed using a cell with a path length of approximately 10 cm. Longer path lengths are obtained by using mirrors to pass the beam of radiation through the sample several times.

Solid samples can be mixed with dry and pressed into a thin disk, which is put into the instrument for analysis.

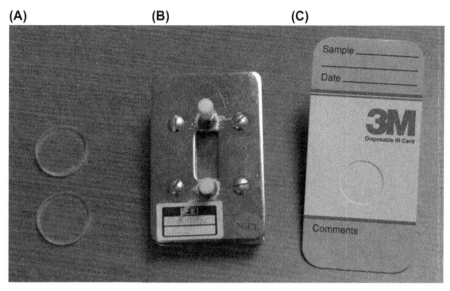

FIG. 7.14

Three examples of IR sample cells: (A) NaCl salts plates; (B) fixed pathlength (0.5 mm) sample cell with NaCl windows; (C) disposable card with a polyethylene window that is IR transparent with the exception of strong absorption bands at 2918 cm^{-1} and 2849 cm^{-1}.

(II) Liquid Samples (Non-volatile)

Liquid Samples (Non-volatile): A liquid samples may be analyzed using a variety of different sample cells (Fig. 7.14). All ordinary solvents have complicated IR spectra, so it is preferable to measure spectra of pure substances rather than a solution, especially a dilute solution.

For non-volatile liquids a suitable sample can be prepared by placing a drop of the liquid sample is often simply put onto a flat plate of NaCl and another similar NaCl flat plate is put on top of it and then clamped. Then a spectrum is taken of the resulting film: This form a thin film that typically is less than 0.01 mm thick.

Liquid Samples (Volatile): Volatile liquids must be placed in a sealed cell to prevent their evaporation.

Liquid Samples (Solution form): The analysis of solution samples is limited by the solvent's IR absorbing properties, with CCl_4, CS_2, and $CHCl_3$ being the most common solvents. Solutions are placed in cells containing two NaCl windows separated by a Teflon spacer. By changing the Teflon spacer, path lengths from 0.015 to 1.0 mm can be obtained.

(III) Solid Samples

Transparent solid samples. This kind of samples can be analyzed directly by placing them in the IR beam.

Opaque solid samples. Opaque samples must be dispersed in a more transparent medium (solvent) before recording the IR spectrum. If a suitable solvent is available, then the solid can be analyzed by preparing a solution and analyzing as described above. When a suitable solvent is not available, solid

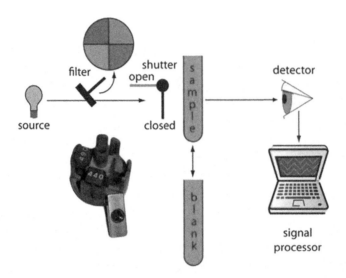

FIG. 7.15

Schematic diagram of a filter photometer. The analyst either inserts a removable filter or the filters are placed in a carousel, an example of which is shown in the photographic inset. The analyst selects a filter by rotating it into place.

samples may be analyzed by preparing a mull of the finely powdered sample with a suitable oil. Alternatively, the powdered sample can be mixed with KBr and pressed into an optically transparent pellet.

7.10.4 Detectors

IR spectroscopy utilizes heat sensing detectors, unlike photon sensing detectors such as photo-diode and photomultiplier tube, in UV–Vis spectroscopy. IR radiation heats the thermocouple by causing its atoms to move more rapidly. A thermocouple is a junction of two different conductors that generates an electric voltage that depends on the temperature difference between the ends of two wires, one of which is maintained at a constant temperature. The intensity of IR radiation falling on this detector causes warming and a change in voltage, thus making it possible to detect the radiation.

7.11 Types of IR spectrophotometers
7.11.1 Filter photometers

The simplest instrument for IR absorption spectroscopy is a filter photometer similar to that shown in Fig. 7.15 for UV/Vis absorption. These instruments have the advantage of portability, and typically are used as dedicated analyzers for gases such as HCN and CO.

Working Principle: Single beam filter photometer consists of a source of light (tungsten filament lamp), lens to make the light beam parallel, filter for wavelength selection, cuvette with sample holder for keeping the solution to be analyzed, detector (photocell) and reading device (galvanometer or micro-ammeter) which is connected to computer for data storage.

The tungsten filament lamp gives the light radiation. This light is incident on the lens which makes it a parallel beam of light. This parallel beam of light is passed through the sample solution after passing through a filter. The sample absorbs some light energy, transmitting the other. This transmitted light falls on the photocell that generates the photocurrent. This photo-current is recorded by the galvanometer or micro ammeter having transmittance scale, since the photometer is directly proportional to the transmitted light, the transmittance scale is linear. The current recording device can also be connected with computer for offline data processing (analysis).

7.11.1.1 Measurement method

(1) With photocell darkened, the meter is adjusted to zero by zero adjustment.
(2) The blank or reference solution is inserted in the path of light beam and light intensity is adjusted by means of rheostat in series with lamp (The blank or reference solution means the solvent used for dissolving analyte or mixture of solvent and reagents which are used for developing color). With this adjustment the meter reading is brought to 100 scale divisions.
(3) Solutions of both standard and unknown samples are inserted in place of blank and the reading of the specimen relative to the blank is recorded.

7.11.2 Double-beam spectrophotometer

Infrared instruments using a monochromator for wavelength selection use double-beam optics similar to that shown in Fig. 7.16. A chopper directs the source's radiation, using a transparent window to pass radiation to the sample and a mirror to reflect radiation to the blank. The chopper's opaque surface

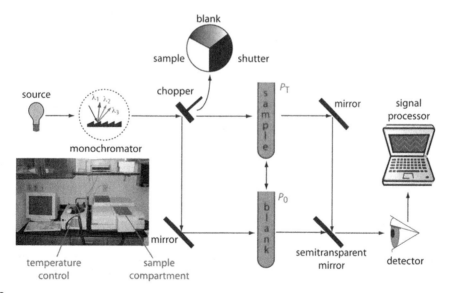

FIG. 7.16

Schematic diagram of a scanning, double-beam spectrophotometer.

serves as a shutter, which allows for a constant adjustment of the spectrophotometer's 0% T. The photographic insert shows a typical instrument. The unit in the middle of the photo is a temperature control unit that allows the sample to be heated or cooled.

Double-beam optics are preferred over single-beam optics because the sources and detectors for infrared radiation are less stable than those for UV/Vis radiation. In addition, it is easier to correct for the absorption of infrared radiation by atmospheric CO_2 and H_2O vapor when using double-beam optics. Resolutions of $1-3$ cm^{-1} are typical for most instruments (Fig. 7.16).

7.11.3 Fourier transforms spectrometer

Fourier-transform infrared spectroscopy (FTIR) is a technique used to obtain an infrared spectrum of absorption or emission of a solid, liquid or gas. Instead of separating the wavelengths in space and time, the wavelength dependence of %T is gained by use of a device called interferrometer, which causes positive and negative interference to occur at sequential wavelengths as a moving mirror changes the path length of the light beam. This initial output recorded by the detector does not look like spectrum, but this direct output is transformed into spectrum by application of mathematical process called **"Fourier-transform"**.

The term Fourier-transform infrared spectroscopy originates from the fact that a Fourier transform (a mathematical process) is required to convert the raw data into the actual spectrum. The major advantage of FTIR is in the speed in which the spectrum is obtained, typically in a few seconds. This means large number of spectra can be gathered in a short time. This high rate of data acquisition also makes it possible to combine a large number of spectra to help to remove random fluctuations in the signal or "noise". The more spectra what are averaged, the better the signal to noise ratio will become. This approach, in turn, means that a good spectrum can be achieved for a small concentration of analyte and that a lower limit of detection for measurement of the analyte can be obtained.

One difficulty in using IR spectrometry (including FTIR) is that atmospheric CO_2 and H_2O vapor in air both absorb IR radiation considerably and obscure the spectrum of the desired sample. For this reason, IR spectrophotometers are often double beam devices. However, in an FT-IR includes only a single optical path, therefore, it is necessary to collect a separate spectrum to compensate for the absorbance of atmospheric CO_2 and H_2O vapor (in which the spectrum of air subtracted from the spectrum of sample). This is done by collecting a background spectrum without the sample and storing the result in the instrument's computer memory. The background spectrum is removed from the sample's spectrum by ratioing the two signals. In comparison to other instrument designs, an FT−IR provides for rapid data acquisition, allowing an enhancement in signal-to-noise ratio through signal-averaging. Therefore, Fast Fourier- Infrared (FT-IR) spectroscopy has been one of the powerful tool in material science research, in particular for identifying the prevailing functional group(s) and also monitoring reaction intermediates (Fig. 7.17).

7.11.4 Attenuated total reflectance (ATR) FT−IR instrument

The analysis of an aqueous sample is complicated by the solubility of the NaCl cell window in water. One approach to obtaining infrared spectra on aqueous solutions is to use attenuated total reflectance instead of transmission. Fig. 7.18 shows a diagram of a typical attenuated total reflectance (ATR) FT−IR instrument. The ATR cell consists of a high refractive index material, such as ZnSe or

(A)

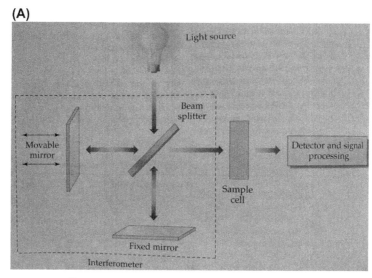

(B)

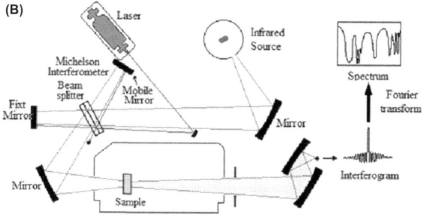

FIG. 7.17

(A) The general design of FTIR spectroscopy. As one mirror is moved in this device, different wavelengths of light from the original source will have constructive interference and make it onto the sample. By moving this mirror it is possible to have different sets of wavelengths pass on to the sample. The absorption of light for each set of wavelengths and at each position of the mirror is measured and converted through the process of Fourier transform into a spectrum. Besides light sources discussed in previous section, a laser beam has also been used as a light source. A section of the instrument that is n the dashed box and that is used for wavelength selection is known as an interferometer. (B) FTIR spectrophotometer.

diamond, sandwiched between a low refractive index substrate and a lower refractive index sample. Radiation from the source enters the ATR crystal where it undergoes a series of total internal reflections before exiting the crystal. During each reflection the radiation penetrates into the sample to a depth of a few microns. The result is a selective attenuation of the radiation at those wavelengths

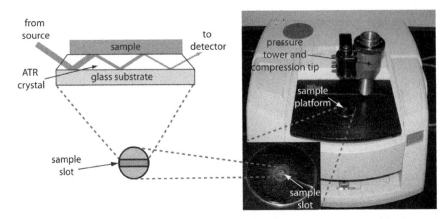

FIG. 7.18

FT-IR spectrometer equipped with a diamond ATR sample cell. The inserts show a close-up photo of the sample platform, a sketch of the ATR's sample slot, and a schematic showing how the source's radiation interacts with the sample. The pressure tower is used to ensure the contact of solid samples with the ATR crystal.

where the sample absorbs. ATR spectra are similar, but not identical, to those obtained by measuring the transmission of radiation.

Solid samples also can be analyzed using an ATR sample cell. After placing the solid in the sample slot, a compression tip ensures that it is in contact with the ATR crystal. Examples of solids that have been analyzed by ATR include polymers, fibers, fabrics, powders, and biological tissue samples. Another reflectance method is diffuse reflectance, in which radiation is reflected from a rough surface, such as a powder. Powdered samples are mixed with a non-absorbing material, such as powdered KBr, and the reflected light is collected and analyzed. As with ATR, the resulting spectrum is similar to that obtained by conventional transmission methods.

7.12 Applications of IR spectroscopy

The general use of mid IR spectroscopy by chemists for identifying organic compounds got popularity when scientist developed an inexpensive and easy-to-use double-beam spectrometers (the discussed above) that produce spectra in the region of 5000 to 670 cm^{-1} (2—15 μm). Indeed, NMR, Mass and IR spectroscopies are the backbone tools for identifying organic, inorganic and bio-molecules species.

Although it is rather difficult to extract precise information about the vibrations of complex molecules, IR spectroscopy provides useful information, by simply comparing the measured spectrum with the standard one. As mentioned above, often, mass and NMR spectra are also required to make accurate interpretation of the measured IR spectrum.

Sections discussed so far in this chapter covered basic concepts on IR spectroscopy, origin of peaks and bands in an IR spectrum, major types of IR spectroscopy and components associated with it. Below we discuss how to interpret the IR spectra recorded by these devices.

7.12.1 Interpretation of IR spectra

Absorption features resembling functional groups present in complex organic molecules show variations band in a wide range because the band arise from interaction of various modes (complex) of vibrations prevailing in the molecule. Even in this case, the band may resemble mainly a single vibrational mode. Certain absorption bands such as those arising from the C$-$H, O$-$H, and C$=$O stretching modes, remain within fairly narrow regions of the spectrum. Very important information about the molecular structure may be achieved from the exact position of absorption band within that narrow regions. Shift (change) in absorption peak positions and changes in band shape, molecular environment may also suggest important structural details.

However, following consideration are required before an attempt to interpret a spectrum.

1. The spectrum must be adequately resolved (well separation between neighboring peaks) and of adequate intensity (clearly distinguished peak height with respect of background with ratio of signal to noise ratio of at least 3).
2. The spectrum should be that of a reasonably pure compound.
3. The spectrophotometer should be calibrated so that the bands are observed at their proper frequencies or wavelengths. Proper calibration must be made with reliable standards such as poly(styrene) film.
4. The method of sample handling must be specified. If a solvent is employed, the solvent, concentration, and the cell thickness should be indicated.

Fig. 7.19 shows two typical spectra obtained with an inexpensive double-beam instrument. Identification of an organic compound from a spectrum of this kind is a two step process. The first step involves determining what functional groups are most likely present by examining the *group frequency region (see below in the consecutive paragraph)*, which covers the radiation frequency in the range about 3600 cm^{-1} to approximately 1250 cm^{-1} (Fig. 7.19 top). In the second step we compare the spectrum of unknown in detail with the spectra of pure compounds that contain all of the functional groups found in the first step. Also, the fingerprint region (Fig. 7.19 bottom) from 1200 to 600 cm^{-1} is particularly useful because small differences in the structure and constitution of the molecule result in significant changes in the appearance and distribution of absorption bands in this region. A close match between two spectra in the fingerprint region and group frequency region provides solid evidence that the two compounds are identical.

(I) **Functional Group Region:** Two important areas for a preliminary examination of a spectrum are the regions from 4000 cm^{-1} to 1300 cm^{-1} and 900 - 650 cm^{-1}. The high frequency region is called functional group region. The characteristic stretching frequencies for important functional groups such as OH, NH, and C$=$O occur in this portion of the spectrum. The presence of absorption in the assigned ranges for a particular functional group provides evidence to confirm the presence of such group in the molecule.

Weak bands in the high frequency region, resulting from fundamental absorption by functional group such as S$-$H and C$=$C are extremely important in the determination of structure.

(II) **Skeletal Bands:** Moreover, prevailing of strong band in 1600$-$1300 cm^{-1} regions (are called skeletal bands) indicates presence of aromatic and hetero-aromatic molecules. Aromatic and hetero-aromatic compounds display strong out-of-plane C$-$H bending and ring bending absorption bands in 900$-$650 cm^{-1} region that can be correlated with the substitution pattern. Broad, moderately strong absorption in the low-frequency region suggests the presence of carboxylic acid dimmers, amines, or amides, all of which show out-of-plane bending in this region.

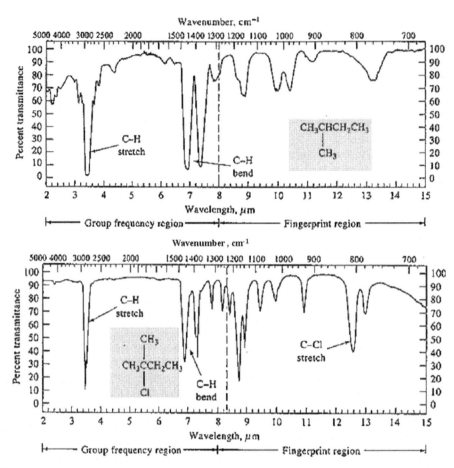

FIG. 7.19

Group frequency and fingerprint regions of the mid-IR spectrum. Examples of two different molecules: $(CH_3)_2CHCH_2CH_3$ (upper) and $(CH_3)_2CClCH_2CH_3$ (lower).

John D. Robert and Marjorie C. Caserio (1977). Basic Principles of Organic Chemistry, second edition. Adapted with the permission from W. A. Benjamin, Inc., Menlo Park, CA. ISBN 0-8053-8329-8.

Also, lack of strong absorption band in the region 900 -650 cm^{-1} generally indicates a non-aromatic structure. If the band of the absorption spectrum extended to 1000 cm^{-1}, it indicates the presence of alkene group.

(III) Finger print region: The intermediate region 1300 - 900 cm^{-1}, is known as finger print region. This portion of the band is extremely important, than that of other regions. For example, if alcoholic or phenolic O—H stretching absorption appears in the high frequency region of the spectrum, the position of C—C—O absorption band in the 1260 - 100 cm^{-1} region frequently make it possible to assign the O—H absorption to alcohols and phenols with highly specific structures. Absorption in this intermediate region is probably unique for every molecular species.

Any conclusion reached after examination of the particular band should be confirmed where possible by examination of other portion of the spectrum. For example, the assignment of carbonyl bond to an aldehyde should be confirmed by the appearance of a band or a pair of bands in the 2900 - 2695 cm^{-1} region of the spectrum, arising from C−H stretching vibrations of the aldehyde group. Similarly assignment of carbonyl bond to an ester should be confirmed by observation of a strong band in the C−O stretching region, 1300 - 1100 cm^{-1}.

Finally, in the "finger print" comparison of spectra, or any other situation in which the shapes of peaks are important, we should also check the appearance of the spectrum both in wavenumber (cm^{-1}) and wavelength (μm): There should be substantial difference between two spectra (Fig. 7.20).

Table: The following table lists **infrared spectroscopy absorptions** by frequency regions.

(A)

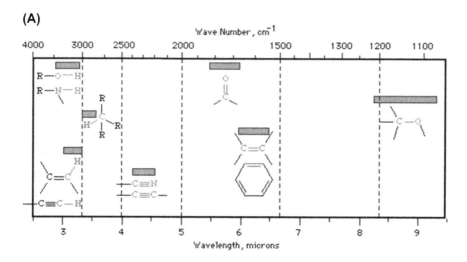

(B)

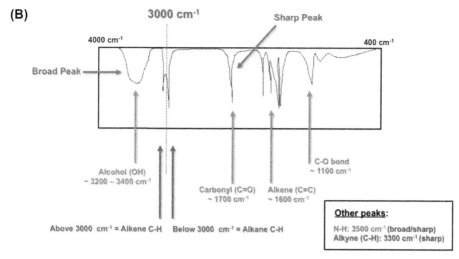

FIG. 7.20

Most commonly used IR values in organic chemistry.

4000-3000 cm^{-1}						
Wavenumber	**Intensity**	**Peak width**	**Group**	**Vibration type**		**Free/bonded**
3700–3584	Medium	Sharp	O–H	Stretching	Alcohol	Free
3550–3200	Strong	Broad	O–H	Stretching	Alcohol	Intermolecular bonded
3500 3400	Medium		N–H	Stretching	Primary amine	
3400–3300 3330–3250	Medium		N–H	Stretching	Aliphatic primary amine	
3350–3310	Medium		N–H	Stretching	Secondary amine	
3300–2500	Strong	Broad	O–H	Stretching	Carboxylic acid	Usually centered on 3000 cm^{-1}
3200–2700	Weak	Broad	O–H	Stretching	Alcohol	Intramolecular bonded
3000–2800	Strong	Broad	N–H	Stretching	Amine salt	
3000-2500 cm^{-1}						
3333–3267	Strong	Sharp	C–H	Stretching	Alkyne	
3100–3000	Medium		C–H	Stretching	Alkene	
3000–2840	Medium		C–H	Stretching	Alkane	
2830–2695	Medium		C–H	Stretching	Aldehyde	Doublet
2600–2550	Weak		S–H	Stretching	Thiol	
2400-2000 cm^{-1}						
2349	Strong		O=C=O	Stretching	Carbon dioxide	
2275–2250	Strong	Broad	N=C=O	Stretching	Isocyanate	
2260–2222	Weak		C≡N	Stretching	Nitrile	
2260–2190	Weak		C≡C	Stretching	Alkyne	Disubstituted
2175–2140	Strong		S-C≡N	Stretching	Thiocyanate	
2160–2120	Strong		N=N=N	Stretching	Azide	
2150			C=C=O	Stretching	Ketene	
2145–2120	Strong		N=C=N	Stretching	Carbodiimide	
2140–2100	Weak		C≡C	Stretching	Alkyne	Monosubstituted
2140–1990	Strong		N=C=S	Stretching	Isothiocyanate	

—cont'd							
4000-3000 cm^{-1}							
Wavenumber	**Intensity**	**Peak width**	**Group**	**Vibration type**			**Free/bonded**
2000−1900	Medium		C=C−C	Stretching	Allene		
2000			C=C=N	Stretching	Ketenimine		
2000-1650 cm^{-1}							
2000−1650	Weak		C−H	Bending	Aromatic compound		Overtone
1870-1540 cm^{-1}							
1818 1750	Strong		C=O	Stretching	Anhydride		
1815−1785	Strong		C=O	Stretching	Acid halide		
1800−1770	Strong		C=O	Stretching	Conjugated acid halide		
1775 1720	Strong		C=O	Stretching	Conjugated anhydride		
1770−1780	Strong		C=O	Stretching	Vinyl/phenyl ester		
1760	Strong		C=O	Stretching	Carboxylic acid		Monomer
1750−1735	Strong		C=O	Stretching	Esters		6-Membered lactone
1750−1735	Strong		C=O	Stretching	δ-lactone		γ: 1770
1745	Strong		C=O	Stretching	Cyclopentanone		
1740−1720	Strong		C=O	Stretching	Aldehyde		
1730−1715	Strong		C=O	Stretching	α,β-unsaturated ester		Or formates
1725−1705	Strong		C=O	Stretching	Aliphatic ketone		Or cyclohexanone or cyclopentenone
1720−1706	Strong		C=O	Stretching	Carboxylic acid		Dimer
1710−1680	Strong		C=O	Stretching	Conjugated acid		Dimer
1710−1685	Strong		C=O	Stretching	Conjugated aldehyde		
1690	Strong		C=O	Stretching	Primary amide		Free (associated: 1650)
1690−1640	Medium		C=N	Stretching	Imine/oxime		
1685−1666	Strong		C=O	Stretching	Conjugated ketone		
1680	Strong		C=O	Stretching	Secondary amide		Free (associated: 1640)

Continued

—cont'd

				4000-3000 cm^{-1}			
Wavenumber	**Intensity**	**Peak width**	**Group**	**Vibration type**			**Free/bonded**
1680	Strong		C=O	Stretching	Tertiary amide		Free (associated: 1630)
1650	Strong		C=O	Stretching	δ -lactam		γ: 1750−1700 β: 1760-1730

1670-1600 cm^{-1}

1678−1668	Weak		C=C	Stretching	Alkene		Disubstituted (trans)
1675−1665	Weak		C=C	Stretching	Alkene		Trisubstituted
1675−1665	Weak		C=C	Stretching	Alkene		Tetrasubstituted
1662−1626	Medium		C=C	Stretching	Alkene		Disubstituted (cis)
1658−1648	Medium		C=C	Stretching	Alkene		Vinylidene
1650−1600	Medium		C=C	Stretching	Conjugated alkene		
1650−1580	Medium		N−H	Bending	Amine		
1650−1566	Medium		C=C	Stretching	Cyclic alkene		
1648−1638	Strong		C=C	Stretching	Alkene		Monosubstituted
1620−1610	Strong		C=C	Stretching	α,β-unsaturated ketone		

1600-1300 cm^{-1}

1550−1500 1372−1290	Strong		N−O	Stretching	Nitro compound		
1465	Medium		C−H	Bending	Alkane		Methylene group
1450 1375	Medium		C−H	Bending	Alkane		Methyl group
1390−1380	Medium		C−H	Bending	Aldehyde		
1385−1380 1370−1365	Medium		C−H	Bending	Alkane		Gem dimethyl

1400-1000 cm^{-1}

1440−1395	Medium		O−H	Bending	Carboxylic acid		
1420−1330	Medium		O−H	Bending	Alcohol		
1415-1380 1200-1185	Strong		S=O	Stretching	Sulfate		
1410−1380 1204−1177	Strong		S=O	Stretching	Sulfonyl chloride		
1400−1000	Strong		C−F	Stretching	Fluoro compound		

—cont'd						
4000-3000 cm^{-1}						
Wavenumber	**Intensity**	**Peak width**	**Group**	**Vibration type**		**Free/bonded**
1390—1310	Medium		O—H	Bending	Phenol	
1372—1335 1195—1168	Strong		S=O	Stretching	Sulfonate	
1370-1335 1170-1155	Strong		S=O	Stretching	Sulfonamide	
1350—1342 1165—1150	Strong		S=O	Stretching	Sulfonic acid	Anhydrous hydrate: 1230-1120
1350-1300 1160-1120	Strong		S=O	Stretching	Sulfone	
1342—1266	Strong		C—N	Stretching	Aromatic amine	
1310—1250	Strong		C—O	Stretching	Aromatic ester	
1275-1200 1075-1020	Strong		C—O	Stretching	Alkyl aryl ether	
1250—1020	Medium		C—N	Stretching	Amine	
1225-1200 1075-1020	Strong		C—O	Stretching	Vinyl ether	
1210—1163	Strong		C—O	Stretching	Ester	
1205—1124	Strong		C—O	Stretching	Tertiary alcohol	
1150—1085	Strong		C—O	Stretching	Aliphatic ether	
1124—1087	Strong		C—O	Stretching	Secondary alcohol	
1085—1050	Strong		C—O	Stretching	Primary alcohol	
1070—1030	Strong		S=O	Stretching	Sulfoxide	
1050—1040	Strong	Broad	CO—O—CO	Stretching	Anhydride	
1000-650 cm^{-1}						
995-985 915-905	Strong		C=C	Bending	Alkene	Monosubstituted
980—960	Strong		C=C	Bending	Alkene	Disubstituted (trans)
895—885	Strong		C=C	Bending	Alkene	Vinylidene
850—550	Strong		C—Cl	Stretching	Halo compound	
840—790	Medium		C=C	Bending	Alkene	Trisubstituted
730—665	Strong		C=C	Bending	Alkene	Disubstituted (cis)
690—515	Strong		C—Br	Stretching	Halo compound	
600—500	Strong		C—I	Stretching	Halo compound	
900-700 cm^{-1}						
880 ± 20 810 ± 20	Strong		C—H	Bending		1,2,4-Trisubstituted

Continued

—cont'd

4000-3000 cm^{-1}

Wavenumber	Intensity	Peak width	Group	Vibration type		Free/bonded
880 ± 20 780 ± 20 (700 ± 20)	Strong		C−H	Bending		1,3-Disubstituted
810 ± 20	Strong		C−H	Bending		1,4-Disubstituted or 1,2,3,4-tetrasubstituted
780 ± 20 (700 ± 20)	Strong		C−H	Bending		1,2,3-Trisubstituted
755 ± 20	Strong		C−H	Bending		1,2-Disubstituted
750 ± 20 700 ± 20	Strong		C−H	Bending		Monosubstituted Benzene derivative

7.12.2 Qualitative application: spectral searching method (for identifying unknown or new compound)

A variety of rules have been developed to correlate infrared absorption bands to chemical structure. For example a carbonyl's C=O stretch is sensitive to adjacent functional groups, occurring at 1650 cm^{-1} for acids, 1700 cm^{-1} for ketones, and 1800 cm^{-1} for acid chlorides.

With the availability of computerized data acquisition and storage it is possible to build digital libraries of standard reference spectra. The identity of an unknown compound can often be determined by comparing its spectrum against a library of reference spectra, a process is known as **spectral searching**. Comparisons are made using an algorithm that calculates the cumulative difference between the sample's spectrum and a reference spectrum. For example, one simple algorithm uses the following equation:

$$D = \sum_{i=1}^{n} \left| (A_{sample})_i - (A_{reference})_i \right|$$

where D is the cumulative difference, A_{sample} is the sample's absorbance at a particular wavelength or wavenumber i, $A_{reference}$ is the absorbance of the reference compound at the same wavelength or wavenumber, and n is the number of digitized points in the spectra. The cumulative difference is calculated for each reference spectrum. The reference compound with the smallest value of D provides the closest match to the unknown compound. The accuracy of spectral searching is limited by the number and type of compounds included in the library, and by the effect of the sample's matrix on the spectrum.

Another advantage of computerized data acquisition is the ability to subtract one spectrum from another. When coupled with spectral searching it may be possible, by repeatedly searching and subtracting reference spectra, to determine the identity of several components in a sample without the need of a prior separation step. An example as shown in Fig. 7.21 in which the composition of a two-component mixture is determined by successive searching and subtraction.

Fig. 7.21A shows the spectrum of the mixture. A search of the spectral library selects cocaine · HCl (Fig. 7.21B) as a likely component of the mixture. Subtracting the reference spectrum for cocaine HCl from the mixture's spectrum leaves a result (Fig. 7.21C) that closely matches mannitol's reference spectrum (Fig. 7.21D). Subtracting the reference spectrum for leaves only a small residual signal (Fig. 7.21E). This searching and subtracting method confirms the presence of cocaine and mannitol.

7.12.3 Quantitative applications

Although infrared absorption based analysis for quantitative analysis has not been frequently used, it is useful for the analysis of chemical species such as organic vapors, including HCN, SO_2, nitrobenzene, methyl mercaptan, and vinyl chloride. Frequently, these analyses are accomplished using portable, dedicated infrared photometers. However, this kind of analysis is less frequently encountered than those for UV/Vis absorption. The reasons for less suitability are due to the following reasons

1. One reason is the greater tendency for instrumental deviations from Beer's law when using infrared radiation. Because an infrared absorption band is relatively narrow, any deviation due to the lack of monochromatic radiation is more pronounced.
2. In addition, infrared sources are less intense than UV/Vis sources, making stray radiation more of a problem.
3. Differences in pathlength for samples and standards when using thin liquid films or KBr pellets are a problem, although an internal standard can be used to correct for any difference in pathlength.
4. Finally, establishing a 100% T ($A = 0$) baseline is often difficult because the optical properties of NaCl sample cells may change significantly with wavelength due to contamination and degradation. However, we can minimize this problem by measuring absorbance relative to a baseline established for the absorption band. Fig. 7.18 shows how this is accomplished. *Note:* Another approach is to use a cell with a fixed path length, such as that shown in Fig. 7.22.

Note: A precise estimation of the vibrations of a complex molecule is not feasible; thus the IR spectrum must be interpreted from empirical comparison of spectra and extrapolation of studies of simpler molecules.

7.12.4 Application of FTIR in material characterization

A key requirement of material science research is the ability to identify and fully characterize all the members of a product's family and their applications, both at the lab- and at mass-production scale. However, to be appealing, a characterization tool must be non-destructive, fast, with high resolution and give the maximum information on structural and electronic behaviors. Beside their applications in

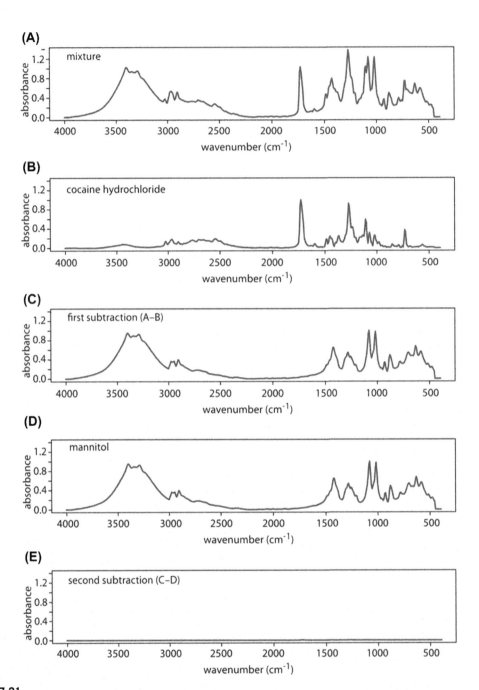

FIG. 7.21

An example of determination of two-component mixture, by successive searching and subtraction shows the spectrum of the mixture.

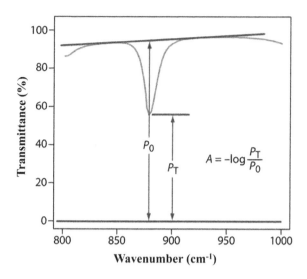

FIG. 7.22

Method for determining absorbance from an IR spectrum.

studying the fundamental properties in molecular level Raman-, FTIR-, UV–Vis-, photo-luminescence- spectroscopy are some of the most widely used tools in material science research as well. Below, we discuss the application of FTIR spectroscopic techniques to characterize some of the versatile materials which have tremendous possibilities in developing new devices for future technology such nanomaterial based solar cells, light emitting diodes, transistors etc.

To note, as in the case of gas and liquid phase, FTIR spectroscopy has been employed to confirm the presence/absence of functional groups prevailing in solid materials as well. The position and nature of a peak (peak height, width and shape) in FTIR spectrum indicates the presence of certain functional group and effect of surrounding on it, respectively.

Indeed, graphene family has already many members with unique properties such as graphene oxide (GO), reduced graphene oxide (rGO), carbon nanotubes (CNTs), graphene based quantum dots (GQDs), metal/inorganic element doped GO, and graphene based composite materials (GCMs). Actually graphene and GO, rGO, CNTs, GQDs, GCMs have already been tested for their potential applications in solar cell (e.g., organic dye, methyl ammonium leadhalide perovskite, PSC), light emitting diode, optoelectronics. Knowing fundamental characteristics of this versatile class of carbon is very crucial for both the fundamental and application points of views.

Here, as examples, we discuss about understanding the properties with FTIR of graphene and its derivatives (products that are synthesized/produced, considering graphene as a starting material) and also of metal halide perovskite thin films, as these materials found to be quite useful to apply in devising future technology such as solar cell, LEDs and electronic devices.

7.12.4.1 Highlights of properties of some promising materials and their applications: graphene based materials (GBMs)

Before, discussing the information/features that we could extract of GBMs with FTIR, lets get an overview of their characteristic features and directions of applications in devising new technologies.

The GBMs family members, such as graphene, GO and rGO, stand out for their good electronic properties, processability, low-costs, possibility to use green-chemistry methods for its production, and suitability for large scale applications. Particularly, these materials have been considered serious alternatives to the organic charge transport materials in above mentioned devices: For example, evidences have shown that pristine graphene itself found to be appropriate for stable transparent conductor (as an electrical contact), whereas the GO and rGO found to be effective for electron and hole transport purposes, respectively, in thin film based organo-metallic solar cells and light emitting diode. The reasons for adaption of these graphene derivative as a charge carrier is mainly due to their possession of unique electronic, optical (high optical transparency: transmittance above 90%, tunable band gap, from zero to insulator level) and electrical properties (sheet resistance of undoped graphene films is in the range of $150-500$ Ω/sq. and high conductivity (10^4 S/cm).

In this context, studies from last 5 years inferred us that substitution of organic charge transport materials with inorganic one (both the ETL and HTL) could be the most appropriate approach to address the underlined issues of PSCs. Because, charge transport layers from inorganic materials (for both the n- and p-type semiconductors) found to possess high chemical stability, high charge mobility and are low cost, when compared with organic counterparts. As highlighted below, some of these inorganic charge carrying materials considered to be promising are of course GBMs.

Recently published review article highlights possibilities of integration of GBMs in perovskite solar cells. For example, rGO found to be quite effective as HTL (hole receiving layer and electron blocking layer) in perovskite solar cells. Although the observed efficiency of PSCs with this material as HTL ($\eta = $ ca.16%) is less by about 30% than that of PSCs (ca. 23%) employing PEDOT:PSS as HTL. Besides, rGO based solar cells can withstand against environmental parameters and also protect perovskite layer (act as an overcoat layer). Below, we discuss investigation of this versatile family with FTIR spectroscopy. Later at the end of this section, we will also give signature of perovskite in FTIR.

7.12.4.2 FTIR spectroscopy of GBMs family

As illustrated in Fig. 7.23, graphene oxide layers deposited on a substrate consists of the attachment of hydroxyl (O—H), carboxyl (C=O) alkoxy (C—O) groups and (carbon-carbon double bonding (C=C) due to which it is highly soluble in water, allowing exfoliation into individual sheets. Besides, there are also possibilities of integrating (attachment by the route of chemical bond) other functional group (known as functionalization) for modifying chemical/physical properties of graphene layers (see Fig. 7.23: A). The availability of these oxygen-containing functional groups on the basal plane and the sheet edge allows GO to interact with a wide range of organic and inorganic materials in non-covalent, covalent and/or ionic manner so that functional hybrids and composites with unusual properties can be readily synthesized. Also to note, as shown in Fig. 7.23B, due to presence of covalently bound oxygen and the displacement of the sp^3-hybridized carbon atoms slightly above and below the original graphene plane, the graphene oxide sheets are expected to be 'thicker' than the pristine graphene sheet (which is atomically flat with a well-known van der Waals thickness of about 0.34 nm).

Fig. 7.24 (a: left) represents the FT-IR spectrum of graphene oxide in solution phase. Apparently, one can clearly observe the peaks in FT-IR spectrum of GO at the wave numbers ~ 1720 cm^{-1}, 1410 cm^{-1}, 1226 cm^{-1} and 1050 cm^{-1} which are related to the stretching vibration of C=O, C—O—H, C—O (epoxy) and C—O (alkoxy) functional groups, respectively. Also, reported were other two peaks observed at about ~ 1558 cm^{-1} and 3600 cm^{-1} which are assigned to C=C and O—H stretching, respectively. After reduction process, the loss (partially) of oxygen-containing functional

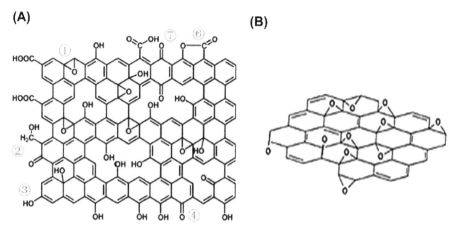

FIG. 7.23

(A) Functionalization possibilities for graphene, including varieties of oxygen containing functional groups: (1) edge-functionalization, (2) basal-plane-functionalization, (3) non-covalent adsorption on the basal plane, (4) asymmetrical functionalization of the basal plane, and (5) self-assembling of functionalized graphene sheets. (B) graphene Oxide sheet (only with epoxy functional group lying at the edge and above and below the graphene sheet.

(A) Dreyer et al., The chemistry of graphene oxide, Chem. Soc. Rev., 2010, 39, Adapted with the permission from The Royal Society of Chemistry 2010, and (B) Szabó et al, et al., Chem. Mater. 2006, 18. Adapted with the permission © 2006 American Chemical Society.

groups is expected, as depicted in Fig. 7.23 (A: right). FTIR spectrum of graphite possesses OH strong OH vibration, which mainly expected to come from humidity and/or adsorbed water and this OH signal changes its shape and becomes wider upon graphite oxidation. But, becomes narrower after reduction of GO. Besides, this technique has also been utilized to study the degree of reduction of GO films by analyzing the number of oxygen-containing functional groups present on the surface of rGO by observing the results for various modes of vibration.

Fig. 7.24B compares the FTIR spectra of graphene (red spectrum) and GO (blue spectrum). Mainly, three strong peaks associated with $C{-}O$ (1087 cm^{-1}), $C{-}O{-}H$ (1404 cm^{-1}), $-O{-}H$ (3410 cm^{-1}) are clearly visible in the spectrum of GO, which are missing or are very week in graphene, as expected.

7.12.4.3 Overview of unique properties and applications of methylammonium leadhalide (MALH) perovskite in thin film based solar cells

MALHs perovskite are solid compounds with perovskite structure with a chemical formula of $CH_3NH_3PbX_3$ (MAPbX$_3$), where X = I, Br or Cl. In the $CH_3NH_3PbX_3$ crystal structure the methyl-ammonium cation ($CH_3NH_3^+$) is surrounded by PbX$_6$ octahedra.

Solar cell technology based on thin film of organic$-$inorganic metal halide perovskite materials is considered to be one of the most promising for harnessing solar energy to fulfill the energy needs in a sustainable manner. One of the main reasons to consider perovskite based photocells promising, specifically, is due to more intensely absorption of solar radiation over a broader region (covers all of the visible portion, all the way to about 750 nm) of the solar spectrum than other light absorbers

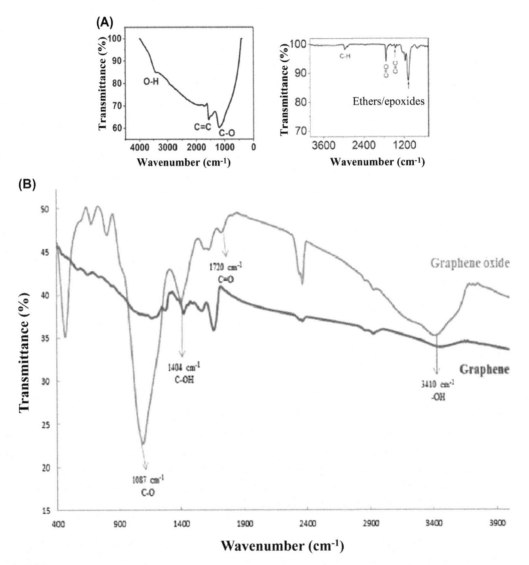

FIG. 7.24

(A) FTIR spectrum of GO (left: Solution) and rGO (right: Thin film). (B) Comparison of FTIR spectra of graphene oxide and graphene.

(A): (Left): [A. J. S. Ahammad et al., J. Electrochem. Soc. 165(5), B174-B183 (2018), With permission from © 2018 The Electrochemical Society. (Right)]: M. Savchak et al., N. Borodinov, et al., ACS Appl. Mater. Interfaces, 10 (4), 3975–3985 (2018).With permission from American Chemical Society. (B): Jafer Ciplak, et al., Fullerenes, Nanotubes and Carbon Nanostructures (2014) 23, 361–370, With permission from Taylors & Francis Group, LLC.

commonly used in solar cells. Besides, it exhibits very long electron hole diffusion length so that most of the photo-generated charge carriers are collected and the solar cells of this kind can be fabricated with inexpensive precursor materials with the simple solution-processing at low temperatures (Fig. 7.25).

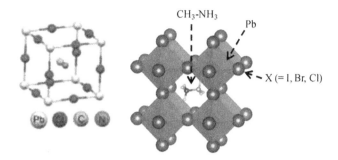

FIG. 7.25

Left; Simple schematic arrangements of atoms and functional groups in perovskite molecule. Right: Crystal Structure of Perovskite. Here, Methylammonium cation ($CH_3NH_3^+$) occupies the central site surrounded by 12 nearest-neighbour iodide ions in corner-sharing PbI_6 octahedra.

Taking advantage of above indicated properties, indeed, achievement of high efficiency of perovskite solar cells (PSCs) even with the thinner than other types of solar cells (due to which material cost is reduced) has already been made. With the use of this crystalline material as a sensitizer, the photo conversion efficiency (PCE) increased from of 3.8% in 2009 to 23.3% in late 2018 in single-junction architectures, which is even higher than the maximum efficiency achieved in single-junction silicon solar cells. Perovskite solar cells are therefore the fastest-advancing solar technology to date within the time period of 10 year.

In the PSCs, primarily, a thin film of the MALH perovskite, $CH_3NH_3PbX_3(X = Cl, Br, I)$, molecule is sandwiched between electron transport layer (ETL), and hole transport layer (HTL). When perovskite thin film is exposed to solar radiation, as elucidated **in** Fig. 7.26, it absorbs the solar radiation and the electron-hole pairs (known as excitons) are generated, which then separated and migrated to opposite directions: the photo-electron migrates toward ETL and hole migrates toward HTL. To collect electrons from these electron−hole pairs created by photon interaction at a perovskite/

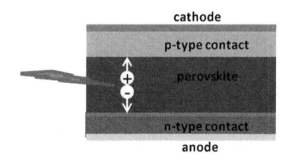

FIG. 7.26

Schematic view of a thin-film perovskite solar cell having a flat layer of perovskite sandwiched between n-type and p-type charge selective contacts. The generation of electron and hole pairs and extraction process is also demonstrated.

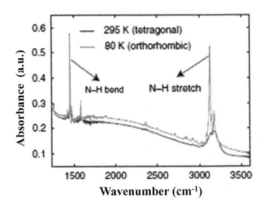

FIG. 7.27

Experimental infrared absorbance at 295 K (tetragonal phase) and 80 K (orthorhombic phase).

Peijun Guo et al., Nature Communications, Springer Nature (2018).

ETL interface, usually mesoporous TiO_2 nanoparticles (a versatile inorganic compound semi-conductor with a wide bandgap) have been used. While on the other hand, mostly, the holes are transported via organic p-type semiconducting mediators (e.g., spiro-MeOTAD) for collection at counter electrode.

FTIR spectroscopy of MALH perovskite: Fig. 7.27 shows the infrared absorbance of a 680-nm thick $MAPbI_3$ film in the tetragonal (295 K) and orthorhombic (80 K) phases, which reveal phonon absorption by the MA^+ cations. The two strong peaks centered between 3100 and 3200 cm^{-1} are assigned to the N–H stretching modes which exhibit strong light-induced changes in dipole moments. For the orthorhombic phase of $MAPbI_3$, these two strong peaks arise from the asymmetric N–H stretching motions with the lower and higher-frequency modes, respectively (Fig. 7.28).

The experimental spectra are stacked for better viewing, and the intensities are scaled by the factors indicated on the graphs. For the DFT calculations (orthorhombic phase), researchers included both the computed normal modes of vibrations with their relative intensities (shown in bars) and the theoretical spectrum obtained with a Lorentzian broadening (shown in solid black line) with an FWHM of 5 cm^{-1}.

Single crystal of the $MAPbX_3$ have been investigated using infrared spectroscopy with the aim of analyzing structural and dynamical aspects of processes that enable the ordering of the MA molecule in the orthorhombic crystal structures of these hybrid perovskites. Researchers' temperature-dependent study was focused on the analysis of the CH/NH rocking, C–N stretching, and CH/NH bending modes of the MA molecule in the 800-1750 cm^{-1} frequency range. Fig. 7.29 clearly shows the behavior of these three kinds of halides on crossing the orthorhombic-tetragonal phase transition in MA lead halide crystals. Specifically, drastic changes of all vibrational modes close to the phase transition can be clearly seen. For low temperatures, it can be stated that the iodide is more strongly influenced by hydrogen bonding than the bromide and the chloride.

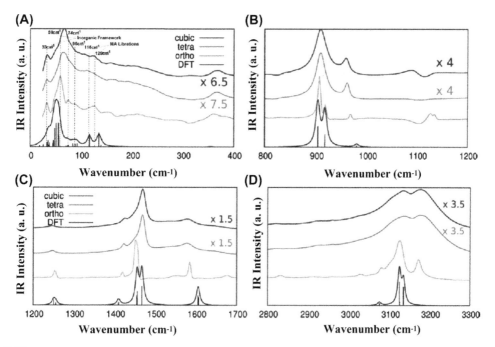

FIG. 7.28

Experimental IR spectra of MAPbI$_3$ recorded at four different frequency regions for the orthorhombic phase (ortho) at 80 K, the tetragonal phase (tetra) at 293 K, and the cubic phase (cubic) at 340 K, compared with the DFT spectra (DFT) calculated for the orthorhombic phase. (A) IR spectrum below 400 cm^{-1}; (B) IR spectrum between 800 and 1200 cm^{-1}; (C) IR spectrum between 1200 and 1700 cm^{-1}; (D) IR spectrum between 2800 and 3300 cm^{-1}.

Dr. T. Ivanovska et al., ChemSusChem, 9 (2016) 1–12, Adapted with permission from © 2016 Wiley-VCH Verlag GmbH & Co. KGaA, Weinheim.

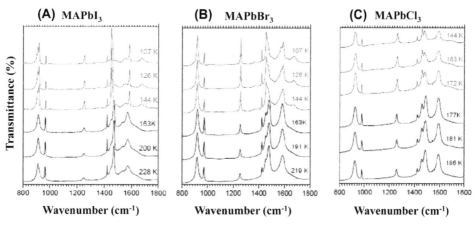

FIG. 7.29

Temperature-dependent infrared spectra of methyl ammonium lead halides perovskites between 800 cm^{-1} and 1800 cm^{-1}.

Schuck et al., J. Phys. Chem. C, 2018, 122 (10), 5227–5237, Adapted with permission from © 2018, American Chemical Society.

7.13 Problems

1. Another historically important series of spectral lines is the Balmer series of emission lines form hydrogen. One of the lines has a wavelength of 656.3 nm. What are the frequency and the wavenumber for this line?

2. What types of energy transitions in a molecule are involved in the absorption of light in IR spectroscopy? How are these energy transitions different from those that are used in UV−Vis spectroscopy?

3. Describe what happens to the motions within a molecule when this molecule absorbs infrared radiation.

4. How is IR spectroscopy typically used for chemical analysis? How does the typical application of IR spectroscopy differ from UV−Vis spectroscopy?

5. Why a glass or quartz prism cannot be used in IR spectroscopy?

6. What types of materials are used for sample holders in IR spectroscopy?

7. Describe how do you prepare liquid for IR spectroscopic analysis. Also discuss the method for solid samples preparations.

8. Describe how IR spectroscopy can be used for the identification of chemicals. What features of an IR spectrum are useful for this type of application?

9. What is a "correlation chart"? Explain how you can use this type of chart for chemical identification in IR spectroscopy.

10. A student receives an unknown organic compound that is either cyclohexane or cyclohexene. Explain how the student could use the IR spectroscopy for identifying the molecule.

11. A compound is known to have either a carbon-carbon double bond or triple bond in its structure. An IR spectrum for this chemical has a sharp peak at 2200 cm^{-1}, but nothing at 1650 cm^{-1}. Determine whether a double or triple carbon-carbon bond is present in this compound.

12. A can of paint solvent that is found in the scene of an arson attempt is believed by a forensic laboratory to be either a mixture of hydrocarbons or acetone, $(CH3)2C{=}O$. An IR spectrum for a sample of this solvent is found to show no appreciable absorbance in the reason of 1700 cm^{-1}. Of the given possibilities, what is the most likely identity for this paint solvent?

Further reading

[1] R.M. Silverstein, F.X. Webster, Spectroscopic Identification of Organic Compounds, sixth ed., John Wiley & Sons, Inc., New Delhi, 1976. Reprint by Wiley India Pvt. Ltd.

[2] Chemistry LibreText, 2019 visited on 26/2/2019, <https://chem.libretexts.org/Bookshelves/Analytical_ Chemistry/Book%3A_Analytical_Chemistry_2.0_(Harvey)/10_Spectroscopic_Methods/10.3%3A_UV%2F %2FVis_and_IR_Spectroscopy>.

[3] P.W. Atkins, Molecular Quantum Mechanics, Oxford University Press, 1970.

[4] W. Guillory, Introduction to Molecular Structure and Spectroscopy, Allyn & Bacon, Inc., Boston, MA, 1977.

[5] W. Struve, Fundamentals of Molecular Spectroscopy, Wiley, 1989.

[6] D. Harris, M. Bertolucci, Symmetry and Spectroscopy, Oxford University Press, 1978.

[7] F.A. Cotton, F. Albert, Chemical Applications of Group Theory, 3rd., Wiley, 1990.

[8] M. Horak, A. Vitek, Interpretations and Processing of Vibrational Specta, Wiley, 1978.

[9] M. Diem, Introduction to Modern Vibrational Spectroscopy, Wiley, 1993.

[10] P. Kondratyuk, Analytical formulas for Fermi resonance interactions in continuous distributions of states, Spectrochim. Acta A Mol. Biomol. Spectrosc. 61 (4) (2005) 589−593 (Web).

[11] S. Dragon, A. Fitch, Infrared spectroscopy determination of lead binding to ethylenediaminetetraacetic acid, J. Chem. Educ. 75 (1998) 1018−1021.

[12] H. Frohlich, Using infrared spectroscopy measurements to study intermolecular hydrogen bonding, J. Chem. Educ. 70 (1993) A3−A6.

[13] N. Garizi, A. Macias, T. Furch, R. Fan, P. Wagenknecht, K.A. Singmaster, Cigarette smoke analysis using an inexpensive gas-phase IR cell, J. Chem. Educ. 78 (2001) 1665−1666.

[14] R. Indralingam, A. I Nepomuceno, The use of disposable IR cards for quantitative analysis using an internal standard, J. Chem. Educ. 78 (2001) 958−960.

[15] L.J. Mathias, M.G. Hankins, C.M. Bertolucci, T.L. Grubb, J. Muthiah, Quantitative analysis by FTIR: thin films of copolymers of ethylene and vinyl acetate, J. Chem. Educ. 69 (1992) A217−A219.

[16] J.D. Schuttlefield, V.H. Grassian, ATR-FTIR spectroscopy in the undergraduate chemistry laboratory. Part I: fundamentals and examples, J. Chem. Educ. 85 (2008) 279−281.

[17] J.D. Schuttlefield, S.C. Larsen, V.H. Grassian, ATR-ftir spectroscopy in the undergraduate chemistry laboratory. Part II: a physical chemistry laboratory experiment on surface adsorption, J. Chem. Educ. 85 (2008) 282−284.

[18] M.B. Seasholtz, L.E. Pence, O.A. Moe Jr., Determination of carbon monoxide in automobile exhaust by FTIR spectroscopy, J. Chem. Educ. 65 (1988) 820−823.

[19] A. Skoog, F.J. Holler, S.R. Crouch, Instrumental Analysis, Cengage Learning, New Delhi, 2007, p. 420 (Reprint).

[20] D.S. Hage, J.D. Carr, Analytical Chemistry and Quantitative Analysis, Pearson Publication (Int. Ed.), 2011, p. 442.

[21] A.J.S. Ahammad, T. Islam, M.M. Hasan, M.N.I. Mozumder, R. Karim, N. Odhikari, Poly rani pal, subrata sarker, dong min kim, J. Electrochem. Soc. 165 (5) (2018) B174−B183.

[22] N.A. Kumar, S. Gambarelli, F. Duclairoir, G. Bidan, L. Dubois, J. Mater. Chem. 1 (2013) 2789.

[23] J.I. Paredes, S. Villar-Rodil, A. Martínez-Alonso, J.M.D. Tascón, Langmuir 24 (2008) 10560.

Raman spectroscopy

8

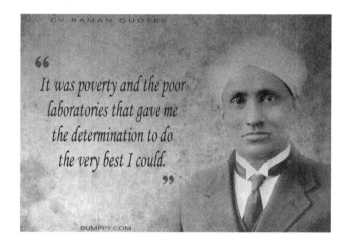

Raman spectroscopy is named after Indian physicist Sir C. V. Raman who was awarded Nobel prize in physics in 1930. C. V. Raman who, with K. S. Krishnan studied an effect, the difference in energy between the light from the original source and the scattered light is equal to the difference in energy of vibrational levels in a molecule. Such scattering of radiation was first experimentally observed in liquids in 1928, coinciding with the theoretical prediction by A. Smekal in 1923. However, due to its very low scattering efficiency, Raman spectroscopy did not become popular until powerful laser systems were available after the 1960s. Now, Raman spectroscopy has become one of the most popular approaches to study the vibrational structures of molecules together with an infrared spectrum.

8.1 Introduction

Raman spectroscopy is a spectroscopic technique commonly used in chemistry and material science to provide a structural fingerprint by which molecules can be identified. It is considered a complementary analytical tool of IR spectroscopy as it also provides information about vibrational transitions in a molecule. It is based on the induced molecular polarizability, after interaction of the molecule with laser light.

Chemical Analysis and Material Characterization by Spectrophotometry. https://doi.org/10.1016/B978-0-12-814866-2.00008-7

When a beam of monochromatic light interacts with a molecule, a portion of the radiation energy is scattered. This scattering process takes only about 10^{-14} s or less. During this brief period of time, the molecule is temporarily raised to a higher energy level called a ***"virtual state or excited state"*** and the molecule returns to a lower energy state after the light is scattered. And the energy of scattered radiation will consist almost entirely of the frequency of incident radiation (i.e., most of these molecules return to their original energy level which gives the incoming light and scattered light the same wavelength (a process known as "Rayleigh scattering"). But, in addition, *certain discrete frequencies above and below that of the incident beam will also be scattered.* This is because a molecule will undergo a change in vibrational and rotational levels during the scattering process. This change means the scattered light and incoming light will now have a difference in energy that is equal to the energy involved in the vibrational transition. This effect provides these two types of light with slightly different wavelengths. This phenomenon is referred to as *Raman scattering*. The general process of Raman scattering is illustrated in Fig. 8.1. These changes are detected as lines falling both above and below the wavelength of the incident light and are quite small. Therefore, can only be seen when using incoming light that is monochromatic. Sir Raman conducted his experiments using an intense mercury discharge lamp, but modern instruments for Raman spectroscopy use laser light source.

In other words, *Raman effect* is observed as a result of photons being "captured" momentarily by molecules in the sample and giving up (or gaining) small increments of energy through changes in the molecular vibrational and rotational energies before being emitted as scattered light.

From the concept of classical mechanics, the Raman effect arises when a photon is incident on a molecule and interacts with the electric dipole of the molecule which causes to bring changes in polarizability. The vibrational mode will be Raman active only when it changes the ***polarizability*** of the molecule (see below to understand the concept of polarizability) and is observed as a peak in the Raman spectra.

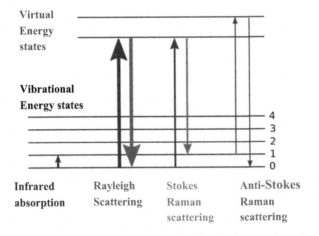

FIG. 8.1

An illustration of the type of energy transitions that occur in Rayleigh scattering versus Raman scattering.

8.2 Quantum theory of Raman effect

Raman scattering can be understood easily with the help Quantum theory. This theory considers radiation of frequency v as a stream of particles (called photons) having energy hv, where h is Planck's constant. Photons can be imagined to undergo collisions with molecules and, if the collision is perfectly elastic, they will be deflected unchanged. A detector placed to collect radiation energy at right angles to the incident beam will thus receive photons of energy hv (that is, the energy of the scattered light also remains the same).

Virtual energy state: At room temperature, most molecules are present in the lowest energy vibrational level. When materials is irradiated with intense electromagnetic radiation of a single frequency it interacts with the molecule and distort (polarize) the electron cloud around the nuclei to form a short-lived high energy state called 'virtual state'. Energies of these virtual states are determined by the frequency of the light source used. These high energy virtual states are very unstable.

In the example shown in Fig. 8.1, a photon of incoming light in each type of scattering has energy equal to E_1. In Rayleigh scattering, the scattered light has also the same energy as the incoming light (Such collision is called elastic collision). In Raman scattering, however, scattered light differs by a value of ΔE from the energy of the incoming light (such interaction between photon and molecule is called inelastic collision). The molecule can gain or lose an amount of energy only in accordance with the quantum laws: its energy change, ΔE, must represent a change in the vibrational/or rotational energy of the molecule. For example, in the type of Raman scattering that is shown in the middle of above Fig. 8.1, the molecule ends at a higher vibrational energy (e.g., molecule initially was at electronically and vibrationally ground state, $v = 0$, then after interaction with photon it transits to *virtual energy state* and comes back to initial electronic state but at higher vibrational state $v = 1$) and therefore the photon loses a very small portion of its energy, is known as "Stokes scattering". That is the photon will have new energy:

$$E_s = hv - \Delta E. \tag{8.1}$$

Conversely, it is also possible for a molecule to lose energy by ending at a lower vibrational state (level), in which case the scattered light gains energy, this process is called "anti-Stokes scattering". In this case, the scattered photon will have energy

$$E_{a-s} = hv + \Delta E(v + \Delta E)/h. \tag{8.2}$$

Here, E_s and E_{a-s}, respectively, represent the energy of emitted photon associated with stokes scattering and anti-Stokes scattering.

8.3 Molecular polarizability and classical theory of the Raman effect
8.3.1 Basic concept of molecular polarizability (polarizability under static electric field)

The classical concept of Raman spectroscopy is based on the molecular polarizability, therefore, before discussing this theory; let us understand the concept of the polarizability and its sources.

Polarizability is a measure of how easily an electron cloud is distorted by an electric field (Typically, the electron cloud will belong to an atom or molecule or ion). The electric field could be caused, for example, by an electrode or a nearby cation or anion. If an electron cloud is easy to distort, we say that the species it belongs to is polarizable. Polarizability, is represented by the Greek letter alpha, α *and* is experimentally measured as the ratio of induced dipole moment μ to the electric field E that induces it:

$$\alpha = \mu/E. \tag{8.3}$$

The units of α are $C\ m^2\ V^{-1}$. Where C, m and V represent the coulomb, meter and V.

Large, negatively charged ions, such as I^- and Br^-, are highly polarizable. Small ions with a high positive charge, such as Mg^{2+} and Al^{3+} have low polarizability, but they have a high ability to polarize polarizable species, such as I^- and Br^-.

In ordinary usage, polarizability refers to the "mean polarizability", over the x,y, z-axes of the molecule. Polarizabilities in different directions (e.g., along with the bond in Cl_2, called "longitudinal polarizability", and in the direction perpendicular to the bond, called "transverse polarizability") can be distinguished, at least in principle.

Below in Fig. 8.2, we show a simple pictorial view of induced electric dipole in bulk materials and simple molecules due to the presence of external electric. Specifically, in panel (a) application of electric field causes to produce net electric dipole (indicated by (+) and (-) sign with red color). Similarly, in the panel (b), in the case of no electric filled molecules are randomly oriented and thus net charge is zero. However, upon application of the external field E_o, it induces certain charge dipole. Similarly, the induction of charge dipole due to the presence of field E_o in benzene and H_2O is shown in the panel (c) and (d), respectively.

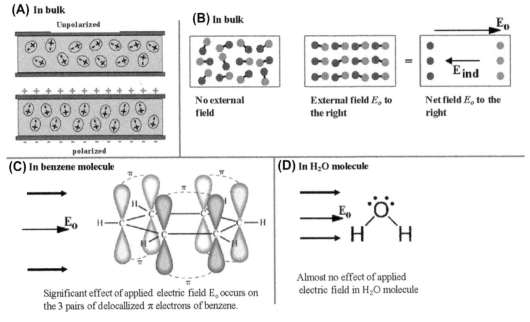

FIG. 8.2

(A) and (B): Upper panel: Representation of induced electric dipole in bulk materials and in molecules
(C) and (D): Lower in the presence of an external electric field.

8.3.2 Sources of polarizability

As indicated above, the central quantity of Raman spectroscopy is the induction of polarization P in the chemical species (or material). *The polarization P is defined as the dipole moment μ per unit volume.* Mathematically, the dipole moment of a system of charges is given by

$$\mu = \sum_i q_i r_i. \tag{8.4}$$

where r_i is the position vector of charge q_i. The simplest case of an electric dipole is a system consisting of a positive and negative charge (see Fig. 8.3) so that the dipole moment is equal to $q*a$. That is:

$$\mu = q * a. \tag{8.5}$$

where a is a vector connecting the two charges (from negative to positive).

This analogy can also be applied to the molecule as well, which is the building block of all substances, as each molecule is composed of both positive charges (nucleus) and negative charges (electrons). When a field acts on a molecule, the positive charges are displaced along the field, while the negative charges are displaced in a direction opposite to that of the field. The effect is therefore to pull the opposite charges apart, i.e., to polarize the molecule (see Fig. 8.2A–D).

There are different types of polarization processes, depending on the structure of the molecules which constitute the solid. If the molecule has a permanent moment, i.e., a moment even in the absence of an electric field, we speak of a dipolar molecule, and a dipolar substance (e.g., water). An example of a dipolar molecule is the H_2O molecule (Fig. 8.3A). The dipole moments of the two OH bonds add vectorially to give a non-vanishing net dipole moment.

Some molecules are non-dipolar, possessing no permanent moments; a common example is a CO_2 molecule in Fig. 8.3 (right). The moments of the two CO bands cancel each other because of the rectilinear shape of the molecule, resulting in a zero net dipole moment. The water molecule has a permanent dipole moment because the two OH bands do not lie along the same straight line, as they do in the CO_2 molecule. The moment thus depends on the geometrical arrangement of the charges, and by measuring the moment one can, therefore, gain information concerning the structure of the molecule. Despite the fact that the individual molecules in a dipolar substance have permanent moments, the net polarization vanishes in the absence of an external field because the molecular moments are randomly oriented, resulting in a complete cancellation of the polarization. Similar to the case depicted in Fig. 8.2, when a field is applied to the substance, however, the molecular dipoles tend to align with the field, which results in a net non-vanishing polarization. This leads to the so-called *dipolar polarizability.*

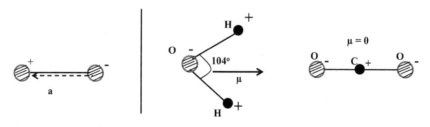

FIG. 8.3

Left: Representation of electric dipole of a system consisting of positive and negative charges. Right: The H_2O molecule and CO_2 molecule.

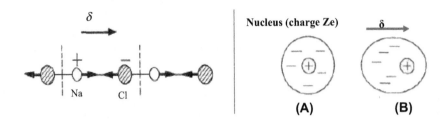

FIG. 8.4

Left: Ionic polarization in NaCl. The field displaces Na$^+$ and Cl$^-$ ions in opposite directions, changing the bond length. Right: Electronic polarization in an individual atom: (A) Unpolarized atom, (B) Atom polarized as a result of the field.

On the other hand, if the molecule contains ionic bonds, then the field tends to stretch the lengths of these bonds. This occurs in NaCl, for instance, because the field tends to displace the positive ion Na$^+$ to the right (see Fig. 8.4), and the negative ion Cl$^-$ to the left, resulting in stretching in the length of the bond. The effect of this change in length is to produce a net dipole moment in the unit cell where previously there was none. Since the polarization here is due to the relative displacements of oppositely charged ions, we speak of *ionic polarizability*.

Ionic polarizability exists whenever the substance is either ionic, as in NaCl, or dipolar, as in H_2O, because in each of these classes there are ionic bonds present. But in substances in which such bonds are missing - such as Si and Ge - ionic polarizability is absent.

The third type of polarizability arises because the individual ions or atoms in a molecule are themselves polarized by the field. In the case of NaCl, each of the Na$^+$ and Cl$^-$ ions are polarized. Thus the Na$^+$ ion is polarized because the electrons in its various shells are displaced to the left relative to the nucleus, as shown in Fig. 8.4B. Such a phenomenon is called *electronic polarizability*.

Electronic polarizability arises even in the case of a neutral atom, again because of the relative displacement of the orbital electrons. Therefore, in general, the total polarizability is given by

$$\alpha_e + \alpha_i + \alpha_d. \tag{8.6}$$

which is the sum of the α_e, α_i, α_d are the electronic, ionic, and dipolar polarizabilities (α_d, respectively. The electronic contribution is present in any type of substance, but the presence of the other two terms depends on the material under consideration.

The relative magnitudes of the various contributions in Eq. (8.6) are such that in non-dipolar, ionic substances the electronic part is often of the same order as the ionic. In dipolar substances, however, the greatest contribution comes from the dipolar part. This is the case for water, as we saw in Fig. 8.3 (middle).

8.3.3 Polarizability due to the interaction of light with a molecule

Another important distinction between the various polarizabilities emerges when one examines the behavior of the ac polarizability that is induced by an alternating field. Fig. 8.5 shows a typical dependence of this polarizability on frequency over a wide range, extending from the static all the way up to the ultraviolet region. It can be seen that in the range from $\omega = 0$ to $\omega = \omega_d$, where ω_d (for dipolar) is some frequency usually in the microwave region, the polarizability is essentially constant.

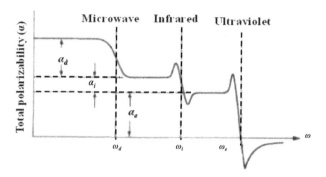

FIG. 8.5

Frequency dependence of the several contributions to the polarizability. One can notice the abrupt change of polarizability value in the vicinity of frequencies $= \omega_d$, ω_i, and ω_e. The values of α_d, α_i and α_e are also shown in the figure, which determines the new contribution of the polarizability of each kind. Moreover, the frequencies ω_d and ω_i, characterizing the dipolar and ionic polarizabilities, respectively, depending on the substance considered.

In the neighborhood of ω_d, however, the polarizability decreases by a substantial amount. This amount corresponds precisely, in fact, to the dipolar contribution α_d. The reason for the disappearance of ω_d in the frequency range $\omega > \omega_d$ is that the field now oscillates too rapidly for the dipole to follow, and so the dipoles remain essentially stationary (That is, the frequency of the dipole (a molecule with dipole) is very small compared to the frequency of radiation.

The polarizability remains similarly unchanged in the frequency range from ω_d to ω_i and then drops down at the higher frequencies. The frequency ω_i lies in the infrared region and corresponds to the frequency of the transverse optical phonon in the crystal. For the frequency range ω_d to ω_i the ions with their heavy masses are no longer able to follow the very rapidly oscillating field, and consequently, the ionic polarizability α_i, vanishes, as shown in Fig. 8.5. Thus in the frequency range above the infrared, only the electronic polarizability remains effective, because the electrons, being very light, are still able to follow the field even at the high frequency. This range includes both the visible and ultraviolet regions.

At still higher frequencies (above the electronic frequency ω_e, however, the electronic contribution vanishes because even the electrons are too heavy to follow the field with its very rapid oscillations.

The frequencies ω_d and ω_i, characterizing the dipolar and ionic polarizabilities, respectively, depending on the substance considered, and vary from one substance to another. However, their orders of magnitude remain in the regions indicated above, i.e., in the microwave and infrared, respectively. The various polarizabilities may thus be determined by measuring the substance at various appropriate frequencies.

8.3.4 Classical (wave) model of Raman and Rayleigh scattering

As discussed in the section, the size of the induced dipole moment (μ) created on a molecule depends on the field E and on the ease with which the molecule can be distorted. In this case, polarizability α of the molecule can be defined as the induced dipole moment *per* unit electric field (*i.e.*, $\alpha = \mu/E$).

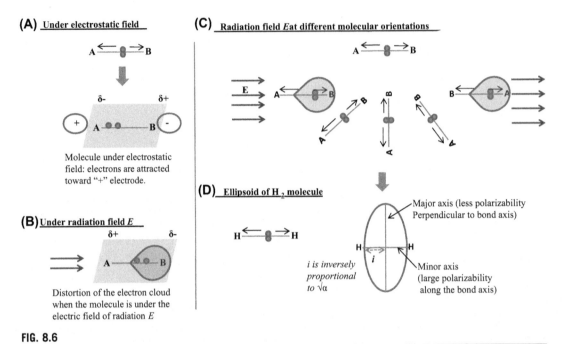

(A) Under electrostatic field

(B) Under radiation field E

(C) Radiation field E at different molecular orientations

(D) Ellipsoid of H_2 molecule

Molecule under electrostatic field: electrons are attracted toward "+" electrode.

Distortion of the electron cloud when the molecule is under the electric field of radiation E

i is inversely proportional to $\sqrt{\alpha}$

Major axis (less polarizability Perpendicular to bond axis)

Minor axis (large polarizability along the bond axis)

FIG. 8.6

(A) (left: top) represents molecule with no induced dipole moment and (B) (left: bottom) the same molecule under the influence of electric field E. (C) (right: top) represents molecule under the influence of radiation field E at different orientations, (D) (right: bottom) represents the polarizability in various directions is conveniently represented by drawing a polarizability ellipsoid.

Fig. 8.6 shows a simple example of induction and variation of polarizability under the static field and EM field, considering a simple diatomic molecule of type AB (or AA, e.g., H_2). The polarizability is anisotropic (of an object or substance having a physical property which has a different value when measured in different directions). This means that the electrons forming the bond are more easily displaced by the external electric field applied along the bond axis than one across this bond axis (when the molecular bond axis is perpendicular to the direction of the electric field, the polarizability is almost zero) (see Fig. 8.6B and C).

Besides, as in Fig. 8.6D, we can represent the polarizability in various directions most conveniently by drawing a polarizability ellipsoid: The polarizability ellipsoid is a three dimensional surface whose distance from the electrical center of the molecule (i.e. center of gravity) is inversely proportional to $\sqrt{\alpha_i}$ where α_i is the polarizability along the line joining points on the ellipsoid with the electrical center. Thus, where the polarizability is greatest the axis of the ellipsoid is least and vice versa.

Since the polarizability of a diatomic molecule is the same for all directions at right angles to the bond axis, the ellipsoid has a circular cross-section in this direction.

When a sample of diatomic molecules is subjected to a beam of radiation (which comprises oscillating electric field) of frequency ν_{rad}, the electric field experienced by each molecule varies as follows:

$$E = E_o \sin(2\pi \vartheta_{rad}.t). \tag{8.7}$$

As the light interacts with the matter, the electron orbits of matter are perturbed periodically by the incident electric field (see Fig. 8.6). The oscillation of the electron cloud results in a periodic separation of charge within the molecules which causes to induce dipole μ due to radiation, (see Eq. 8.3), and thus the μ also oscillates with a frequency of $\vartheta_{rad.}$

$$\mu = \alpha E = \alpha E_o \sin(2\pi\vartheta_{rad.}t). \tag{8.8}$$

where E_o is the amplitude (strength) of the oscillating applied electric field coming from radiation. Such an oscillating dipole emits radiation of its own oscillation frequency and we have immediately in Eq. (8.8) the classical explanation of Rayleigh scattering.

If, in addition, the molecule undergoes some internal motion, such as vibration or rotation, which changes the polarizability periodically, then the oscillating dipole will have superimposed upon it the vibrational or rotational oscillation. The oscillating dipole has frequency v_{vib} as well as the exciting frequency v_{rad}. It should be noted that if the vibration or rotation does not change the polarizability of the molecule the dipole oscillates only at the frequency of the incident radiation.

On the other hand, in the case of change of polarizability, we can write

$$\alpha = \alpha_o + \beta \sin(2\pi\vartheta_{vib.}t). \tag{8.9}$$

where α is the polarizability of the molecule at equilibrium, $\beta(= \partial\alpha/dQ$, where ∂Q is the physical displacement of atoms; $\partial\alpha$ the changes in polarizability) is the rate of change of polarizability with the vibration.

$$\mu = (\alpha_o + \beta \sin(2\pi\vartheta_{vib.}t)) \cdot E_o \sin((2\pi\vartheta_{rad.}t)). \tag{8.10}$$

due to inner motion due to changing electric field Further solving Eq. (8.10) we get:

$$\mu = \alpha_o E_o \sin(2\pi\vartheta_{rad.}t) + \beta E_o \sin(2\pi\vartheta_{vib.}t)\sin(2\pi\vartheta_{rad.}t). \tag{8.11}$$

Now using the concept of trigonometric relation:

$$\sin A.\sin B = 1/2\{\cos(A - B) - \cos(A + B)\}$$

we have,

$$\mu = \alpha_o E_o \sin(2\pi\vartheta_{rad.}t) + \frac{1}{2}\beta E_o\{\cos 2\pi(\vartheta_{rad.} - \vartheta_{vib.})t - \sin(2\pi\vartheta_{rad.}t)\cos 2\pi(\vartheta_{rad.} + \vartheta_{vib.})t\}.$$

$$\tag{8.12}$$

Thus the oscillating dipole has frequency components $\vartheta_{rad.}\pm\vartheta_{vib.}$ As well as exciting frequency $\vartheta_{rad.}$.

It should be carefully noted, however that if the vibration does not alter the polarizabity of the molecule, then $\beta = 0$ and the dipole oscillates only at the frequency of the incident radiation; the same is true of a rotation. Thus we have the general rule:

In order to be Raman active a molecule's rotation or vibration must cause some change in a component of the molecular polarizability. A change in polarizability, is of course, reflected by a change in either the magnitude or the direction of the polarizability ellipsoild.

In summary, the polarizability is a material property that depends on the molecular structure and nature of the bonds. What we see in Raman scattering is due to the changes in polarizability, when the atoms (or charged particles) moves.

This rule should be contrasted with that for infra-red and microwave activity, which is that the molecular motion must produce a change in the electric dipole of the molecule.

8.4 Selection rule

The gross selection rule for the observation of vibrational Raman transitions is that the polarizability of the molecule should change as the molecule vibrates. Both homo-nuclear and hetero-nuclear diatomic molecules fulfill this requirement, so both molecules are vibrationally Raman active.

If we approximate the potential energy curve of a vibrating bond as a parabola (i.e. assume that the vibration is harmonic) then the specific selection rule for vibrational Raman transitions is $\Delta v = \pm 1$. Just as with IR spectroscopy, much weaker overtone transitions appear at $\Delta v = \pm 2$.

Transitions for which $\Delta v = -1$ lie to the high frequency of the incident radiation, and are called the Stokes lines: For example, as shown in Fig. 8.1, if the electron starts from ground state at $v = 0$, the electron excites to virtual state and comes back to $v = 1$ of the ground state (same electronic level) to for transition to higher vibrational level, for example, from $v = 0$ to $v = 1$, for this molecule should absorb part of incident radiation and transfer to vibrational motion. Therefore, the scattered radiation will have lower energy than that of incident radiation.

Transitions for which $\Delta v = +1$ lies on low frequency of the incident radiation, and are called the anti-Stokes lines (See in Fig. 8.1): transition to lower vibrational level, for example, from $v = 1$ to $v = 0$, the molecule releases part of its vibrational energy, therefore, the scattered radiation will have lower energy than that of incident radiation.

In gas-phase spectra, the lines of the vibrational spectrum have a branch structure similar to that of the lines in an infra-red vibrational spectrum. They again arise from rotational transitions simultaneous with the vibrational transitions (**see** Fig. 8.7).

In this case, the selection rules are $\Delta J = 0, \pm 2$ (as in pure rotational Raman spectroscopy), and the three branches are termed the O branch ($\Delta J = -2$), the Q branch ($\Delta J = 0$), and the S branch ($\Delta J = +2$).

Note that in Raman spectroscopy, the Q branch is always observed.

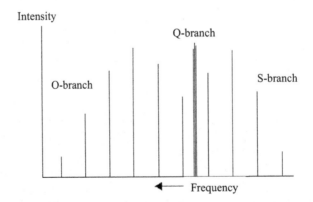

FIG. 8.7

This diagram is a schematic representation of the rotational structure within a vibrational line in a high-resolution Raman spectrum. Note that in this diagram, unlike that for the infra-red rotation-vibration spectrum; the frequency (and thus the wavenumber) increases from right to left.

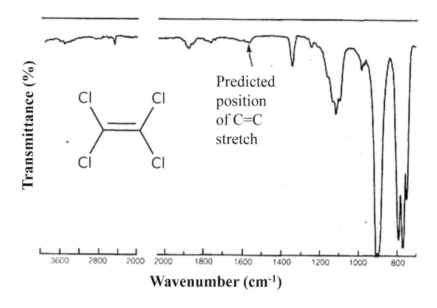

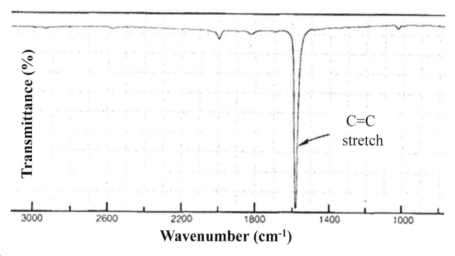

FIG. 8.8

Infrared spectra (top) and Raman spectra (bottom) of tetrachloroethene (notice that the spacing and alignment of the horizontal scales are not the same).

-J. Roberts, Basic Principles Organic Chemistry, second ed., ©1977. Reprinted by permission of Pearson Education, Inc.,

New York, New York.

For a given bond, the energy shifts observed in a Raman experiment should be identical to the
energies of its IR absorption bands, provided the vibrational modes involved are both the IR and
Raman active. Fig. 8.8 illustrates the similarity of the two types of spectra; it is seen that there are
several bands with identical v and Δv values for the two compounds. We should also note, however,
that the relative intensities of the corresponding bands are frequently quite different.

If a molecule has little or no symmetry, it is usually correct to assume that all its vibrational modes will be Raman active. But when the molecule has considerable symmetry it is not always easy to make the decision, since it is sometimes not clear, without detailed consideration, whether or not the polarizability changes during the vibration.

8.5 Comparison/contrast between IR spectroscopy and Raman spectroscopy

Raman spectroscopy is similar to IR spectroscopy (discussed in the previous chapter) in that both techniques can be used for chemical identification by examining vibrational transitions in molecules. However, Raman spectroscopy relies on inelastic scattering of monochromatic light, usually from a laser in the visible, near-infrared, or near-ultraviolet range. In particular, in this kind of probing method, the laser light interacts with molecular vibrations, phonons or other excitations in the system, resulting in the energy of the laser photons being shifted up or down. The shift in energy gives information about the vibrational modes in the system.

Infrared spectroscopy yields similar, but complementary, information using IR radiation as a light source. Besides vibrational information of a molecule, Raman spectra also provide information about rotational transitions of the molecule under investigation.

Comparison: Although Raman and FTIR Spectroscopy gives complementary information and are often interchangeable, there are some practical differences that influence which one will be optimal for a given experiment. Below we discuss similarities and contrasts between these analytical tools.

1. The information provided by Raman spectroscopy results from a light scattering process, whereas IR spectroscopy relies on the absorption of IR radiation.
2. Raman spectroscopy uses laser light in visible and near UV light range as an excitation source, whereas IR spectroscopy uses IR radiation. Indeed, the ability to use a laser as a light source and use of visible and near UV light instead of infrared light for these measurements are two important advantages of Raman spectroscopy.
3. While Infrared absorption requires a change in dipole moment or charge distribution during the vibration, Raman scattering involves a momentary distortion of the electrons distributed around a bond in a molecule, followed by re-emission of radiation as the bond returns to its normal state. In its distorted form, the molecule is temporarily polarized; i.e. it develops momentarily an induced dipole that disappears on relaxation and re-emission. In particular, one general rule is that functional groups that have large changes in dipoles are strong in the IR, whereas functional groups that have weak dipole changes or have a high degree of symmetry will be better seen in Raman spectra.
4. As indicated in (3), although changes in wavelength in Raman scattering correspond to absorption or emission of infrared radiation, infrared and Raman spectra are not always identical. Utilizing these unique features of both the analytical tools, indeed, valuable information about molecular symmetry may be obtained by comparison of infrared and Raman spectra.
5. Both the Raman spectroscopy and IR spectroscopy provide a spectrum characteristic of the specific vibrations of a molecule ("molecular fingerprint') and are valuable for identifying a substance. However, Raman spectroscopy can give additional information about lower frequency modes, and vibrations that give insight into the crystal lattice and molecular backbone structure.

6. Most molecular symmetry will allow for both Raman and IR activity. One special case is if the molecule contains a center of inversion. In a molecule that contains a center of inversion, Raman bands and IR bands are mutually exclusive, i.e. the bond will either be Raman active or it will be IR active but it will not be both.

7. Because of the fundamental difference in mechanism pointed in (3−4) the IR activity of a given vibrational mode may differ significantly from its Raman activity. For example, homo-nuclear diatomic molecules (of which bond is electrically symmetrical) such as H_2, O_2, Cl_2 has no dipole moment either in its equilibrium position or when a stretching vibration causes a change in the distance between the two nuclei. Thus the absorption of IR radiation associated with the molecule's vibrational frequency cannot occur (i.e., do not give infrared absorption spectra). On the other hand, however, excitation of symmetrical vibrations does occur in Raman scattering: Specifically, after irradiation with visible or UV light from laser source, the polarizability of the bond between the two atoms of such a molecule varies periodically in phase with the stretching vibrations, reaching a maximum at the greatest separation and a minimum at the closest approach.

For example, in a molecule such as ethene, $CH_2{=}CH_2$, the double-bond stretching vibration is symmetrical, because both ends of the molecule are the same. As a result, the double-bond stretching absorption is not observable in the infrared spectrum of ethene and is weak in all nearly symmetrically substituted ethenes. Nonetheless, this vibration appears strongly in the Raman spectrum and provides evidence for a symmetrical structure for ethene.

Therefore, as a general conclusion, a molecule has no important symmetry if all its infrared bands have counterparts in Raman scattering. To further illustrate these effects, the Raman and infrared spectra of tetrachloroethene (left: C_2Cl_4) and cyclohexene (C_6H_{10}) are discussed below.

In Fig. 8.8, we compare the IR spectra (top) and Raman spectra (bottom) of tetrachloroethene. To note, absorption due to the stretching vibration of the double bond in tetrachloroethene (at 1570 cm^{-1}) is strong in the Raman and absent in the infrared.

Similarly, Fig. 8.9 compares the IR spectrum with that of Raman of cyclohexane. One can clearly notice the difference that the peak arising from the less symmetrical C=C double bond of cyclohexene (at 1658 cm^{-1}) is weak in the infrared and slightly stronger in the Raman (Fig. 8.10).

Similarly, it is of interest to compare the IR and Raman activities of coupled vibrational modes such as those described earlier for the CO_2 molecule. In the symmetric mode, no change in the dipole moment occurs as the two oxygen atoms move away from or toward the central carbon atom; thus, this mode is IR inactive. However, molecular polarizability, which is associated with Raman activity, fluctuates in phase with the vibration because the distortion of bonds becomes easier as they shorten.

In contrast, the dipole moment of CO_2 fluctuates in phase with the asymmetric vibrational mode. Thus, an IR absorption band arises from this mode. On the other hand, as the polarizability of one of the bonds increases as it lengthens, the polarizability of the other decreases, resulting in no net change in the molecular polarizability. Thus, the asymmetric stretching vibration is Raman inactive. For molecules with a center of symmetry, such as CO_2, no IR active transitions are in common with Raman active transitions. This is often called the mutual exclusion principle. Therefore, for such homo-nuclear molecules, IR and Raman spectra are considered complementary to each other-each being associated with a different set of vibrational modes within a molecule.

However, for non-centrosymmetric molecules, many vibrational modes may be both Raman and IR active. For example, all of the vibrational modes of SO_2 result in both Raman and IR bands. The intensities of the bands differ, however, because the probability for the transitions are different for the two mechanisms.

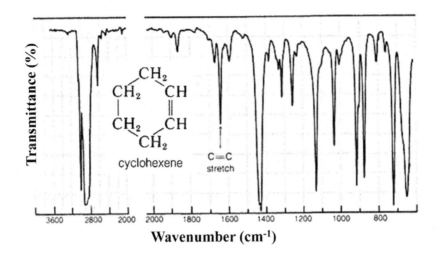

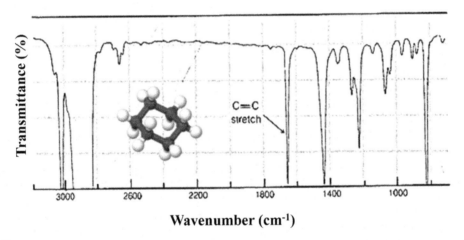

FIG. 8.9

Infrared spectra (left) and Raman spectra (right) of cyclohexane.

-J. Roberts, Basic Principles Organic Chemistry, second ed., ©1977. Reprinted by permission of Pearson Education, Inc., New York, New York.

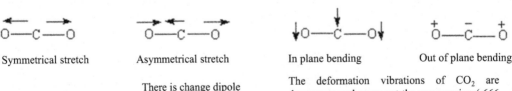

Symmetrical stretch	Asymmetrical stretch	In plane bending	Out of plane bending
No change in dipole moment therefore IR inactive There is change in polarizability therefore Raman active	There is change dipole moment therefore IR active There is no change in polarizability therefore Raman inactive	The deformation vibrations of CO_2 are degenerate and appear at the same region (666 cm^{-1}) in the IR spectrum of CO_2. There is no change in polarizability therefore these vibrations are Raman inactive.	

FIG. 8.10

Vibrational modes in CO_2 molecules and their features in IR and Raman spectroscopy.

Note: Raman spectra are often simpler than IR spectra because the occurrence of overtone and combination bands is rare in Raman spectra.

8.6 Intensity of normal Raman bands

The intensity or radiant power of a normal Raman band depends, mainly, on the polarizability of the molecule, the intensity of the source, and the concentration of the active group, as well as other factors. *If there is no absorption, the power of Raman emission increases with the fourth power of the frequency of the source.* However, this relation is not applicable, if the energy of the radiation causes to break the bond.

Raman intensities are usually directly proportional to the concentration of the active species. In this regard, Raman spectroscopy more closely resembles fluorescence than absorption, in which the concentration-intensity relationship is logarithmic.

8.7 Experimental setup

In principle, the experimental arrangement for Raman spectroscopy is quite simple (see Fig. 8.11). A modern compact Raman system typically consists of four major components:

1. Radiation (Excitation) source
2. Sample illumination system and light collection optics
3. Wavelength selector (Grating and Filter)
4. Detector (Photodiode array, CCD or PMT)

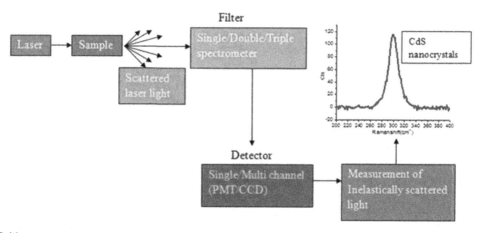

FIG. 8.11

Block diagram of a Raman spectrophotometer, with general alignment of components (light source, laser transmitting filter, sample holder, laser blocking filter and detector). The laser radiation is directed into a sample cell. Then, the Raman scattering is usually measured at perpendicular to the incoming radiation beam to avoid signal (un-scattered radiation) from the source radiation. A brief overview of each component used in Raman spectroscopy is given below.

8.7.1 Radiation source (laser)

Monochromatic radiation is passed through a sample, and the light scattered at right angles to the incident beam is analyzed by an optical spectrometer. In modern Raman instruments, typically, solid-state lasers are used with popular wavelengths of 532 nm, 785 nm, 830 nm and 1064 nm. The shorter wavelength lasers have higher Raman scattering cross-sections so the resulting signal is greater, however, the incidence of fluorescence also increases at shorter wavelengths. For this reason, many Raman systems feature the 785 nm laser.

8.7.2 Filters/grating

Since Raman scattering (the Stokes and Anti-Stokes signal) is very weak the main difficulty of Raman spectroscopy is separating it from the intense Rayleigh scattering. More precisely, the major problem here is not the Rayleigh scattering itself, but the fact that the intensity of stray light from the Rayleigh scattering may greatly exceed the intensity of the useful Raman signal in the closeness, proximity to the laser wavelength. Optical filters are used to prevent the undesired light from reaching the spectrometer and while allowing the relatively weak Raman signal. Below we briefly discuss the types of filters and their applications (purposes). Types of commonly used filters in Raman spectroscopy are:

(i) Longwave pass (LWP) edge filter
(ii) Short wave pass (SWP) edge filter
(iii) Notch filter and
(iv) Laser line filter (each shown below).

In many cases, the problem is resolved by simply using a "Laser line filter" which transmits only the laser and block all other light (see Fig. 8.12).

While notch filter blocks only the laser line while passing both long and shorter wavelengths (see also a very simple pictorial representation of the arrangement of these filters in Figs. 8.12F). By using these two filters (Laser nine filter and Notch filter) together, both Stokes and Anti-Stokes Raman scattering can be measured simultaneously.

Generally, commercially available interference (notch) filters are used, which cut-off spectral range of $\pm$ 80-120 cm^{-1} from the laser line. This method is efficient in stray light elimination but it does not allow detection of low-frequency Raman modes in the range below 100 cm^{-1}.

An edge filter can provide a superior alternative, as they offer the narrowest transition to see Raman signals extremely close to the laser line.

The examples in Fig. 8.12 show how different filter types can be used in a Raman system. The blue lines represent the filter transmission spectra, the green lines represent the laser spectrum, and the red lines represent the Raman signal.

Alternatively, the grating can also be used for both the wavelength selection (filters radiation of unwanted frequency). Moreover, it is highly likely to produce stray light inside the spectrophotometer due to reflection by the optical devices and, actually, stray light is generated in the spectrometer mainly upon light dispersion on gratings and strongly depends on grating quality.

To overcome this problem, Raman spectrometers typically use *holographic gratings* which normally have much less manufacturing defects in their structure then the ruled once. Stray light produced by *holographic gratings* is about an order of magnitude less intense than from ruled gratings of the same groove density.

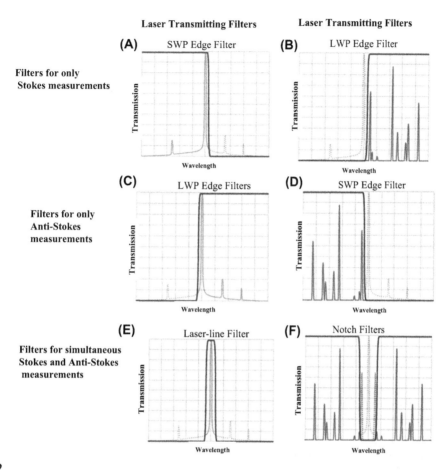

FIG. 8.12

Combinations of optical filters used in Raman spectroscopy with their respective radiation transmission windows. Combination of SWP Edge filter (A) and LWP Edge filter (B) are combined for only Stokes measurements. LWP Edge filter (C) and SWP Edge filter (D) are combined for only anti-Stokes measurements. Similarly, Laser-line filter (E) and Notch filter (F) are combined for simultaneous Stokes and Anti-Stokes measurements.

-IH&S Semrock, a unit of IDEX Health & Science.

8.7.3 Detectors (transducer)

The transducer converts the Raman signal into a proportional electrical signal that is processed by the computer data system. For detecting weak signals produced by Raman scattering, it is most important that high-quality, optically well-matched components are used. Previously, single-point detectors such as photon-counting Photomultiplier Tubes (PMT) have been primarily used. However, a single Raman spectrum obtained with a PMT detector in wavenumber scanning mode used to take a substantial period of time, slowing down any research or industrial activity based on Raman analytical technique. Nowadays, more and more often researchers use multi-channel detectors like Photodiode Arrays (PDA) or, more commonly, a Charge-Coupled Devices (CCD) to detect the Raman scattered light.

Sensitivity and performance of modern CCD detectors are rapidly improving. In many cases, CCD is becoming the detector of choice for Raman spectroscopy.

8.7.3.1 Solvent

Water is regarded as good solvents for the study of inorganic compounds in Raman spectroscopy. In fact, a major advantage of sample handling in Raman spectroscopy compared to IR arises because water is a weak Raman scatterer but a strong absorber of IR radiation. Thus samples prepared in aqueous solution can be studied with Raman spectroscopy. This advantage is particularly important for biological and inorganic systems and in studies dealing with water pollution.

8.7.3.2 Sample-illumination system

Unlike in IR spectroscopy, glass can be used for windows, lenses, and other optical components instead of more fragile and environmentally less stable crystalline halides. In addition, the radiation (laser light) can be easily focused on a small sample area and the emitted radiation efficiently focused o a slit or entrance aperture of a spectrophotometer. Due to this arrangement, very small samples can be investigated.

8.7.3.3 Sample holder

Glass and silica tubes: A common sample holder for non-absorbing liquid samples is an ordinary glass capillary tube.

8.7.3.4 Gas samples

Gaseous samples are usually filled in glass tubes with $1-2$ cm in diam. and about 1 mm thick.

8.7.3.5 Liquid samples

Liquids can be sealed in ampoules, glass tubes or capillaries. The capillaries can be as small as $0.5-0.1$ mm bore and 1 mm long. The spectra of even nanoliter volumes of the sample can be obtained with capillary cells. Relatively larger cylindrical cells can be used to reduce the local heating particularly for absorbing samples. Moreover, the laser beam is focused on an area near the wall to minimize absorption of the incident beam.

8.7.3.6 Solid samples

Raman spectra of solid samples are often acquired by filling a small cavity or capillary with the sample after grinding it to a fine powder. Polymers can usually be examined directly with no sample pre-treatment. In some cases, KBr pellets similar to those used in IR spectroscopy are used. Dilution with KBr can reduce decomposition of the sample produced by local heating.

   As an example, in Fig. 12.12 below we show compact sample holders produced by Inphotonics company. Raman measurement of routine samples is easy and safe with such compact sample holders. A spring-loaded cap prevents unnecessary exposure to laser radiation and blocks stray light from contributing to the spectrum. The sample holder is supplied with four inserts for three sizes of vials and a 1-cm pathlength cuvette. The complete holder can also be turned upright for the measurement of neat solids and films in small sampling cups.

   The sample holder has been designed for use with a Raman Probe with various working distances (For example, Inphotonics' holders are designed to use at either 5, 7.5 mm or 10 mm working distance (see for example in Fig. 8.13). Fine focusing of the probe can be done for maximizing the Raman signal on specific samples.

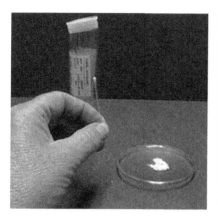 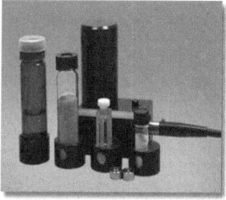

FIG. 8.13

Ordinary glass-melting-point capillary tube (left). Compact sample holder with four inserts.

8.8 Application of Raman spectroscopy
8.8.1 Raman spectra of inorganic species

Application of Raman spectroscopy for analyzing inorganic species may be summarized as follows:

1. IR and Raman spectroscopy plays a crucial role to elucidate vibrational information of inorganic systems. In particular, combining the vibrational information with other spectroscopic information, such as X-ray diffraction (XRD) NMR, electron diffraction, or UV−Visible) help us to determine the structure of small inorganic molecules and solids in inorganic systems. Raman spectroscopy plays a crucial role in determining structures of small molecules. For example, some small reactive molecules only exist in the gas phase and XRD can only be applied for solid state. Also, XRD cannot distinguish between the following bonds: CN versus −NC, −OCN versus −NCO, −CNO versus −ONC, −SCN versus −NCS. Furthermore, Raman and IR are fast and simple analytical methods and are commonly used for the first approximation analysis of an unknown compound.

2. Raman spectroscopy has considerable advantages over IR in inorganic systems due to two reasons. First, since the laser beam used in Raman spectroscopy and the Raman-scattered light are both in the UV and visible region, glass (Pyrex) tubes can be used in RS. On the other hand, glass absorbs infrared radiation and cannot be used in IR. However, some glass tubes, which contain rare earth salts, will give rise to fluorescence or spikes. Thus, using glass tubes in Raman spectroscopy still, need to be careful. Secondly, since water is a very weak Raman scatter but has a very broad signal in IR, the aqueous solution can be directly analyzed using Raman spectroscopy.

3. The vibrational energies of metal-ligand bonds are generally in the range of $100-700$ cm^{-1}, a region of the IR that is experimentally difficult to study. However, these vibrations are frequently Raman active and lines with $\Delta \bar{u}$ values in this range are readily observed. Raman studies are potentially useful sources of information concerning the composition, structure and stability of coordination compounds. For example, numerous halogens and halogenated complexes produce

Raman spectra and thus are acceptable to investigate, by this method. Metal-oxygen bonds such as VO_3^{4-}, $Al(OH)_4^-$, $Si(OH)_6^{2-}$ $Sn(OH)_6^{2-}$ are also Raman active and spectra from such species have been obtained. For example, in perchloric acid solutions, vanadium (IV) appears as $VO^{2+}(aq)$ rather than as $V(OH)_2^{2+}$ (aq).

4. As Raman Spectroscopy and IR have different selection rules (Raman spectroscopy detects the polarizability change of a molecule, while IR detects the dipole momentum change of a molecule). Thus, some vibrational modes that are active in Raman may not be active in IR, vice versa. As a result, both of Raman and IR spectra are provided in the structure study, complementing each other. As an example, in the study of Xenon Tetrafluoride, there are 3 strong bands in IR and solid Raman shows 2 strong bands and 2 weaker bands. This indicates that Xenon Tetrafluoride is a planar molecule and has a symmetry of D_{4h}.

8.8.2 Raman spectra of organic species

Both the Raman and IR spectra are useful for studying the functional group detection and fingerprint regions that permit the identification of specific compounds. Thus, Raman and IR spectroscopy is widely used in organic systems. Characteristic vibrations of many organic compounds both in Raman and IR are widely studied and summarized in many literatures. Qualitative analysis of organic compounds can be done based on the characteristic vibrations of the functional group(s) in a given analyte molecule (see Table 8.1).

Raman spectra yield more information about certain types of organic compounds than do their IR counterparts. For example, the double-bond stretching vibration for olefins results in weak and sometimes undetected IR absorption. On the other hand, the Raman band (which is like the IR band, occurs at about 1600 cm^{-1}) is intense, and its position is sensitive to the nature of substituents as well as to their geometry. Thus Raman studies are likely to yield useful information about the olefinic functional group that may not be revealed by IR spectroscopy. This applies to cycloparaffin derivatives as well: These compounds have a characteristic Raman band in the region of $700-1200 \text{ cm}^{-1}$. This band has been attributed to a breathing vibration in which the nuclei move in and out symmetrically with respect to the center of the ring. The position of the band decreases continuously from 1190 cm^{-1} for cyclopropane to 700 cm^{-1} for cyclo-octane. Raman spectroscopy thus appears to be an excellent diagnostic tool for the estimation of the ring size in paraffins. The IR band associated with this vibration is weak or nonexistent.

Table 8.1 Functional groups and associated frequency, and nature of Raman and IR signals.

Vibration	Region (cm^{-1})	Raman intensity	IR intensity
ν (O—H)	3650–3000	Weak	Strong
ν (N—H)	3500–3300	Medium	Medium
ν (C=O)	1820–1680	Strong $\sim$ weak	Very strong
ν (C=C)	1900–1500	Very strong $\sim$ medium	0 $\sim$ weak

8.9 Application of Raman spectroscopy in material science

Indeed, among spectroscopic methods, Raman spectroscopy has been becoming one of the standard tools in the fast-growing field of materials science and engineering. It is due to the fact that this technique is a fast, non-destructive, and is a high-resolution tool for characterization of the lattice structure, electronic properties, optical properties and phonon properties of nanomaterial-based materials (including graphene, organo-metal halide based perovskite family).

Below we discuss the application of Raman spectroscopy in material characterization, considering graphene and derivatives as an example.

(1) Raman spectra enable us to identify the vibrational modes, using laser excitation. For example, the external perturbations with radiation to graphene flakes could affect their lattice vibrations and also band structures.

(2) Through Raman imaging, we can extract spatially distributed information concerning the number and quality of layers, local stress within such a multilayer graphene structure, doping level and nature of defects in graphene (point defects, line defects, and edges on the D, G, and 2D modes) and growth mechanism of graphene flakes.

(3) Besides, this tool also enables to monitor the property modification of three-dimensional (3d) diamond, graphite, 2d graphene, carbon dots, epitaxial graphene on SiC, GO, and rGO prepared by various methods.

For example, in the process of making graphene, be it from mechanical cleavage, epitaxial growth, chemical vapor deposition, chemical exfoliation, all kinds of carbon species can in principle prevail. Moreover, while shaping graphene and graphene derivatives into devices, unwanted by-products and structural damage can also be created. Therefore, it is necessary to have a structural reference for which Raman spectroscopy has been employed as a common technique.

As an example, here we describe the Raman scattering processes of the entire first- and second-order modes in the intrinsic graphite, GO and rGO. In Fig. 8.14, we present the Raman spectra measured by three different research laboratories. As demonstrated in the panels of (a), (b) and (c), we can clearly notice differing features of graphite, GO, rGO and the GO films of monolayer and multilayer graphene.

Specifically, as depicted in Raman spectra in Fig. 8.14A recorded by Stankovich et al., the significant structural changes were observed during the chemical processing from pristine graphite to GO, and then to the rGO: Raman spectrum of the pristine graphite in Fig. 8.14A: displays a prominent G peak as the only feature at $1581 \ cm^{-1}$ which corresponds to the first-order scattering of the E_{2g} mode of graphite. Whereas in the Raman spectrum of GO (a: central panel), the G band was broadened and slightly shifted to $1594 \ cm^{-1}$. In addition, the D band at $1363 \ cm^{-1}$ becomes prominent, indicating the reduction in the size of the in-plane sp^2 domains, which is understood to be due to the oxidation of the graphite. Moreover, the Raman spectrum of the rGO (a: bottom panel) also contains both G and D bands (at 1584 and $1352 \ cm^{-1}$, respectively) which were prevailed in GO spectrum; however, with an increased intensity ratio of D peak to G peak (D/G ratio) compared to that in GO. This change suggests a decrease in the average size of the sp^2 domains upon reduction of the exfoliated GO, and this is due to creation new graphitic domains that are smaller in size to the ones present in GO before reduction, but more numerous in number.

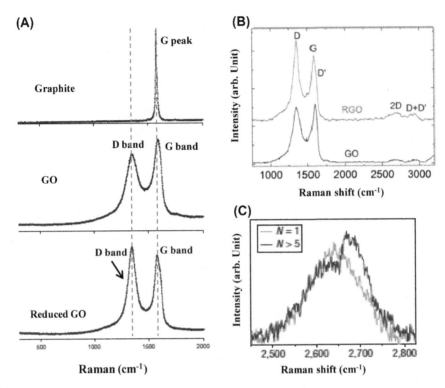

FIG. 8.14

Panel (A): left shows the Raman spectra of graphite (top), GO (middle) and the reduced GO (bottom). (B): right-upper: Raman spectrum of GO (*black line*) and RGO (*red line* [light gray in print version]). The main peaks are labeled. (C): right-bottom: Raman spectra of single-layered (*N=1*) and multilayered (*N> 1*) GO.

-(A) S. Stankovich, D.A. Dikin, R.D. Piner, K.A. Kohlhaas, A. Kleinhammes, Y. Jia, Y. Wu, S.T. Nguyen, R.S. Ruoff, Carbon 45 (2007) 1558, Adapted with permission from Elsevier. (B) A.L. Palma, L. Cinà, S. Pescetelli, A. Agresti, M. Raggio, R. Paolesse, F. Bonaccorso, AldoDiCarlo, Reduced graphene oxide as efficient and stable hole transporting material in mesoscopic perovskite solar cells, Nano Energy 22 (2016) 349–360, Adapted with permission from Elsevier. (C) G. Eda, G. Fanchini, M. Chhowalla, Nat. Nanotechnol. 3 (2008) 270, Adapted with permission from Springer Nature.

Moreover, the spectra presented in Fig. 8.14B were measured by Palma et al.(2008) from GO (black line) and rGO (red line) layers in the range of 500 cm^{-1} to 3500 cm^{-1} of GO. Similar to spectra recorded by Stankovich et al. in Fig. 8.14A, the spectrum of rGO (red line in Fig. 8.14(B)) shows D peak at $\bar{\upsilon}(D) = 1351$ cm^{-1}, and G peak $\bar{\upsilon}$ (G) $= 1581$ cm^{-1}. It is understood that the narrower FWHM of D peak (FWHM (D) $= 115$ cm^{-1}) and downshifted frequency ($\bar{\upsilon}(G) = 1581$ cm^{-1}) with respect to GO (FWHM(D) $= 160$ cm^{-1} and $\bar{\upsilon}$ (G) $= 1591$ cm^{-1}, respectively, is an indication of partial sp^2 restoration.

Moreover, the ratio of the intensity of D peak and G peak provides us with information about the fraction of defect level in the crystal. For example, in Fig. 8.14A, there is an increase in the I(D)/I(G) ratio passing from GO(I(D)/I(G) $= 0.95$) to RGO (I(D)/I(G) $= 1.18$) indicating a medium level of defects and presence of both crystalline and amorphous carbons. The G can be roughly considered constant as a function of disorder, be in related to the relative motion of sp^2 carbons, while an increase of I(D) is directly linked to the presence of sp^2 rings. Thus an increase of the I(D)/I(G) ratio is considered as the restoration of sp^2 rings. Moreover, it is also reported that 2D peak (not shown here) is highly intense in single layer defectless graphene but with the stacking of the layers, interaction increases, and this 2D peak splits into multiple peaks making it shorter and wider.

Fig. 8.14 (c-bottom) also represents Raman spectrum recorded by Eda et al. in (2009) rGO layers (number of layer, N = 1 and N = 5) in the range of 2500 cm^{-1} to 2800 cm^{-1}. One can clearly notice the shift of the wavenumber in the peak position and variation in the peak intensity.

8.10 Questions

1. What is virtual state?
2. What is polarizability? How do we relate the induced dipole moment and the polarizability?
3. Why does the ratio of anti-Stokes to Stokes intensities increase with sample temperature?
4. The Raman spectra of N_2 is obtained by using excitation radiation of wavelength 54 nm. Predict the wavelength of three stokes, and anti-stokes lines in the pure rotational Raman spectra of N_2.
5. Under what circumstances would a helium-neon laser be preferable to an argon-ion laser Raman spectroscopy?
6. A molecule A_2B_2 has infra-red absorptions and Raman spectral lines as in the following table

Cm^{-1}	Infra-red	Raman
3374	Absent	Strong
3287	Very strong	Absent
1973	Absent	Very strong
729	Very strong	Absent
612	Absent	Weak

7. Deduce what you can about the structure of the molecule and assign the observed vibrations to particular molecular modes as far as possible.
8. Identify the modes of vibrations of graphene, graphene oxide and reduced graphene oxide that are Raman active. Compare and contrast the Raman spectra of graphite, graphite oxide, graphene, graphene oxide and reduced graphene oxide.
9. Why 2D peak is very important in the Raman spectrum of graphene oxide Hint: Its intensity is associated with number of layers of graphene oxide).

Further reading

[1] D.A. Skoog, F. J Holler, S.R. Crouch, Instrumental Analysis, 11th Indian Reprint, Brooks/Cole, Cengage Learning India Private Limited, Delhi, 2012 (chapter 18), pg. 533.
[2] H. Skoog, Nieman, Principles of Instrumental Analysis, fifth ed., Brooks Cole, 1997.
[3] K. Nakamoto, Infrared and Raman Spectra of Inorganic and Coordination Compounds, sixth ed., John Wiley & Sons, Inc., 2008.
[4] D.C. Harris, M.D. Bertolucci, Symmetry and Spectroscopy an Introduction to Vibrational and Electronic Spectroscopy, Courier Corporation, 1989.
[5] P. Bisson, G. Parodi, D. Rigos, J.E. Whitten, Chem. Educ. 11 (2) (2006).
[6] B. Schrader, Infrared and Raman Spectroscopy, VCH, 1995.

[7] S.A. Borman, Analytical Chem,, Vol. 54, No. 9, 1982 (1021A-1026A).

[8] K. Nakamoto, Infrared Spectra of Inorganic and Coordination Compounds, third ed., Wiley Intrsc John Wiley & Sons, New York London Sydney Toronto, 1978.

[9] J.D. Robert, M.C. Caserio, Basic Principles of Organic Chemistry, second ed., W. A. Benjamin, Inc., Menlo Park, CA, 1977, ISBN 0-8053-8329-8.

[10] D.S. Hage, J.D. Carr, Analytical Chemistry and Quantitative Analysis, Pearson Education, Inc., Publishing as Prentice-Hall, Upper Saddle River, New Jersey, 2011 (chapter 18), pg. 447.

[11] M.C. Gupta, Atomic and Molecular Spectroscopy,, New Age International (P) Limited, New Delhi, 2001 (chapter 9), pg. 241.

[12] C.N. Banwell, Fundamentals of Molecular Spectroscopy, 15th Reprint, Tata McGraw-Hill Book Co. (UK) Ltd., New Delhi, 1994 (chapter 4), pg. 124.

[13] S. Stankovich, D.A. Dikin, R.D. Piner, K.A. Kohlhaas, A. Kleinhammes, Y. Jia, Y. Wu, S.T. Nguyen, R.S. Ruoff, Carbon 45 (2007) 1558.

[14] A.L. Palma, L. Cinà, S. Pescetelli, A. Agresti, M. Raggio, R. Paolesse, F. Bonaccorso, AldoDiCarlo, Reduced graphene oxide as efficient and stable hole transporting material in mesoscopic perovskite solar cells, Nano Energy 22 (2016) 349−360.

[15] G. Eda, G. Fanchini, M. Chhowalla, Nat. Nanotechnol. 3 (2008) 270.

[16] A.C. Ferrari, Raman spectroscopy of graphene and graphite: disorder, electron-phonon coupling, doping and nonadiabatic effects, Solid State Commun. 143 (2007) 47−57.

[17] G. Eda, C. Mattevi, H. Yamaguchi, H. Kim, M. Chhowalla, Insulator to semi-metal transition in graphene oxide, J. Phys. Chem. C 113 (2009) 15768−15771.

[18] P. Venezuela, M. Lazzeri, F. Mauri, Phys. Rev. B 84 (2011) 035433.

[19] N.A. Kumar, S. Gambarelli, F. Duclairoir, G. Bidan, L. Dubois, Synthesis of high quality reduced graphene oxide nanosheets free of paramagnetic metallic impurities, J. Mater. Chem. A. 1 (2013) 2789.

[20] J.I. Paredes, S. Villar-Rodil, A. Martínez-Alonso, J.M.D. Tascón, Graphene oxide dispersions in organic solvents, Langmuir 24 (2008) 10560−10564.

[21] M.F. Islam, E. Rojas, D.M. Bergey, A.J. Yodh, Nano lett. 3 (2003) 269.

Molecular luminescence spectroscopy

9.1 Introduction

When a molecule absorbs a photon, the molecule is promoted to a more energetic excited state. However, the molecule cannot remain in the excited state for long and, therefore, it releases the absorbed energy in the form of a photon. Therefore, when a molecule of an analyte in the excited state possesses energy (E_2) that is greater than its energy when it was in a lower energy state (E_1). When the analyte returns to its lower energy state (a process we call *relaxation*) the excess energy (ΔE) must be released. The released energy can be related as $\Delta E = E_2 - E_1$. This process of emission of light by an excited chemical substance is known as Luminescence (or photoluminescence).

In other words, photoluminescence occurs when an electron returns to the electronic ground state from an excited state and loses its excess energy as a photon. Time periods between absorption and emission may vary from femtosecond (such fast emission occurs in inorganic semiconductors) up to milliseconds for Phosphorescence processes in molecular systems.

Depending on the time duration of radiation emission by the excited molecule, luminescence is divided into two categories: (1) fluorescence, (2) phosphorescence. In general, both of these kinds of photoluminescence are observed when a molecule is excited by electromagnetic radiation. There is also another kind of luminescence called chemiluminescence. Unlike fluorescence, phosphorescence processes, chemiluminescence is observed when a molecule of interest reacts with another reactant to form an excited state. A brief introduction of each process is given below.

9.2 Mechanism for fluorescence and phosphorescence

In order to understand the origin of fluorescence and phosphorescence, we must consider what happens to a molecule (in electronic level) following the absorption of a photon. Here, we assume a case of a rather simple molecule, formaldehyde.

9.2.1 Electronic states of formaldehyde

Fig. 9.1 shows electronic configuration (distribution of electrons) in a formaldehyde molecule. Electronic labels σ_1, σ_2, σ_3 and σ_4, are four low-lying orbitals which are each occupied by a pair of electrons with opposite spins (spin quantum numbers $= +\frac{1}{2}$ and $-\frac{1}{2}$, respectively, represented by ↑ and ↓. At higher energy level is a π bonding orbital, made of the p_y atomic orbitals of carbon and oxygen. The highest-energy occupied molecular orbital (HOMO) is a nonbonding orbital (n),

Chemical Analysis and Material Characterization by Spectrophotometry. https://doi.org/10.1016/B978-0-12-814866-2.00009-9

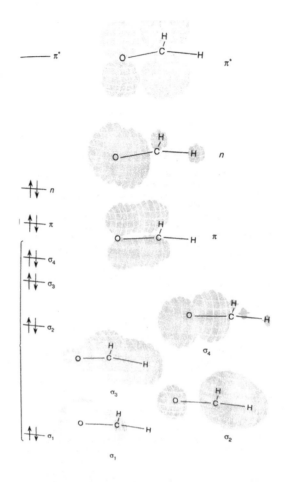

FIG. 9.1

Electronic configuration of a formaldehyde molecule. The n orbital is the highest occupied molecular orbital (HOMO) and π^* is the lowest unoccupied molecular orbital (LUMO). The shaded area represents the shape of the molecular orbital.

From W.L. Jorgensen and L. Salem, The Organic Chemist's Book of Orbitals *(Adapted and reprinted with permission from New York: Academic Press, 1973).*

composed principally of the oxygen $2p_x$ atomic orbital. The lowest-energy unoccupied molecular orbital (LUMO) is anti-bonding orbital π^*. An electron in this orbital produces repulsion rather than attraction, between the carbon and oxygen atoms.

9.2.1.1 What happens after absorption of light energy?

In an electronic transition, an electron moves from one orbital to another. The lowest-energy electronic transition (from HOMO to LUMO) of formaldehyde promotes a nonbonding *n* orbital to $\pi*$ orbital.

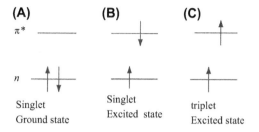

FIG. 9.2

Electron configurations for (A) a singlet ground state; (B) a singlet excited state; and (C) a triplet excited state.

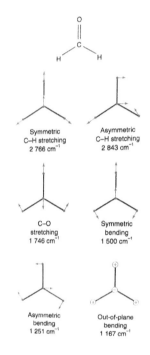

FIG. 9.3

The molecular structure (at the top) and six kinds of vibrations of CH_2CO molecule: (1) Symmetric C-H stretching, (2) Asymmetric C-H stretching, (3) C-O stretching, (4) Symmetric bending, (5) Asymmetric bending, and (6) Out-of-plane bending.

From D. C. Harris, Quantitative Chemical Analysis, *second Ed, W. H. Freeman and Company, New York (2001).*

Depending on the spin quantum numbers in the excited state, there are actually two possible transitions. As shown in Fig. 9.2B, the molecular state in which the spins of two electrons, one situated in n orbital and other in π^* orbital, are opposed is called a *singlet state*. If the spins are parallel, the excited state is a *triplet state* (Fig. 9.3).

9.2.1.2 How do molecules release energy?

Let's assume that the molecule initially occupies the lowest vibrational energy level of its electronic ground state, which is a singlet state labeled S_0 (see in Fig. 9.4). Absorption of a photon excites the molecule to one of several vibrational energy levels in the first excited electronic state, S_1, or the second electronic excited state, S_2, both of which are singlet states. Similar to other molecules, the lowest-energy excited singlet and triplet states of the formaldehyde molecule are represented by symbols S_1 and T_1. In general, the T_1 state has lower energy than the S_1 state. In this molecule, the transition $n \rightarrow \pi^*$ (S_1) takes 355 nm ultraviolet radiation. Whereas, the transition $n \rightarrow \pi*$ (T_1) requires the absorption of visible light with a wavelength of 397 nm. Although formaldehyde is planar in its ground state (S_0), it takes pyramidal shape in both the S_1 and T_1 excited states. Promotion of a nonbonding electron to an antibonding C–O orbital weakens and lengthens the C–O bond and changes the molecular geometry.

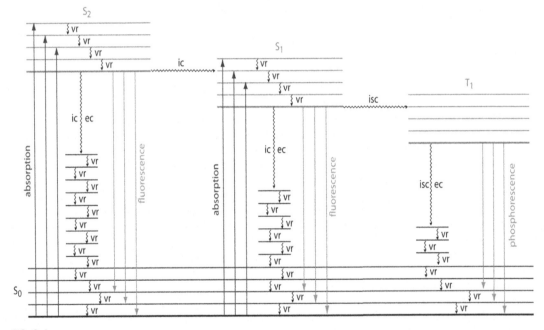

FIG. 9.4

Energy level diagram for a molecule showing pathways for the deactivation of an excited state: 'vr' is vibrational relaxation; 'ic' is internal conversion; 'ec' is external conversion, and 'isc' is an intersystem crossing. The lowest vibrational energy for each electronic state is indicated by the thicker line. The electronic ground state is shown in black and the three electronic excited states are shown in green. The absorption, fluorescence, and phosphorescence of photons also are shown (Note it is also called 'Jablonski Diagram for Absorption', Fluorescence, and Phosphorescence).

With permission from D. Harvey, Modern Analytical Chemistry, Pg. 614, Chapter 9, first, McGraw-Hill, 1999.

9.2.2 Vibrational and rotational states of formaldehyde (CH$_2$CO)

The six ($3*4-6 = 6$) kinds of vibrations of CH$_2$CO molecule are shown in Fig. 9.3. Infrared and microwave radiation are not energetic enough to induce electronic transitions, but they can change the vibrational or rotational motion of the molecule. For example, when CH$_2$CO molecule absorbs an infrared photon with a wavenumber 1746 cm^{-1}, C$-$O stretching is stimulated: Oscillations of the atoms increase in amplitude and the energy of the molecule increases. Also, the rotational energies of a molecule are even smaller than vibrational energies. Absorption of microwave radiation increases the rotational speed of a molecule.

In general, when a molecule absorbs light of sufficient energy to cause an electronic transition, vibrational and rotational transitions (changes in the vibrational and rotational states) occur as well. Formaldehyde can absorb one photon with just the right energy to (1) promote the molecule from ground state S$_0$ $\rightarrow$ S$_1$ (excited electronic state) and (2) increase the vibrational energy from the ground vibrational state of S$_0$ to an excited state of S$_1$; and (3) change from one rotational state of S$_0$ to an excited state of S$_1$. Due to the fact that many different vibrational and rotational levels are excited at slightly different energies, as mentioned above, electronic absorption bands are usually very broad (see for example the UV$-$Vis spectra presented in Figure 6.3 in Chapter 6). Before discussing the emission spectra let us discuss how the excited molecule comes back to the ground state (i.e., methods of releasing the absorbed energy).

9.3 Types of relaxation (deactivation) processes

Relaxation to the ground state occurs by a number of mechanisms, some involving the emission of photons and others occurring without emitting photons. In general, the relaxation process may be classified into four kinds: Radiationless deactivation, fluorescence, phosphorescence, and chemiluminescence. These relaxation processes (mechanisms) are demonstrated in Fig. 9.4. The most likely relaxation pathway is the one with the shortest lifetime for the excited state.

9.3.1 Radiationless deactivation

When a molecule relaxes without emitting a photon we call the process *radiationless deactivation*. The radiationless relaxation is classified into four kinds: (i) Vibrational relaxation (vr), (ii) internal conversion (ic), (iii) external conversion, and (iv) intersystem crossing (isc).

9.3.1.1 Vibrational relaxation, vr

In this kind of radiationless deactivation, a molecule in an excited vibrational energy level loses energy by moving to a lower vibrational energy level in the same electronic state (in Fig. 9.4, vibrational relaxation is represented by "vr"). Suppose our molecule is in the highest vibrational energy level of the second electronic excited state S$_2$. After a series of vibrational relaxations brings the molecule to the lowest vibrational energy level of the same electronic state S$_2$. Vibrational relaxation is very rapid, with an average lifetime of $<10^{-12}$ s (i.e., $<$ picosecond). Because vibrational relaxation is so efficient, a molecule in one of its excited state's higher vibrational energy levels quickly returns to the excited state's lowest vibrational energy level.

9.3.1.2 Internal conversion, ic

It is another form of radiation-less deactivation, in which a molecule in the ground vibrational level of an excited state passes directly into a higher vibrational energy level of a lower energy electronic state of the same spin state (such as the transition from S_2 to S_1 in Fig. 9.4). Again after switching to lower excited state S_1, vibrational relaxations proceeds and bring the molecule to the lowest vibrational energy level of S_1. Following an internal conversion into a higher vibrational energy level of the ground state (S_0), the molecule continues to undergo vibrational relaxation until it reaches the lowest vibrational energy level of S_0 (ground state). Therefore, by a combination of internal conversions and vibrational relaxations, a molecule in an excited electronic state may return to the ground electronic state without emitting a photon.

9.3.1.3 External conversion

Another related form of radiationless deactivation is an **external conversion** in which excess energy is transferred to the solvent or to another component of the sample's matrix. That is, energy is lost to other molecules (solvent, for example) through collision. The net effect is to convert part of the energy of the absorbed photon into heat spread through the entire medium.

9.3.1.4 Intersystem crossing, isc

It is another well-known form of radiationless deactivation, in which a molecule in the ground vibrational energy level of an excited electronic state passes into a higher vibrational energy level of a lower energy electronic state with a different spin state. For example, an intersystem crossing is shown in Fig. 9.4 (**right**) between a singlet excited state, S_1, and a triplet excited state, T_1.

9.3.2 Relaxation by photon emission *(fluorescence and phosphorescence)*

Release of the energy by the excited molecule is carried out by mainly two processes: i) *Fluorescence Emission and ii) Phosphorescence Emission.*

9.3.2.1 Fluorescence emission

A pair of electrons occupying the same electronic ground state have opposite spins and are said to be in a *singlet spin state* (see Fig. 9.2A). When an analyte absorbs an ultraviolet or visible photon, one of its valence electrons moves from the ground state to an excited state without conversion of the electron's spin (Fig. 9.2B). Emission of a photon from the singlet excited state to the singlet ground state—or between any two energy levels with the same spin is called *fluorescence* (Fig. 9.4: **left and middle sketches with red lines**). A method that uses fluorescence to characterize or measure chemicals is called *fluorescence spectroscopy.*

The probability of fluorescence is very high and the average lifetime of an electron in the excited state is only 10^{-5}–10^{-8} s. Fluorescence, therefore, decays rapidly (lasting less than 10 ns) once the source of excitation is removed.

In other words, fluorescence is an emission of light by a sample after it has become electronically excited by the absorption of a photon, with the light emission being due to a "spin-allowed" transition (such as singlet to singlet transition as shown in Fig. 9.2A).

The emitted light in fluorescence is frequently in the visible region, while the original light absorbed by the analyte is often in the UV region but can also occur in the visible range. For low concentrations of a fluorescing compound, the relationship between fluorescence intensity and concentration is nearly linear.

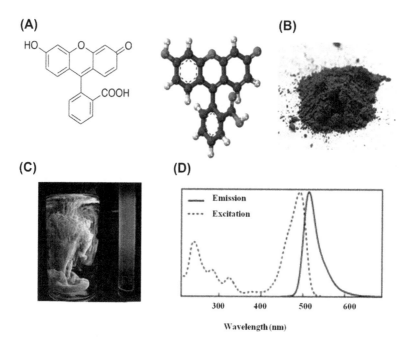

FIG. 9.5

(A) Molecular structure and model of Fluorescein ($C_{20}H_{12}O_5$ with molar mass = 332.311 g mol^{-1}).
(B) Fluorescein in powder form. (C) Fluorescein under UV illumination and (D) Fluorescence excitation
(represents absorption spectrum) and emission spectrum of fluorescein.

From: The World Health Organization (WHO), WHO Model List of Essential Medicines (19th List) (2015).

Similar to the absorbance, the fluorescence is also different for the different analyte (that is, each kind of molecules has its own unique fluorescence emission behavior. Some molecules do not fluorescence efficiently: Instead of releasing most of the energy from their excited state in the form of light, much of this energy is lost as heat to their surroundings. Molecules that have relatively good fluorescence capacity have rigid structures and are often planar, and aromatic groups, as demonstrated by the example in Fig. 9.5 (of Fluorescein. See also in the consecutive subsection, to understand the theoretical concept and parameters that influence the fluorescence emission).

Fluorescein is a manufactured organic compound and dye. As shown in Fig. 9.5B, it is available as a dark orange/red powder and is slightly soluble in water and alcohol. The color of its aqueous solution varies from green to orange as a function of the way it is observed (by reflection or by transmission). It is widely used as a fluorescent tracer for many applications. Fluorescein has an absorption maximum at 494 nm and an emission maximum of 512 nm (in water).

9.3.2.2 Quantitative relation between analyte concentration and fluorescence

As indicated above, fluorescence occurs when a molecule in an excited state's lowest vibrational energy level returns to a lower energy electronic state by emitting a photon. Because molecules return to their ground state by the fastest mechanism, fluorescence is observed only if it is a more efficient means of relaxation than a combination of internal conversions and vibrational relaxation.

The efficiency of fluorescence by a chemical species have been estimated by measuring a *fluorescence quantum yield (Φ_f). It is the ratio of the number of photons observed by the fluorescence process to the absorbed photons.* In other words, it is the fraction of excited state molecules returning to the ground state by fluorescence. A chemical with perfect fluorescence will emit all its absorbed photons, with a maximum value of $\Phi_f = 1$). On the contrary, an analyte that absorbs light but does not produce any fluorescence will have $\Phi_f = 0$. The φ_F values for other molecules will be somewhere in between. That is, fluorescent quantum yields range from 1, when every molecule in an excited state undergoes fluorescence, to 0 when fluorescence does not occur.

The intensity of fluorescence (I_f) is proportional to the amount of radiation absorbed by the sample ($P_0 - P_T$) and the φ_F,

$$I_f = k \cdot \Phi_f (P_0 - P_T) \tag{9.1}$$

where k is a constant accounting for the efficiency of collecting and detecting the fluorescent emission. P_o and P_T are the incident and transmitted light, respectively. From Beer's law, we know that

$$P_T/P_0 = 10^{-\varepsilon bc} \tag{9.2}$$

where c is the concentration of the fluorescing species. Solving Eq. (9.2) for P_T and substituting into Eq. (9.1) gives, after simplifying

$$I_f = k \cdot \Phi_f \cdot P_0 \left(1 - 10^{-\varepsilon bc}\right) \tag{9.3}$$

When $\varepsilon bc < 0.01$, which often is the case when concentration is small, Eq. (9.3) simplifies to:

$$I_f = 2.303k \cdot \Phi_f \cdot \varepsilon.b.c.P_0 = k'P_0 \tag{9.4}$$

where k' is a collection of constants. The intensity of fluorescent emission, therefore, increases with an increase in the quantum efficiency Φ_f, the molar absorptivity ε, the concentration c of the fluorescing species and the incident radiation power P_0. As an example, below in Fig. 9.6 shows the fluorescence of quinine under a UV lamp.

Besides analyte concentration, quantum efficiency, molar absorption coefficient and radiation source's power, a molecule's fluorescence quantum yield is also influenced by external variables, such as temperature and solvent. Increasing the temperature generally decreases Φ_f due to the frequent collisions between the molecule and the solvent increase external conversion. A decrease in the solvent's viscosity decreases Φ_f *for* similar reasons. For an analyte with acidic or basic functional groups, a change in pH may change the analyte's structure and its fluorescent properties.

Fluorescence is generally observed when the molecule's lowest energy absorption is a $\pi \rightarrow \pi*$ transition, although some $n \rightarrow \pi*$ transitions show weak fluorescence. Most unsubstituted, nonheterocyclic aromatic compounds have favorable fluorescence quantum yields, although substitutions on the aromatic ring can significantly affect Φ_f. For example, the presence of an electron-withdrawing group, such as $-NO_2$, decreases Φ_f, while adding an electron-donating group, such as $-OH$, increases Φ_f. Fluorescence also increases for aromatic ring systems and for aromatic molecules with rigid planar structures.

In summary, fluorescence may return the molecule to any of several vibrational energy levels in the ground electronic state. Fluorescence, therefore, occurs over a range of wavelengths. Because the change in energy for fluorescent emission is generally less than that for absorption, a molecule's fluorescence spectrum is shifted to higher wavelengths than its absorption spectrum.

FIG. 9.6

Tonic water, which contains quinine, is fluorescent when placed under a UV lamp.

From: Splarka: http://en.wikipedia.org/wiki/Image:Tonic_water_uv.jpg.

9.3.2.3 Phosphorescence emission

Unlike fluorescence, after excitation of a molecule, the excited electron first undergoes an intersystem crossing into a triplet state. That is, as shown in Fig. 9.2C, in some cases an electron in a singlet excited state is transformed to a triplet excited state (the initial spin of the electron in its ground state is flipped in the opposite direction) in which its spin is no longer paired with the ground state. This means that the release of light from this excited state will now require a "spin-forbidden" transition from this triplet state to the singlet state. Emission between a triplet excited state and a singlet ground state—or between any two energy levels that differ in their respective spin states—is called *phosphorescence*. This type of emission process is much less likely to occur and is slower than the singlet-to-singlet transitions that led to light emission in fluorescence. Because the average lifetime for phosphorescence ranges from $10^{-4} - 10^4$ s (in the range of microseconds to minutes), phosphorescence may continue for some time after removing the excitation source. A spectroscopic technique that utilizes phosphorescence to characterize or measure chemicals is called ***phosphorescence spectroscopy***.

As the measurement of phosphorescence requires low-temperature condition, which is usually maintained by liquid nitrogen, it is much more difficult to measure than fluorescence. The reason for the requirement of low temperature is that the lifetime of an excited triplet state (typically, $10^{-4} - 1$ s or longer) is much greater than an excited singlet state ($10^{-9} - 10^{-8}$ s). This longer lifetime means the probability of energy loss through collisions and heat loss is also much greater in phosphorescence than in fluorescence. Maintaining low temperatures for this measurement will minimize the molecular

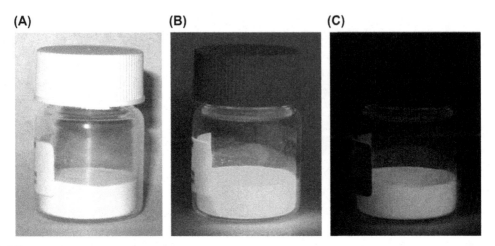

FIG. 9.7

A europium doped strontium silicate-aluminum oxide powder under (A) natural light, (B) a long-wave UV lamp, and (C) in total darkness. The picture taken in total darkness shows the phosphorescent emission.

From: Splarka, commons.wikipedia.org.

motion around the analyte and make its collisions with the solvent or other sample components less likely to occur.

Phosphorescence occurs over a range of wavelengths, all of which are at lower energies than the molecule's absorption band. The intensity of phosphorescence, I_p, is given by an equation similar to *Eq.(9.4)* for fluorescence:

$$Ip = 2.303k \cdot \Phi p \cdot \varepsilon.b.c.P_0 = k'P_0 \tag{9.5}$$

where Φ_p is the phosphorescent quantum yield.

Unlike fluorescence, phosphorescence is the most favorable process for molecules with $n \rightarrow \pi*$ transitions, which have a higher probability for an intersystem crossing than $\pi \rightarrow \pi*$ transitions. For example, phosphorescence is observed with aromatic molecules containing carbonyl groups or heteroatoms. Aromatic compounds containing halide atoms also have higher efficiency for phosphorescence. In general, an increase in phosphorescence corresponds to a decrease in fluorescence.

Because the average lifetime for phosphorescence is very long, the phosphorescent quantum yield is usually quite small. An improvement in Φ_p is realized by decreasing the efficiency of external conversion. As indicated above, this may be accomplished in several ways, including lowering the temperature, using a more viscous solvent, depositing the sample on a solid substrate, or trapping the molecule in solution. Fig. 9.7 shows an example of phosphorescence.

9.3.3 Chemiluminescence (excitation by chemical process and relaxation by photon emission)

Chemiluminescence is the emission of light (luminescence), as the result of a chemical reaction: Specifically, chemiluminescence occurs when a chemical reaction produces an electronically excited species that emits a photon in order to reach the ground state. This may be explained by first breaking down its name and look at the meaning of its pieces. The prefix "chemi" is related to chemicals, and the word "luminescence" is related to giving off light. Putting them together then, chemiluminescence means giving off the light via chemical reaction. Sometimes, the term bioluminescence is also used when the chemical reaction is of biological origin. Fireflies and glowworms are well-known examples of bioluminescence. In the Firefly, an enzyme called luciferase (a name meaning "light-bearing") triggers a reaction that produces energy emitted as light from the insect's lower abdomen.

Another good example of a nonbiological chemical that can undergo chemiluminescence is luminol (5-amino-1, 4-phthalazdione), which reacts with hydrogen peroxide (H_2O_2) to form the excited molecule. The reactions involved in the production of chemiluminescence by luminol is given below:

$$\text{5-amino-1, 4-phthalazdione} + H_2O_2 \rightarrow \text{3-Aminophthalate}* \rightarrow \text{3-Aminophthalate} + \text{light}.$$

The luminol is first reacted under basic conditions with the oxidizing agent H_2O_2 and in the presence of a catalyst. This reaction results the production of 3-Aminophthalate in an excited state (indicated with an '$*$' sign). Some of the excited product release extra energy in the form of light.

The number of chemical reactions which produce chemiluminescence is small. However, some of the compounds which react to produce this phenomenon are significant in terms of environment and health. For example, a good example of chemiluminescence is the determination of nitric oxide. Nitric oxide is a molecule that is produced naturally by our body, and it is important for many aspects of our health. Its most important function is vasodilation (meaning it relaxes the inner muscles of the blood vessels, causing them to widen and increase circulation). In order to measure nitric oxide, NO, it is reacted with ozone, which produces nitrogen dioxide and oxygen.

$$NO + O_3 \rightarrow NO_2^* + O_2$$

$$NO_2^* \rightarrow NO_2 + hv \ (\lambda = 600-2800 \text{ nm}).$$

The following graph (Fig. 9.8) shows the spectral distribution of radiation emitted by the above reaction:

9.4 Instrumentation for luminescence measurements
9.4.1 Brief overview of components for fluorescence measurements

When a filter is used the instrument is called a *fluorimeter*, and when a monochromator is used to select the excitation and emission wavelengths, the instrument is called a *spectrofluorimeter*.

The design of the *fluorometer* is simpler than that of a *spectrofluorometer* because this kind of device only uses filters to select the wavelength of light that is used for excitation as well as to select the wavelength of the fluorescence. Although it does not allow measuring continuous spectrum, it does make it possible to measure the emission intensity at a given set of wavelengths for the quantitative analysis of a particular analyte that undergoes fluorescence at these wavelengths.

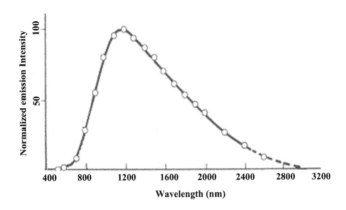

FIG. 9.8

The spectral distribution of radiation emitted by the above reaction.

Adapted with permission from 20th International Symposium on Atmospheric and Ocean Optics: Atmospheric Physics, edited by Gennadii G. Matvienko, Oleg A. Romanovskii, Proc. of SPIE Vol. 9292, 92925I· Copyright © 2014, SPIE.

9.4.2 Light sources

The excitation source for a fluorimeter is usually a low-pressure Hg vapor lamp that provides intense emission lines distributed throughout the ultraviolet and visible region (254, 312, 365, 405, 436, 546, 577, 691, and 773 nm).

With a monochromator, the excitation source is usually a high-pressure Xe arc lamp, which has a continuous emission spectrum. Both of the light sources are appropriate for quantitative work, although only a spectrofluorometer can be used to record an excitation or emission spectrum. To note, a spectrophotometer allows the selection of the exciting wavelength and allows scanning of the spectrum of light that is emitted through fluorescence. As shown in Fig. 9.9, such an instrument has a light source, a monochromator before the sample (hereafter, represented by M1), and another monochromator between the sample and the detector (hereafter, represented by M2).

9.4.3 Excitation versus emission spectra

If we measure the fluorescence by keeping wavelength of the excitation wavelength fixed, varying the wavelength of the second monochromator placed between the sample holder and the detector, the resulting plot of fluorescence intensity versus wavelength is called "*fluorescence spectrum*" (*emission spectrum*).

Alternatively, we can also measure the fluorescence intensity as a function of wavelength by varying the excitation wavelength (wavelength at the second monochromator is fixed). The resultant spectrum is called "*excitation spectrum*". When corrected for variations in the source's intensity and the detector's response, a sample's excitation spectrum is nearly identical to its absorbance spectrum. The excitation spectrum provides a convenient means for selecting the best excitation wavelength for either a quantitative or qualitative analysis.

In an emission spectrum, a fixed wavelength is used to excite the sample and the intensity of emitted radiation is monitored as a function of wavelength. Although a molecule has only a single excitation spectrum, it has two emission spectra, one for fluorescence and one for phosphorescence.

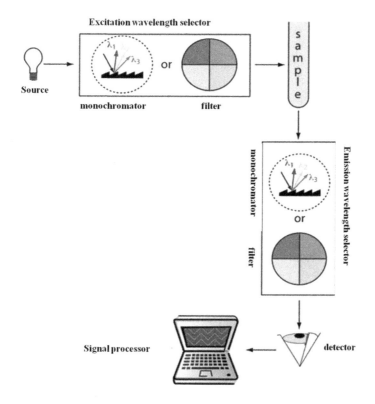

FIG. 9.9

A schematic diagram for measuring fluorescence. Two monochromators (or filters) are used in this instrument. One for wavelength selectors for excitation and another for emission.

From: D. Harvey, Modern Analytical Chemistry, Pg. 619, Chapter 9, first, McGraw-Hill, 1999.

9.4.4 Relationship between absorption and emission spectra

As demonstrated in Fig. 9.10A following absorption, the vibrationally excited S_1 molecule relaxes back to the lowest vibrational level of S_1 prior to emitting any radiation. Emission from S_1 can go to any of the vibrational level of S_0 in Fig. 9.10A. The highest-energy transition comes at wavelength λ_o, with a series of peaks following at longer wavelengths. The absorption and emission spectra will have an approximate mirror image relationship if the spacing between vibrational levels roughly equal and if the transition probabilities are similar.

Fig. 9.10A and B, respectively, represent a simple drawing of an expected excitation (absorption) and emission processes. Whereas, Fig. 9.10B represents excitation and emission spectra. This figure intends to show how these two spectra overlap each other. As an example, the measured UV absorption spectrum and the UV fluorescence emission spectrum for tyrosine and anthracene is given in Fig. 9.11. Note, the first absorption and emission spectrum is measured separately and then analyzed data are displayed in a single figure (as depicted in Fig. 9.11). Note that as expected, in both species, fluorescence is red-shifted.

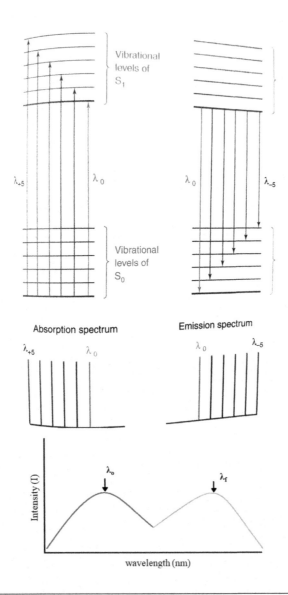

FIG. 9.10

(A) Energy-level diagram showing why structure is seen in the absorption and emission spectra, and why the spectra are rough mirror images of each other. In absorption, wavelength λ_0 comes at the lowest energy, and λ_{+5} is at the highest energy. In emission, wavelength λ_0 comes at the highest energy, and λ_{+5} is at the lowest energy. (B) Excitation/Absorbance Spectrum (Red [dark gray in print version]) coupled with an emission spectrum (Green [gray in print version]). Each spectrum peaks at its respective wavelength. Excitation and emission spectra should overlap and be mirror images of each other.

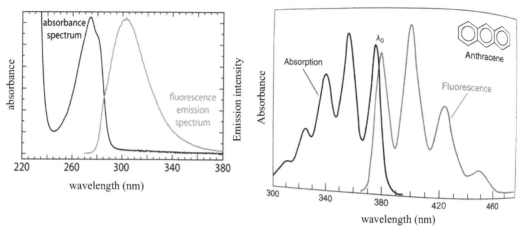

FIG. 9.11

Left: Absorbance spectrum and fluorescence emission spectrum for tyrosine in a pH 7, 0.1 M phosphate buffer. The emission spectrum uses an excitation wavelength of 260 nm. Right: Absorption and emission spectra of anthracene. The spectrum of anthracene shows a typical approximate mirror image relationship between absorption and fluorescence. Fluorescence comes at a lower energy (longer wavelength) than absorption.

Left: Adapted with permission from Mark Somoza, commons.wikipedia.org. Right: Adapted with permission from C.M. Byron and T.C. Werner J. Chem. Ed.1991, 68,433, Copyright 1991, American Chemical Society.

9.4.5 Lifetime measurements

9.4.5.1 Introduction

Fluorescence lifetime (FLT) is the time a fluorophore spends in the excited state before emitting a photon and returning to the ground state. FLT can vary from picoseconds to hundreds of nanoseconds depending on the fluorophore.

The lifetime of a population of fluorophores is the time measured for the number of excited molecules to decay exponentially to N/e (36.8%) of the original population via the loss of energy through fluorescence or non-radiative processes.

Fluorescence lifetime is an intrinsic property of a fluorophore. FLT does not depend on fluorophore concentration, absorption by the sample, sample thickness, method of measurement, fluorescence intensity, photo-bleaching, and/or excitation intensity. It is affected by external factors, such as temperature, polarity, and the presence of fluorescence quenchers. Fluorescence lifetime is sensitive to internal factors that are dependent on fluorophore structure.

9.4.5.2 Methods to determine fluorescence lifetime of fluorophores

Fluorescence lifetime can be measured in either the frequency domain or the time domain. The time-domain method involves the illumination of a sample (a cuvette, cells, or tissue) with a short pulse of light, followed by measuring the emission intensity against time. The FLT is determined from the slope of the decay curve.

Several fluorescence detection methods are available for lifetime measurements, of which time-correlated single-photon-counting (TCSPC) enables simple data collection and enhanced quantitative photon counting.

The frequency-domain method involves the sinusoidal modulation of the incident light at high frequencies. In this method, the emission occurs at the same frequency as the incident light accompanied by a phase delay and change in the amplitude relative to the excitation light (demodulation).

9.4.6 Phosphorescence measurement device

An instrument designed for phosphorescence measurements is similar to fluorescence but differ in two important respects. First, because phosphorescence is such a slow process, we must prevent the excited state from relaxing by external conversion by keeping the sample at low temperature. Traditionally, this has been accomplished by dissolving the sample in a suitable organic solvent, usually in a mixture of ethanol, isopentane, and diethyl ether. The resulting solution is frozen at liquid-N_2 temperatures, forming an optically clear solid. The solid matrix minimizes *external conversion* (discussed in the previous section) due to collisions between the analyte and the solvent. *External conversion* also is minimized by immobilizing the sample on a solid substrate, making possible room temperature measurements. One approach is to place a drop of the solution containing the analyte on a small disc of filter paper. After drying the sample under a heat lamp, the sample is placed in the *spectrofluorimeter* for analysis.

Second, fluorescence often occurs simultaneously with phosphorescence, therefore an alternate method is needed to differentiate rapid fluorescence from slower phosphorescence. For this, the sample is usually radiated for a very short time (max. A few milliseconds). Fluorescence is then observed only in the nanosecond time window and dies quickly, but the phosphorescence continues to produce at least several milliseconds. Then the measured emission spectrum as a function of wavelength with a photo-array detector can be separated by using time gate during analysis (or measurement itself).

9.4.7 Luminometer (devices for chemiluminescence measurement)

An instrument used to measure chemiluminescence is called *luminometer*. This instrument includes a device to mix the analyte with a reagent that will lead to the formation of a luminescent product (a product that can produce luminance). The mixing device is placed close to a photomultiplier tube to measure the intensity of light given off by the exciting product. A simple device (similar to Fig. 9.9, but that only comprises the mixing system in a sample holder, a filter and a detector) merely measure this intensity. Whereas a more complicated one passes the emitted light (by the process of chemiluminescence) through a monochromator to allow the study of the wavelength of the luminescence.

9.5 Instrument standardization method

Because of variations in experimental conditions such as source intensity, transducer sensitivity and other instrumental variables, it is impossible to obtain exactly the same result from the same fluorometer or spectrofluorometer for a solution or a set of solutions from day to day. For this reason, it is common to calibrate (standardize) an instrument and set it to a reproducible sensitivity level. Standardization is usually carried with a standard solution of a stable luminescence. The most common reagent for this purpose is a standard solution of quinine sulfate having the concentration of about 10^{-5} M. It is generally excited by radiation at 350 nm and emits radiation of 450 nm. Nowadays companies specialized to instrumentation related to spectroscopy like Simatsu and PerkinElmer Corporation offer standard solution and details of measurement protocols. For example, PerkinElmer Corporation offers a set of six fluorescence standards dissolved in a plastic matrix to give stable solid blocks that can be used indefinitely without any special storage. With these, the instrument is easily standardized for the wavelength region to be used for the analysis.

In regard to measurement, from Eqs. (9.4) and Eq. (9.5), we know that the intensity of fluorescent or phosphorescent emission is a linear function of the analyte's concentration provided that the sample's absorbance of source radiation ($A = \varepsilon bc$) is less than approximately 0.01. Calibration curves often are linear over four to six orders of magnitude for fluorescence and over two to four orders of magnitude for phosphorescence. For higher concentrations of analytes, the calibration curve becomes nonlinear because of the assumptions leading to Eqs. (9.4) and Eq. (9.5) no longer apply. Nonlinearity may be observed for small concentrations of analyte due to the presence of fluorescent or phosphorescent contaminants.

9.6 Application of molecular luminescence spectroscopy

Application of luminescence spectroscopy may be summarized as follows.

9.6.1 Analyte's Concentration determination

1. *Concentration determination of fluorescent species:* Fluorescence, and to a small extent phosphorescence, is a valuable tool for measuring analytes at low concentrations (Fluorescence and phosphorescence measurements allow to detect lower concentrations than that of absorbance measurements for molecules. See for example, in the "Example 9.1" below: Determination of quinine in urine).

2. **(i)** *Concentration determination of non-fluorescent species (e.g., amines and amino acids):* It is also possible to use an approach based on fluorescence to examine many types of nonfluorescent chemicals by first reacting these analytes with a reagent that converts them into a fluorescent form. A good example is the detection of amines and amino acids by fluorescence. Most amino acids that do not have good fluorescence emission capacity can be combined with the reagent o-phthaldialdehyde to yield a strongly fluorescent product. Similar reactions are available for chemicals that contain alcohol groups, aldehydes, or ketones as part of their structure.

   (ii) *Concentration determination of metal ions (by forming metal complex):* Although most inorganic ions are not sufficiently fluorescent for a direct analysis (except for a few metal ions, most notably UO_2^{2+}), many metal ions may be determined indirectly by reacting with an organic ligand to form a fluorescent, or less commonly, a phosphorescent metal-ligand complex. One example is the reaction of Al^{3+} with the sodium salt of 2, 4, 3′-trihydroxy azobenzene-5′-sulfonic acid (also known as alizarin garnet R) which forms a fluorescent metal-ligand complex (Fig. 9.12). The analysis is carried out using an excitation wavelength of 470 nm, monitoring fluorescence at 500 nm (Please see the fluorescence spectrum in Fig. 9.13).

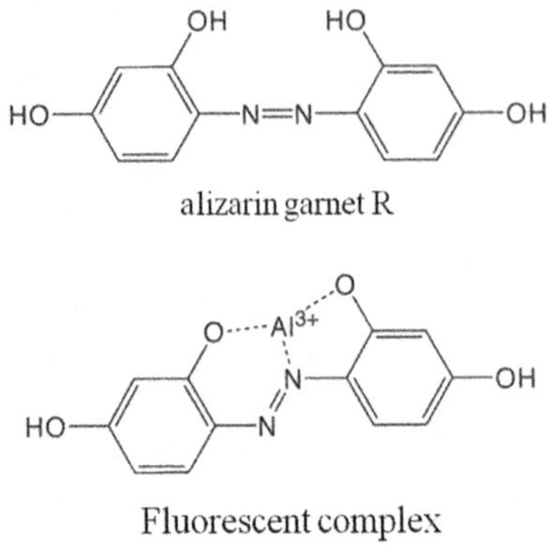

alizarin garnet R

Fluorescent complex

FIG. 9.12

Structure of alizarin garnet R and its complex with Al^{3+}.

From: D. Harvey, Modern Analytical Chemistry, first ed. Chapter10, Pg. 78, McGraw-Hill.

3. *Fluorescent lamp production (most common practical application):* Fluorescent lamp is a glass tube filled with mercury vapor, the inner walls are coated with a phosphor (luminescent substance) consisting of a calcium halophosphate ($Ca_5(PO_4)_3F_{1-x}Cl_x$) doped with Mn^{2+} and Sb^{3+}. The mercury atoms, promoted to an excited state by an electric current passing through the lamp, emit mostly UV radiation at 254 nm and 185 nm. This radiation is absorbed by the Sb^{3+}, and some of the energy is passed to Mn^{2+}. Sb^{3+} emits blue light and Mn^{2+} emits yellow light, with the combined emission white. Fluorescent lamps are more important energy-saving device because they are more efficient than incandescent lamps in the conversion of electricity to light.

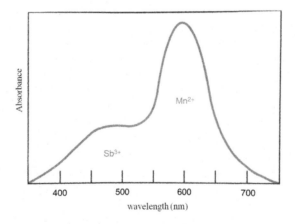

FIG. 9.13

Combined emission spectrum of Sb^{3+} (blue light) and Mn^{2+} (yellow light). D. C.

Harris, Quantitative Chemical Analysis, second Ed, Pg.399, Chapter 18, W. H. Freeman and Company, New York (2001).

Also, a few inorganic nonmetals are determined by their ability to decrease, or quench the fluorescence of another species. One example is the analysis for florine (F)— based on its ability to quench the fluorescence of the Al^{3+}—alizarin garnet R complex.

Below we give an example of glycine on how luminescence can be used to determine the concentration of an analyte.

9.6.1.1 Determination of quinine in urine

Example 9.1. After ingesting 10.0 mg of quinine, a volunteer provided a urine sample 24-h later. Analysis of the urine sample gives a relative emission intensity of 28.16. Report the concentration of quinine in the sample in mg/L and the percent recovery for the ingested quinine.

i. Description of Method

Quinine is an alkaloid used in treating malaria. It is a strongly fluorescent compound in dilute solutions of H_2SO_4 ($\Phi f = 0.55$). Quinine's excitation spectrum has absorption bands at 250 nm and 350 nm and its emission spectrum have a single emission band at 450 nm. Quinine is rapidly excreted from the body in urine and is easily determined by measuring its fluorescence following its extraction from the urine sample.

(See in Fig. 9.6 for the fluorescence of quinine in tonic water.)

ii. Measurement Procedure
1. Transfer a 2.00-mL sample of urine to a 15-mL test tube and adjust its pH to between 9 and 10 using 3.7 M NaOH.
2. Add 4 mL of a 3:1 (v/v) mixture of chloroform and isopropanol and shake the contents of the test tube for 1 min.
3. Allow the organic and the aqueous (urine) layers to separate and transfer the organic phase to a clean test tube. Add 2.00 mL of 0.05 M H_2SO_4 to the organic phase and shake the contents for 1 min.
4. Allow the organic and aqueous layers to separate and transfer the aqueous phase to the sample cell. Measure the fluorescent emission intensity versus concentration of quinine standards at 450 nm using an excitation wavelength of 350 nm [Note: Use distilled water as a blank.]

Table 9.1 Determination of quinine with fluorescence measurement in urine samples.

[quinine] (µg/mL)	Fluorescence, I_f
1	10.11
3	30.2
5	49.84
7	69.89
10	100

5. Determine the concentration of quinine in the urine sample using a calibration curve prepared with a set of external standards in 0.05 M H_2SO_4, prepared from a 100.0 ppm solution of quinine in 0.05 M H_2SO_4.

To evaluate the method described in Representative Method 10.3, a series of the external standard was prepared and analyzed, providing the results shown in the following table. All fluorescent intensities were corrected using a blank prepared from a quinine-free sample of urine. The fluorescent intensities are normalized by setting, if for the highest concentration standard to 100 (Table 9.1).

iii. Analysis

Linear regression of the relative emission intensity versus the concentration of quinine in the standards gives a calibration curve with the following equation.
If = 0.124 + weight of quinine per mL (g/mL)

$$I_f = 0.124 + \text{weight of quinine per mL} \left(\frac{g}{mL}\right)$$

Substituting the sample's relative emission intensity into the calibration equation gives the concentration of quinine as 2.81 µg/mL. Because the volume of urine taken, 2.00 mL, is the same as the volume of 0.05 M H_2SO_4 used in extracting quinine, the concentration of quinine in the urine also is 2.81 µg/mL. The recovery of the ingested quinine is

2.81mL urine × 2.0 mL urine1 mg1000 g × 10010.0 mg quinine ingested = 0.056

$$\frac{\frac{2.81}{ml\ Urine} \times 2.0\ Urine \times \frac{1\ mg}{1000\ g} \times 100}{10.0\ mg\ quinine\ ingested} = 0.056$$

(It can take up 10–11 days for the body to completely excrete quinine.)

The above description of the determination of quinine in urine provides an instructive example of a typical procedure. The description here is based on Mule, S. J.; Hushin, P. L. Anal. Chem. 1971, 43, 708–711, O'Reilly, J. E.; J. Chem. Educ. 1975, 52, 610–612 and https://chem.libretexts.org/courses/Northeastern_University.

Example 9.2. Estimate the concentration of glycine in the following unknown sample based on the fluorescence intensity that is measured for this sample and a series of standards that have each been reacted with an excess of o-phthaldialdehyde to create a fluorescent product.

Concentration of glycine (µM)	Measured fluorescence intensity, I_F
0.0	0.1
0.2	3.4
0.4	6.9
1.0	17.1
2.0	34.3
5.0	83.2
10.0	152
Sample	22.8

As demonstrated in Fig. 9.14B, a plot of the measured fluorescence intensity versus glycine concentration in the standards gives a linear response of the range of concentrations that were examined in this study (to note, Fig. 9.14B was produced from the data given in Table 9.1). The best-fit line for this graph has a slope of 15.3 μM^{-1} and an intercept of 1.77. Using this plot and fluorescence intensity obtained for the sample, the concentration of glycerine in this sample is estimated to be 1.37 μM.

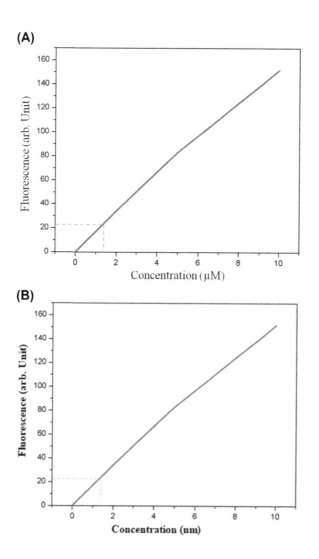

FIG. 9.14

(A and B) A plot of the measured fluorescence intensity versus glycine concentration in the standards.

9.6.2 Material properties (*e.g.*, nature of excitation/absorption, emission/relaxation processes, and determination of bandgap)

Photoluminescence technique has been considered one of the reliable probing methods to understanding material's optical and electronic behaviors. As an example, below we discuss the photoluminescence (PL) measurements of graphene derivatives, specifically graphene oxide (GO) and reduced graphene oxide (rGO).

9.6.2.1 GO and rGO

Both the GO and rGO suspensions found to show unique photoluminescence (PL) features and, therefore, PL measurement technique has been utilized to characterize both of these materials. For example, Fig. 9.15 shows PL spectrum as a function of radiation wavelength under illumination by visible and UV light sources of GO suspensions in water (represented with l-GO) as well as in films (represented with s-GO). To note, in the case of GO two distinct types of PL have been reported so far. The first type is a broad PL covering visible to the near-IR range often exhibiting maximum intensity between 500 and 800 nm (1.55−2.48 eV; Fig. 9.15: left panel). The second type is blue emission, centered around 390−440 nm (2.82−3.18 eV) and is observed upon excitation with UV light. The origin of the two different kinds of PL is still being debated but it has been suggested that the type of PL could be related to the state of dispersion. No absorption features (That is, no PL was observed near the absorption peak position in the PL-energy range, and thus the PL couldn't be directly correlated with the bandgap of the material. Nevertheless, the energies of visible-to-near-IR PL coincide with the bandgap values of graphite oxide estimated from diffuse reflectance measurements, which range from 1.7 to 2.4 eV depending on the degree of oxidation.

Interestingly, the PL spectrum was found to be independent of the GO sheet size and no obvious peak shift was observed even when the GO sheets were cut down to a few nanometers in size. Therefore, on

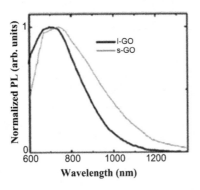 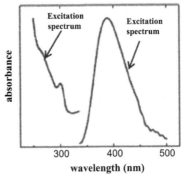

FIG. 9.15

Left panel, Photo-luminescence (PL) spectra of a GO suspension in water (l-GO) and a GO film on a solid substrate (s-GO), excited at 500 nm. Right panel, PL emission spectrum for excitation at 325 nm and excitation spectrum for emission at 388 nm for a GO thin film reduced by hydrazine for 3 min.

Left: Z. Luo, P. Vora, E. J. Mele, A.T. C. Johnson Jr, J. M. Kikkawa, Appl. Phys. Lett. 94 (2009) 111909. Adapted and reprinted with permission from Copyright 2009, American Institute of Physics. Right: Eda et al. Adv. Mater. 2010,22,505, Reproduced with the permission from John Wiley and Sons Inc.

the basis of these observations, the PL in GO has been ascribed to the atomic-scale structure of the material and that the size of the sheet does not define the electron confinement, in contrast to the case of single-walled carbon nanotubes, SWCNTs, where PL wavelength is strongly dependent on the tube diameter. Thus, GO is expected to possess a range of local energy gaps across the sheet, giving rise to the broad PL. The behavior of the PL upon gradual reduction of GO (i.e., conversion to rGO) is also distinctly different in the two systems, suggesting that the origin of the two types of PL is also different. The common trend is that the PL intensity is quenched upon extensive reduction.

9.6.3 Applications for lifetime measurements

As mentioned in the instrumentation section, the fluorescence lifetime of a substance usually represents the average amount of time the molecule remains in the excited state prior to its return to the ground state. Lifetime measurements are frequently necessary for fluorescence spectroscopy. These data can reveal the frequency of collisional with quenching agents, the rate of energy transfer, and the rate of excited state reactions.

The nature of the fluorescence decay can reveal details about the interactions of the fluorophore with its environment. For example, multiple decay constants can be a result of a fluorophore being in several distinct environments or a result of excited state processes. The measurement of fluorescence lifetimes is difficult because these values are typically near 10 nanosec, necessitating the use of high-speed electronic devices and detectors. However, because of the importance of these data, a great deal of effort has been directed toward developing reliable means for the measurement of fluorescence lifetimes. Below, we will discuss practical aspects of lifetime measurements.

9.6.3.1 Fluorescence lifetime assays

The fluorescence lifetime is a robust parameter for use in several biological assays. It has the potential to replace conventional measurement techniques, such as absorption, luminescence, or fluorescence intensity. Any change in the physicochemical environment of the fluorophore leads to changes in the fluorescence lifetime. Lifetime-based assays can be developed by utilizing various mechanisms, such as a simple binding assay that involves the binding of two components (one being fluorescently labeled) to bring about a change in FLT. Another mechanism is a quench-release type assay that involves a quenched species, present in large excess, having low but finite fluorescence. Once the fluorescent compound is released (by an enzymatic reaction or by binding to a complementary DNA), the lifetime of the system changes.

9.6.3.2 Fluorescence lifetime sensing

This technique is based on changes in the lifetime or decay time of the probe. Nanosecond (ns) decay times can be measured by phase-modulation. This technique has been extensively used for sensing of pH, Ca^{2+}, K^+, glucose, and other metabolites. There have been recent developments in the application of lifetime-based sensing in tissues and other random media by using optical probes with excitation and emission spectra in the near-infrared region.

9.6.3.3 Fluorescence lifetime imaging (FLI)

This technique is relatively new and involves the determination of the spatial distribution of fluorescence decay times at every pixel of an image simultaneously. It is based on the fact that the fluorescence lifetime of a fluorophore depends on its molecular environment but not on its concentration. It can be applied in fluorescence microscopy where the local probe concentration cannot be controlled. Fluorescence lifetime imaging microscopy (FLIM) is used in the measurement of molecular environment parameters, protein-interaction by Förster resonance energy transfer (FRET), and the metabolic state of cells and tissue via their autofluorescence. The molecular environment parameters can be measured from lifetime changes induced by fluorescence quenching or conformation changes of the fluorophores. FLIM can be used in biological applications including scanning of tissue surfaces, mapping of tissue type, photodynamic therapy, DNA chip analysis, skin imaging, etc.

Weak emitters have shorter fluorescence lifetimes, while fluorophores with longer lifetimes have low photon turnover rates. They are not very useful for lifetime imaging because of their limited sensitivity and the necessity for long exposition and acquisition time (Based on the reference by J. R. Lakowicz, Principles of Fluorescence Spectroscopy, pp 51−93).

Key Equation:
Fluorescence Intensity.

$I_f = kP_oc$.

I_f = Fluorescence Intensity, k = constant, P_o = radiant power of incident radiation
c = concentration of fluorescing species.

9.7 Questions

1. What is fluorescence? Describe the general process by which light is emitted during fluorescence.
2. What features are often found in molecules that undergo fluorescence?
3. Explain why the wavelength of the light produced from fluorescence by a molecule is longer than the exciting light, but the light fluorescence by an atom is the same wavelength as the exciting light.
4. How is the intensity of light that is emitted by fluorescence related to the concentration of an analyte that is undergoing fluorescence?
5. How is phosphorescence similar to fluorescence? How are these processes different?
6. What is the chemical luminescence? Describe how light is emitted by this process.
7. Explain the difference between a fluorescence emission spectrum and a fluorescence excitation spectrum. Which more closely resembles an absorption spectrum?
8. Define the following terms: (a) Monochromator, Fluorescence, phosphorescence, singlet state, triplet state, vibrational relaxation, internal conversion, intersystem crossing, quantum yield, chemiluminescence.
9. Why do some absorbing compounds fluorescence but others do not?
10. Which compound in each of the pairs below would you expect to have a greater fluorescence quantum yield? Explain.

Phenolphthalein   O,O'-dihydroxyazobenzene

11. The reduced form of nicotinamide adenine dinucleotide (NADH) is an important and highly fluorescent coenzyme. It has an absorption maximum of 340 nm and an emission maximum at 465 nm. Standard solutions of NADH gave the following intensities.

Conc. NADH, μmol/L	Relative intensity
0.1	2.24
0.2	4.52
0.3	6.63
0.4	9.01
0.5	10.94
0.6	13.71
0.7	15.49
0.8	17.91

(a) Construct a spreadsheet and use it to draw a calibration curve for NADH.
(b) Find the least-squares slope and intercept for the plot in (a).
(c) Calculate the standard deviation of the slope.
(d) An unknown exhibits a relative fluorescence intensity of 12.16. Use the spreadsheet to calculate the concentration of NADH.
(e) Calculate the relative standard deviation for the result in part (d).

12. The volumes of a solution containing 1.10 ppm of Zn^{2+} shown in the table were pipetted into separatory funnels, each containing 5.0 mL of an unknown zinc solution. Each was extracted with three 5 mL aliquots of CCl_4 containing an excess of 8-hydroxyquinoline. The extracts were then diluted to 25.0 mL and their fluorescence measured with a fluorometer. The results were the following:

Conc. NADH, μmol/L	Relative intensity
0	6.12
4	11.16
8	15.68
12	20.64

13. Quinine in a 1.664 g anti-malarial tablet was dissolved in sufficient 0.1 M HCl to give 500 mL of solution. A 20.00 mL aliquot was then diluted to 100 with the acid. The fluorescence intensity for the diluted sample at 347.5 nm provided a reading of 245 on an arbitrary scale. A standard 100-ppm quinine solution registered 125 when measured under conditions identical to those for the diluted sample. Calculate the mass in milligrams of quinine in the tablet.

14. Riboflavin emits yellow-green light through fluorescence when this molecule is excited with ultraviolet light. The following data were obtained when measuring the fluorescence intensity of this analyte in a series of standards and a sample. Estimate the concentration of riboflavin in the sample.

Conc. NADH, µmol/L	Relative intensity
1.0E-05	4
2.0E-05	8
4.0E-05	16
8.0E-05	32
1.6E-04	58
3.2E-04	105
6.4E-04	170
Unknown sample	25.8

15. What is the probable color of an aqueous solution that shows a maximum molar absorptivity at (a) 500 nm, or (b) 320 nm?

16. The most abundant substances in unpolluted air are nitrogen, oxygen, argon, carbon dioxide and water. Explain why carbon dioxide and water are considered as greenhouse gases.

References

[1] D.S. Hage, J.D. Carr, Analytical Chemistry and Quantitative Analysis, International ed., Pearson Education, Inc., Publishing as Prentice Hall, Upper Saddle River, New Jersey, 2011 (chapter 18, pg. 450).

[2] D.A. Skoog, F. J Holler, S.R. Crouch, Instrumental Analysis, 11th Indian Reprint, Brooks/Cole, Cengage Learning India Pvt. Ltd., Delhi, 2012 (chap. 15, pg. 443).

[3] D.C. Harris, Exploring Chemical Analysis, second ed., Freeman and Company, USA, 2001 (chap. 18, pg. 401).

[4] W.L. Jorgensen, L. Salem, The Organic Chemist's Book of Orbitals, first ed., Academic Press, New York, 1973.

[5] D. Harvey, Modern Analytical Chemistry, first ed., McGraw-Hill, 1999 (Chap. 9, Pg. 619).

[6] D. Harvey, Molecular Fluorescence, LibreTexts, https://chem.libretexts.org/Courses/University_of_California_Davis/UCD_Chem_115_Lab_Manual/Lab_4%3A_Molecular_Fluorescence, assessed on 5/16/2019.

[7] Primary source, World Health Organization, WHO Model List of Essential Medicines (19th List), 2015. Fluorescein, https://en.wikipedia.org/wiki/Fluorescein.

[8] J.R. Lakowicz, Principles of Fluorescence Spectroscopy, Springer, 2006, ISBN 978-0387-31278-1 (Chap. 2. page 54).

[9] FK1954, File Photo of Red Phosphorescent Pigment, Wikipedia, https://commons.wikimedia.org/wiki/File:Phosphorescent.jpg.

[10] Sheffield Hallam University, UV-visible Luminescence Spectroscopy, Visited on 17/05/2019, https://teaching.shu.ac.uk/hwb/chemistry/tutorials/molspec/lumin1.htm.

[11] S.J. Mule, P.L. Hushin, Anal. Chem. 43 (1971) 708−711, and J. E. O'Reilly, J. Chem. Educ. 52 (1975) 610−612.

[12] F.J. Holler, D.A. Skoog, S. Crouch, Principles of Instrumental Analysis,11th Indian Reprint, Cengage Learning, New Delhi, 2007. Chap. 15, Pg. 443.

[13] J. Mendham, R.C. Denney, J.D. Barnes, M.J.K. Thomas, Vogel's Quantitative Chemical Analysis, sixth ed., Dorling Kindersley (India), 2009. Chap. 17, Pg. 671.

[14] S.J. Mule, P.L. Hushin, Anal. Chem. 43 (1971) 708−711.

[15] J.E. O'Reilly, J. Chem. Educ. 52 (1975) 610−612.

[16] Libretexts, Photoluminescence Spectroscopy. https://chem.libretexts.org/Courses/Northeastern_University/10%3A_Spectroscopic_Methods/10.6%3A_Photoluminescence_Spectroscopy (accessed on 10/05/2019).

[17] C.M. Byron, T.C. Werner, J. Chem. Educ. 68 (1991) 433.

[18] T.A. Amollo, T.M. Genene, O. Nyamori Vincent, High-performance organic solar cells utilizing graphene oxide in the active and hole transport layers, Sol. Energy 171 (2018) 83−91.

[19] J.-S. Yeo, R. Kang, S. Lee, Y.-J. Jeon, N. Myoung, et al., Highly efficient and stable planar perovskite solar cells with reduced graphene oxide nanosheets as electrode interlayer, Nano Energy 12 (2015) 96−104.

[20] A. Agresti, S. Pescetelli, L. Cina', D. Konios, G. Kakavelakis, E. Kymakis, A. Di Carlo, Efficiency and stability enhancement in perovskite solar cells by inserting lithium-neutralized graphene oxide as electron transporting layer, Adv. Funct. Mater. 26 (16) (2016) 2686−2694.

[21] H. Sung, N. Ahn, M.S. Jang, J.-K. Lee, H. Yoon, N.-G. Park, M. Choi, Transparent conductive oxide-free graphene-based perovskite solar cells with over 17% efficiency, Adv. Energy Mater. 6 (3) (2016) 1501873.

[22] Q.-D. Yang, J. Li, Y. Cheng, H.-W. Li, Z. Guan, B. Yu, S.-W. Tsang, Graphene oxide as an efficient hole-transporting material for high-performance perovskite solar cells with enhanced stability, J. Mater. Chem. 5 (2017) 9852−9858.

[23] H. Luo, X. Lin, X. Hou, L. Pan, S. Huang, X. Chen, Efficient and air-stable planar perovskite solar cells formed on graphene-oxide modified PEDOT:PSS hole transport layer, Nano-Micro Lett. 9 (2017) 39.

[24] T. Gatti, S. Casaluci, M. Prato, M. Salerno, F. Di Stasio, A. Ansaldo, E. Menna, A. Di Carlo, F. Bonaccorso, Boosting perovskite solar cells performance and stability through doping a poly-3(hexylthiophene) hole transporting material with organic functionalized carbon nanostructures, Adv. Funct. Mater. 26 (2016) 7443−7453.

[25] T. Gatti, F. Lamberti, P. Topolovsek, M. Abdu-Aguye, R. Sorrentino, L. Perino, M. Salerno, L. Girardi, C. Marega, G.A. Rizzi, M.A. Loi, A. Petrozza, E. Menna, Sol. RRL (2018) 1800013.

[26] B.H. Lee, J.H. Lee, Y.H. Kahng, N. Kim, Y.J. Kim, J. Lee, T. Lee, K. Lee, Adv. Funct. Mater. 24 (2014) 1847−1856.

[27] M.J. Ju, I.Y. Jeon, J.C. Kim, K. Lim, H.J. Choi, S.M. Jung, I.T. Choi, Y.K. Eom, Y.J. Kwon, J. Ko, Adv. Mater. 26 (2014) 3055−3062.

[28] A. Liscio, G.P. Veronese, E. Treossi, F. Suriano, F. Rossella, V. Bellani, R. Rizzoli, P. Samori, V. Palermo, J. Mater. Chem. 21 (2011) 2924.

[29] S. Rafique, S.M. Abdullah, M.M. Shahid, M.O. Ansari, K. Sulaiman, Significantly improved photovoltaic performance in polymer bulk heterojunction solar cells with graphene oxide/PEDOT:PSS double decked hole transport layer, Nature: Sci. Rep. 7 (2017) 1−10, 39555.

[30] H.-W. Cho, W.-P. Liao, W.-H. Lin, M. Yoshimura, Pristine reduced graphene oxide as an energy-matched auxiliary electron acceptor in nanoarchitectural metal oxide/poly(3-hexylthiophene) hybrid solar cell, J.-J. Wu, J. Power Sources 293 (2015) 246.

[31] T. Mahmoudi, Y. Wang, Y.-B. Hahn, Graphene and its derivatives for solar cells application, Nano Energy 47 (2018) 51−65.

[32] S. Bae, H. Kim, Y. Lee, X.F. Xu, J.S. Park, Y. Zheng, J. Balakrishnan, T. Lei, H.R. Kim, Y.I. Song, Y.J. Kim, K.S. Kim, B. Ozyilmaz, J.H. Ahn, B.H. Hong, Iijima, S. Roll-to-Roll production of 30-in. Graphene films for transparent electrodes, Nat. Nanotechnol. 5 (2010) 574−578.

[33] T. Kobayashi, M. Bando, N. Kimura, K. Shimizu, K. Kadono, N. Umezu, K. Miyahara, S. Hayazaki, S. Nagai, Y. Mizuguchi, Y. Murakami, D. Hobara, Production of a 100-m-Long high-quality graphene

transparent conductive film by roll-to-roll chemical vapor deposition and transfer process, Appl. Phys. Lett. 102 (2013) 023112.

[34] G. Eda, Y.-Y. Lin, S. Miller, C.-W. Chen, W.-F. Su, M. Chhowalla, Transparent and conducting electrodes for organic electronics from reduced graphene oxide, Appl. Phys. Lett. 92 (2008) 233305.

[35] G. Eda, M. Chhowalla, Chemically derived graphene oxide: towards large-area thin-film electronics and optoelectronics, Adv. Mater. 22 (2010) 2392−2415.

[36] K.P. Loh, Q. Bao, G. Eda, M. Chhowalla, Graphene oxide as a chemically tunable platform for optical applications, Nat. Chem. 2 (2010) 1015−1024.

[37] C. Hu, D. Liu, Y. Xiao, L. Dai, Functionalization of graphene materials by heteroatom-doping for energy conversion and storage, Prog. Nat. Sci.: Materials International 28 (2018) 121−132.

Index

Note: 'Page numbers followed by "t" indicate tables'.

Printed in the United States
By Bookmasters